ELSEVIER ASTRODYNAMICS SERIES

Modern Astrodynamics

Elsevier Astrodynamics Series

ELSEVIER ASTRODYNAMICS SERIES

Modern Astrodynamics

Edited by

Pini Gurfil

ELSEVIER

AMSTERDAM • BOSTON • HEIDELBERG • LONDON • NEW YORK • OXFORD
PARIS • SAN DIEGO • SAN FRANCISCO • SINGAPORE • SYDNEY • TOKYO

Academic Press is an imprint of Elsevier

Academic Press is an imprint of Elsevier
Linacre House, Jordan Hill, Oxford OX2 8DP, UK
The Boulevard, Langford Lane, Kidlington, Oxford OX5 1GB, UK
84 Theobald's Road, London WC1X 8RR, UK
Radarweg 29, PO Box 211, 1000 AE Amsterdam, The Netherlands

First edition 2006

British Library Cataloguing in Publication Data
Modern astrodynamics. — (Elsevier astrodynamics book series)
1. Astrodynamics
I. Gurfil, Pini
629.4 '1

Library of Congress Catalog Number : 2006930357

ISBN-13: 978-0-12-373562-1
ISBN-10: 0-12-373562-9

For information on all Academic Press publications
visit our web site at http://books.elsevier.com

Printed and bound in the United Kingdom

Transferred to Digital Print 2011

Contents

Foreword

I am honored to have been asked to write a foreword for this new *Elsevier Astrodynamics Series*. The word *Astrodynamics* is often attributed to R.M.L. Baker in the late 1950s. So exactly what is Astrodynamics? An aerospace dictionary from the 1960s defines Astrodynamics as "The practical application of celestial mechanics, astroballistics, propulsion theory and allied fields to the problem of planning and directing the trajectories of space vehicles." The aims and scopes of this series states that: Astrodynamics is a well-recognized, stand-alone field of discipline comprising many auxiliary fields including dynamical systems analysis, optimization, control, estimation, numerical analysis, perturbation methods, Lagrangian and Hamiltonian dynamics, geometric mechanics and chaos. The primary difference between these definitions is the list of the allied fields; the list today is much more extensive, and even includes some that did not exist in 1960, such as chaos. These new allied fields being applied to astrodynamics problems along with some exciting new missions, some of the New Horizons missions and satellite formations, are creating excitement in astrodynamics. With this new series we are trying to present to the aerospace community new results and methods that will help fill the gap between the state-of-the-art and existing books.

Let's look at some of the new problems. The problem of swarms of small satellites flying in precise formation and operating autonomously has probably created more excitement than any problem in years. Hundreds of papers have been published and presented at many conferences in different venues, and its interdisciplinary nature has drawn people from numerous disciplines beyond the traditional dynamics, control and navigation. If some of the discussed future NASA formation flying missions occur, there will be some significant challenges for the astrodynamics community. The application of dynamical systems theory to astrodynamics problems has resulted in the identification of new fuel-efficient interplanetary trajectories and has probably enabled some new missions. The GENESIS mission is an example. From this research we now have the *Interplanetary Super Highway* (from Martin Lo). This is an exciting area with many new results still to be discovered. New missions, small satellites and space control are resulting in new demands in orbit determination. We have the precise orbit determination of a few satellites, such as GRACE, and the maintenance of the space object catalog, which is the less precise orbit determination of thousands of satellites with just a few observations per day. The precise orbit determination is requiring more accurate dynamic models and better estimation. A few years ago small space objects were debris; now they may be micro- or pico-satellites. This, coupled with the need to protect the ISS from collision with small objects, means we need to track objects at least as small as 5 cm in diameter, if not 2 cm in diameter. This would increase the size of the space object catalog from its current size of approximately 14,000 objects to 100,000–300,000 objects. While doing this we need to improve the accuracy of their orbits and provide an accurate covariance. Here, again, the problem is multi-disciplinary. The solution will require improved estimation methods, optimum utilization of sensors, improved atmospheric physics, etc. Here I have tried to identify just some of the new problems creating an exciting challenge in astrodynamics. Even though the demands on astrodynamics from NASA's New Space Exploration Initiative are not yet all clear, one thing is certain: There will be demands. An obvious one is the need for real-time trajectory optimization. As with many fields today, astrodynamics is becoming more interdisciplinary and I believe this is where many of the future advances will be made.

Astrodynamics can be viewed from both an engineering and a mathematical-theoretical standpoint. It is a multi-disciplinary science, drawing largely on systems research and multi-disciplinary design, an area that does not fit very well with current journals, and is not covered in any existing book series. With this new series we are presenting some of the new cutting edge research in Astrodynamics. Topics in this first volume '*Modern Astrodynamics*' include orbital dynamics and perturbations, satellite formation flying, dynamical systems theory, trajectory optimization and novel propulsion systems. Details on the contents are provided in the Introduction.

Kyle T. Alfriend
Department of Aerospace Engineering
Texas A&M University
College Station, TX, USA

Introduction

In recent years, an unprecedented interest in novel and revolutionary space missions has risen out of new NASA and ESA programs. Astrophysicists, astronomers, space systems engineers, mathematicians, and scientists have been cooperating to develop and implement novel, ground-breaking space missions. Recent progress in mathematical dynamics has enabled development of low-energy spacecraft orbits; significant progress in the research and development of electric and propellantless propulsion system promises revolutionary, energy-efficient spacecraft trajectories; and the idea of flying several spacecraft in formation will break the boundaries of mass and size by creating virtual space-borne platforms.

The growing interest in the astrodynamical sciences at large creates a sound need for a new book series solely devoted to astrodynamics. The purpose of the *Elsevier Astrodynamics Series* is, therefore, to give scientists and engineers worldwide an opportunity to publish their works utilizing the high professional and editorial standards of Elsevier Science under the supervision and guidance of a superb editorial board comprised of world-renowned scientists, engineers, and mathematicians.

The first volume in the series, *Modern Astrodynamics*, reviews emerging topics in astrodynamics. The book is designed as a stepping stone for the exposition of modern astrodynamics to students, researchers, engineers, and scientists, and covers the main constituents of the astrodynamical sciences in a comprehensive and rigorous manner.

Modern Astrodynamics deals with the following key topics: Orbital dynamics and perturbations; low-energy orbits, chaos, and Hamiltonian methods; trajectory optimization; novel propulsion systems and spacecraft formation flying.

This volume will be of value to research and graduate students, for its clear and comprehensive portrayal of state-of-the-art astrodynamics; to aerospace and mechanical engineers, for its discussion of advanced trajectory optimization and control techniques, spacecraft formation flying and solar sail design; for mathematicians, for its discussion of Hamiltonian dynamics, chaos and numerical methods; for astronomers, for its presentation of perturbation methods and orbit determination schemes; and for astrophysicists, for its discussion of deep-space and libration point orbits suitable for observational science missions. It is also a must-read for commercial and economic policymakers, as it presents the forefront of space technology from the broad perspective of astrodynamics.

Modern Astrodynamics is a multi-authored volume comprising invited technical contributions written by some of the world's leading researchers: David Vallado (Analytical Graphics Inc., USA), Michael Efroimsky (United States Naval Observatory), Vincent Guibout and Daniel Scheeres (University of Michigan, USA), Edward Belbruno (Princeton University, USA), Oliver Junge and Michael Dellnitz (Paderborn University, Germany), Michael Ross (Naval Postgraduate School, USA), Colin McInnes and Matthew Cartmell

(University of Strathclyde, UK), and Louis Breger, Gokhan Inalhan, Michael Tillerson, and Jonathan How (Massachusetts Institute of Technology, USA).

David Vallado opens this volume with an informative chapter on orbital dynamics and perturbations, including the classical distinction between secular, short- and long-periodic motions, Keplerian orbits, and the quantitative and qualitative effects of gravitational and non-gravitational perturbations on satellite orbits.

Michael Efroimsky continues the discussion on orbital dynamics by presenting one of the most remarkable recent discoveries of theoretical astrodynamics: Gauge freedom. If the inertial Keplerian solution in a non-perturbed setting is expressed via time and some six adjustable constants called elements, then under perturbations this expression is used as *ansatz* and the 'constants' are endowed with time dependence. The perturbed velocity will consist of a partial derivative with respect to time and a so-called convective term, one that includes the time derivatives of the variable 'constants'. Out of sheer convenience, the so-called Lagrange constraint is often imposed. It nullifies the convective term and, thereby, guarantees that the functional dependence of the velocity upon the time and 'constants' stays, under perturbation, the same as it used to be in the undisturbed setting. The variable 'constants' obeying this condition are called osculating elements. Efroimsky shows that it is sometimes convenient, however, to deliberately permit deviation from osculation, by substituting the Lagrange constraint with an essentially arbitrary condition. Moreover, each such condition will then give birth to an appropriate family of non-osculating elements, and the freedom of choosing such conditions will be analogous to the gauge freedom in electrodynamics.

Vincent Guibout and Daniel Scheeres embark on a quest for solving well-known mathematical problems, with important applications in astrodynamics: Two-point boundary value problems. The Hamilton–Jacobi theory for dynamical systems predicts the existence of functions that transform Hamiltonian systems to ones with trivial solutions. These functions, called generating functions, have been widely used to solve a variety of problems in fields ranging from geometric optics to dynamical systems. Guibout and Scheeres' recent work has applied generating functions to solve problems in astrodynamics, with applications to targeting, formation flight, and optimal control. Their chapter defines an algorithm which solves the Hamilton–Jacobi equation for the generating functions associated with the canonical transformation induced by the phase flow. A new algorithm for computing the generating functions, specialized to two-point boundary value problems, is developed.

Ed Belbruno discusses, from the theoretical standpoint, the fascinating applications of Chaos Theory in astrodynamics. Prior to 1985, the Hohmann transfer was viewed as the only way to get a spacecraft from Earth to another planet of the solar system, e.g., the Moon. Is there a better way? Belbruno shows that, indeed, the answer is positive: Have the spacecraft arrive at the Moon with a lesser velocity than the Hohmann transfer, and let the subtle interactions of the gravitational fields of the Earth and Moon gradually slow it down with no fuel required. This theory is called *weak stability boundary theory*. It estimates a region about the Moon where the motion of a spacecraft is chaotic in nature and approximately feels the gravitational pulls of the Earth and Moon almost equally—so that, like a surfer trying to ride a wave, the spacecraft can arrive at the Moon balancing itself on the transition boundaries of the gravity fields of the Earth and the Moon. This yields a capture of the spacecraft into lunar orbit requiring no fuel at all.

Oliver Junge and Michael Dellnitz continue the study of chaos and low-energy orbits by dwelling upon pertinent numerical aspects. They extend recent studies of energy-efficient trajectories for space missions based on the circular restricted three-body problem model. In their chapter, Junge and Dellnitz develop numerical methods for computing approximations to invariant manifolds, which are important for the design of low-energy trajectories. They show how to detect connecting orbits as well as pseudo-trajectories that might serve as initial guesses for the solution of more complex optimal control problems.

Michael Ross continues the quest for energy-efficient orbits by considering spacecraft trajectory optimization problems, or, stated differently, Ross is developing methods for enabling new space missions by reducing the amount of consumed fuel. This problem, as well as its concomitant mathematical modelling and solution, are of prime importance to modern space missions. In particular, Ross provides a well-thought distinction among several options for defining what is meant by "optimal" and "energy efficient", and concludes that these terms are dependent on the particular propulsion system in use.

Colin McInnes and Matthew Cartmell complete the discussion of efficient space travel by a comprehensive study of propellantless mass systems, enabling the breaking of boundaries of currently conceived space missions. Conventional spacecraft are limited in their ability to deliver high-energy missions by a fundamental reliance on reaction mass. However, this basic constraint can be overcome by a class of propulsion systems which either extract momentum from the environment (solar sails), or balance momentum through payload exchanges (tethers). This chapter provides an introduction to the physics of solar sail and tether propulsion systems, along with a review of the recent development of the technologies. This chapter also suggests an outlook for future innovation, including some practical applications of highly non-Keplerian orbits for solar sails and performance optimization for interplanetary tether transfers using motorized momentum exchange principles.

Louis Breger, Gokhan Inalhan, Michael Tillerson, and Jonathan How conclude this volume by addressing an important emerging topic in space systems: Spacecraft formation flying. Efficient execution of precise spacecraft formation flying relies on having accurate descriptions of the fleet dynamics and accurate knowledge of the relative states. However, there are numerous sources of error that exist in real-time as a result of perturbations and differential disturbances. Breger, Inhalan, Tillerson and How analyze the impact of key perturbations on formation flying control. The main point is that analyzing the closed-loop system gives a common framework for comparing both navigation and modeling errors.

Finally, a few debts of gratitude. I would like to acknowledge the dedicated members of the Editorial Board, who diligently and professionally reviewed the contributed chapters for this volume. Special thanks to Professor Terry Alfriend, member of the National Academy of Engineering, for his encouragement, endorsement and support. Finally, great many thanks to Isabelle Kandler, who prepared the infrastructure for this project, and Graham Hart and Jonathan Simpson, the Commissioning Editors, for their professionalism, far-reaching vision and good will.

Pini Gurfil
Elsevier Astrodynamics Series Editor
Technion – Israel Institute of Technology
Haifa 32000, Israel

1 Perturbed Motion

David A. Vallado
Analytical Graphics Inc.

Contents

1.1 Basic definitions

Perturbations are deviations from a normal, idealized, or undisturbed motion. The actual motion will vary from an ideal undisturbed path (two-body) due to perturbations caused by other bodies (such as the Sun and Moon) and additional forces not considered in Keplerian motion (such as a non-spherical central body and drag).

It is important to know about *gradients*, *accelerations* (*specific forces*), and *functions*. A ***gradient*** is really a directional derivative which gives the rate of change of a *scalar function* in a particular direction (Kreyszig, [7]). It's a vector quantity and the ***del operator***,∇, designates the gradient process. The gradient gives an acceleration if the scalar function is a potential function related to a specific potential energy, such as the potential function of a central body's gravity field. I distinguish a potential function as the negative of the potentil energy. Two conventions are "standard" in this area because many schools of thought have evolved over the last few decades. Brouwer and Clemence [3], Battin [2], Long et al. [10], and others express one of the two main approaches, in which the acceleration is the negative gradient of the potential function. This implies that positive work is done as the potential decreases. The other approach, used mainly by the geophysical community, holds an acceleration to be the positive gradient of the potential function [Lambeck [9], Kaula [6], Moritz and Mueller [13], Kaplan [5], Roy [14], and others]. Of course, both methods use potential functions that differ only by a minus sign; therefore, the results are identical! We'll follow the second method and place the sign change between the potential energy and the potential function. I'll also refer to the potential function instead of simply the *potential*, to avoid confusion with *potential energy*.

The distinction between a specific force (often used interchangeably with acceleration) and a potential is important because analysis of perturbations typically uses both concepts. It's common to analyze perturbations using a *disturbing function* and a *disturbing force*. The *disturbing force* simply expresses (in some coordinate system) the specific force (acceleration) that is perturbing the satellite's orbit. Non-conservative forces, such as the perturbing effects of drag and solar-radiation pressure, are usually modeled as a specific force ***Disturbing functions*** are simply the difference between perturbed and unperturbed potential functions. They model conservative forces that perturb the orbit, such as the central body's non-sphericity and third-body attractions.

A ***potential function*** is one way to mathematically characterize a conservative force, such as the gravitational potentialof a *spherical* central body ($U_{2\text{-}body} = \mp\mu/r$). Some people distinguish a disturbing function from a disturbing potential by a minus sign. As mentioned earlier, considering the two to be equal is just as correct, as long as we maintain the correct sign convention. The potential function for an *aspherical* central body, U(sometimes referred to as the *anomalous potential*) includes the spherical potential ($U_{2\text{-}body}$) as the first term. The term *geopotential* is often used for this aspherical potential when the central body is the Earth.

Because we wish to examine the effect of perturbations on the orbital elements, we must characterize how they vary over time. Perturbations on orbital motion result in *secular* and *periodic* changes.

Secular changes in a particular element vary linearly over time, or in some cases, proportionally to some power of time, such as a quadratic. Secular terms grow with time, and errors in secular terms produce unbounded error growth. Secular terms are the primary contributor to the degradation of analytical theories over long time intervals. Although the dominant perturbing force for the Earth, J_2, results in all three types of effects, we can do a first-order approximation and approximate the main variations. We can also develop some higher-order solutions. ***Periodic*** changes are either *short-* or *long*-periodic, depending on the length of time required for an effect to repeat. Because so many definitions exist in the literature and in practice, I'll define each type.

Short-periodic effects typically repeat on the order of the satellite's period or less. ***Long-periodic*** effects have cycles considerably longer than one orbital period—typically one or two orders of magnitude longer. These long-periodic effects are often seen in the motion of the node and perigee and can last from a few weeks to a month or more. This means a short-periodic effect for a satellite at an altitude of 400 km could vary with periods up to about 100 minutes, whereas a short-periodic effect for a geosynchronous satellite would be up to about 24 hours. Also, short-periodic variations occur when a fast variable (true anomaly, for instance) is present in the contributing perturbational effect.

We also distinguish certain orbital elements as either *fast* or *slow* variables, depending on their relative rate of change. ***Fast variables*** change a lot during one orbital revolution, even in the absence of perturbations. Examples are the mean, true, and eccentric anomalies, which all change 360°, or the cartesian coordinates, which also change dramatically in a single revolution. ***Slow variables*** (semimajor axis, eccentricity, inclination, node, argument of perigee) change very little during one orbital revolution. Perturbations cause these changes. Without perturbations, all the slow elements would remain constant. Fast variables would continue to change.

We can describe the perturbed motion of a satellite by an ordered series of position and velocity vectors. Consequently, at each point in time, we can use these vectors to find the orbital elements using two-body techniques. The corresponding position and velocity vectors define these ***osculating elements*** at any instant in time. "Osculate" comes from a Latin word meaning "to kiss." Thus, the osculating orbit kisses the trajectory at the prescribed instant. We define an ***osculating ellipse*** as the two-body orbit the satellite would follow if the perturbing forces were suddenly removed at that instant. Therefore, *each* point on the trajectory has a corresponding set of osculating elements. Osculating elements are the true time-varying orbital elements, and they include all periodic (long- and short-periodic) and secular effects. They represent the high-precision trajectory and are useful for highly accurate simulations, including real-time pointing and tracking operations.

In contrast, ***mean elements*** are "averaged" over some selected time (or an appropriate angle such as true anomaly), so they are relatively smoothly varying and do not chase the short-periodic variations. Notice that mean elements depend on some unspecified averaging interval of the time; the true, eccentric, or mean anomaly; or the longitude. Because there are many kinds of mean elements, it is important to understand how they are defined and used. Mean elements are most useful for long-range mission planning because they approximate the satellite's long-term behavior.

1.2 Forces

The accuracy of orbit determination largely depends on modeling of all physical forces affecting the motion of the Earth satellite or spacecraft in its orbital path through space. By far the largest effect is due to gravitation, usually followed by atmospheric drag, third body perturbations, solar radiation pressure effects and a suite of smaller effect such as tides, and several others. Vallado [16] shows the relative effect of various forces on several satellites at two different satellite altitudes. Figure 1.1 shows these quantitative effects of all physical forces in terms of positional differences for a 500 km altitude, 97.6° inclined satellite. Note that most of the effects like tides, third body forces and relativity are very small, but need to be taken into account when high precision is of

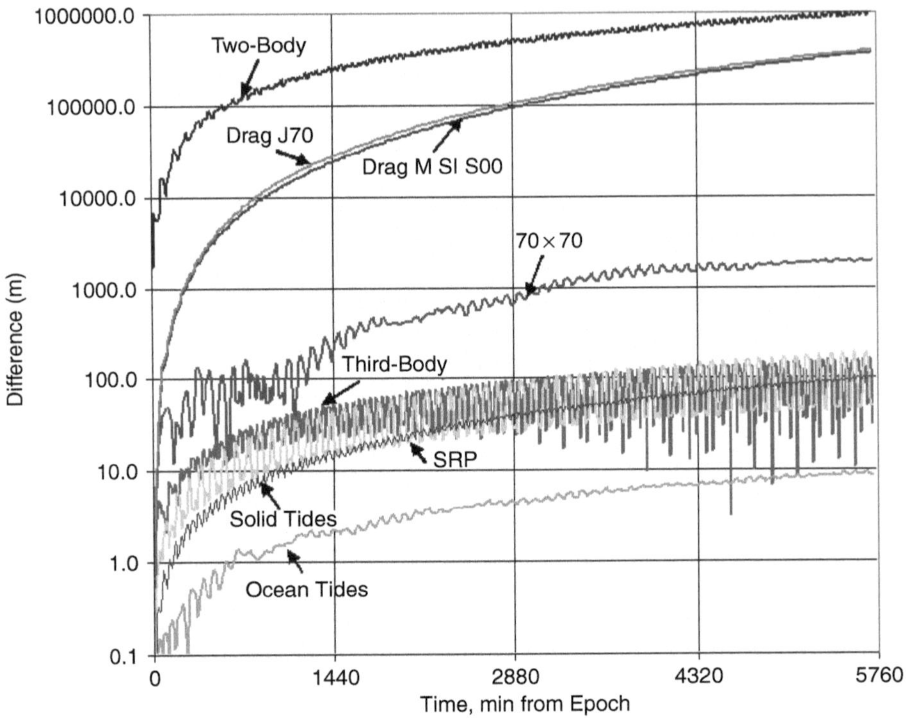

Fig. 1.1. Force Model Comparisons: This figure shows the positional difference over time (four days) when using various force models on the same initial state.

importance. The satellite parameters were chosen to illustrate force model effects. The coefficient of drag $c_{\mathrm{D}} = 2.2$, coefficient of solar radiation pressure $c_{\mathrm{R}} = 1.2$, and area to mass ratio $A/m = 0.04\,\mathrm{m}^2/\mathrm{kg}$. The simulation time, January 4, 2003, was chosen as the epoch to propagate as this was a moderate period of solar activity (solar flux $F_{10.7} \sim 140$).

1.3 Gravity

The general equation for the gravitational attraction uses a spherical harmonic potential equation in an Earth-centered, Earth-fixed reference frame of the form. The fundamental expression for Earth's gravitational potential acting on a satellite is usually given in the familiar form of Earth's geopotential with the origin at Earth's center of mass:

$$V = \frac{\mu}{r}\left[1 + \sum_{n=2}^{\infty}\sum_{m=0}^{n}\left(\frac{R_\oplus}{r}\right)^n P_{nm}(\sin\phi_{gcsat})(C_{nm}\cos m\lambda_{sat} + S_{nm}\sin m\lambda_{sat})\right] \qquad (1.1)$$

Where $\mu =$ gravitational parameter, r is the satellite radius magnitude, ϕ_{gcsat} and λ_{sat} are the geographic coordinates of the satellite, $R_{\oplus}$ is the Earth radius, and C_{nm} and S_{nm} are the gravitational coefficients. Notice the presence of Legendre polynomials.

$$\begin{aligned} P_{nm}\left(\sin \phi_{gcsat}\right) &= \left(\cos \phi_{gcsat}\right)^{m} \frac{\mathrm{d}^{m}}{\mathrm{d}^{m}(\sin \phi_{gcsat})} P_{n}\left(\sin \phi_{gcsat}\right) \\ P_{n}\left(\sin \phi_{gcsat}\right) &= \frac{1}{2^{n}\, n!} \frac{\mathrm{d}^{m}}{\mathrm{d}^{m}(\sin \phi_{gcsat})} \left(\sin^{2} \phi_{gcsat} - 1\right)^{n} \end{aligned} \tag{1.2}$$

For computational purposes, this expression is often used in the *normalized* form. This results from replacing P_{nm}, C_{nm}, and S_{nm} with $\bar{P}_{nm}$, C_{nm}, and $\bar{S}_{nm}$ where

$$\bar{P}_{nm} = \left[\frac{(2n+1)k(n-m)!}{(n+m)!}\right]^{\frac{1}{2}} P_{nm},$$

and

$$\begin{Bmatrix} \bar{C}_{nm} \\ \bar{S}_{nm} \end{Bmatrix} = \left[\frac{(n+m)!}{(2n+1)k(n-m)!}\right]^{\frac{1}{2}} \begin{Bmatrix} C_{nm} \\ S_{nm} \end{Bmatrix}, \tag{1.3}$$

with $k = 1$ for $m = 0$, and $k = 2$ for $m \neq 0$.

A Legendre function (polynomial or associated function) is referred to as a zonal harmonic when $m = 0$, sectorial harmonic when $m = n$, and tesseral harmonic when $m \neq n$.

When normalized coefficients are used, they must be used with the corresponding normalized associated Legendre function:

$$P_{nm} = \frac{\bar{P}_{nm}}{\Pi_{nm}} \tag{1.4}$$

such that $\bar{C}_{nm}\bar{P}_{nm} = C_{nm}P_{nm}$ and $\bar{S}_{nm}\bar{P}_{nm} = S_{nm}P_{nm}$ and the standard model is preserved. Computer software programs generally all use double precision values when converting these coefficients.

1.3.1 Earth Gravitational Models

The first attempts to standardize models of the Earth's gravitational field and the shape of the Earth were begun in 1961. A series of gravitational constants in the form of low degree and order spherical harmonic coefficients were published based on Sputnik, Vanguard, Explorer, and Transit satellite tracking data by special investigators within their respective sponsoring organizations. The first gravity models differed greatly primarily due to observational and computational limitations. As satellite tracking has become more commonplace and computing power has increased, there are still several gravitational models, but their differences are minimal for most applications. Currently there are several prevailing gravitational models being used within the scientific community for a variety of purposes. These models were determined from a wide range of measurement types,

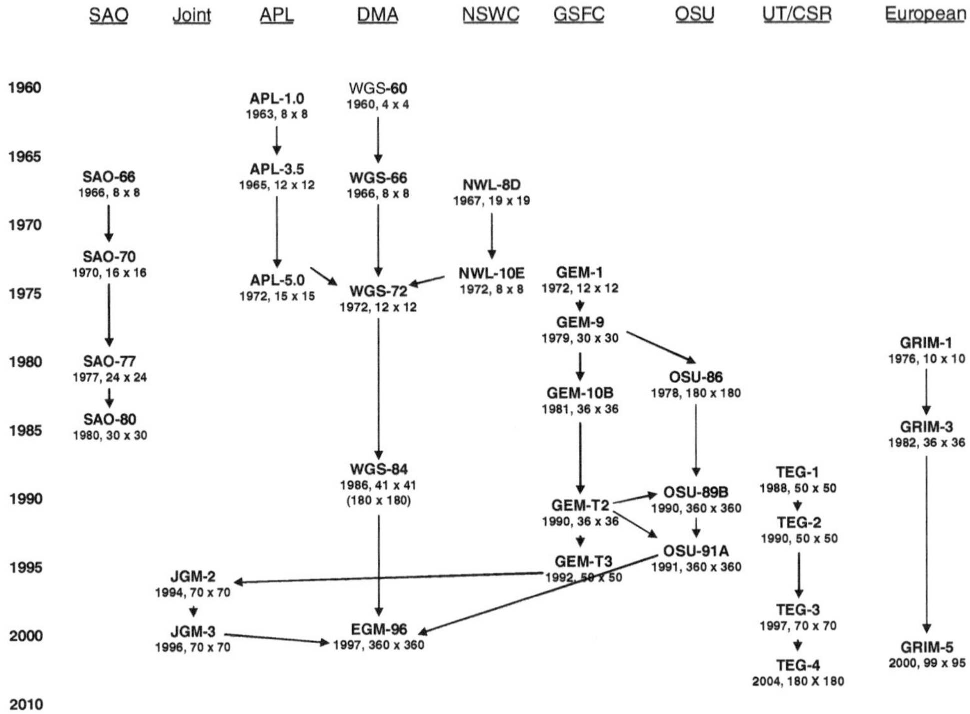

Fig. 1.2. Gravitational models: The *Joint Gravity Models* (JGM) come from Goddard Space Flight Center (GSFC), Ohio State University (OSU), University of Texas at Austin (UT), and the European communities. The *Earth Gravity Model* (EGM) combines the JGM work with Defense Mapping Agencies efforts. The *Goddard Earth Model*s (GEM) were produced annually beginning with GEM-1 in 1972. Even numbered models contain satellite and surface gravity data. Odd numbered models contain only satellite data. *Standard Earth* (SAO) and *Applied Physics Laboratory* (APL) models were among the first models. The basic information is from Vetter [18] and [19].

satellite inclinations and altitudes including surface gravity measurements and satellite altimetry data (Figure 1.2). From Vallado [15],

Many computational applications choose to truncate the gravitational field. While the rigorous approach requires the complete field, many applications use reduced gravity field orders to speed computational processing. Historically, there was some interest to truncate the gravity field for computational or program limitations. While this is often overlooked, some operational systems (AFSPC) often use a blanket 24×24 (for example) field for LEO orbits, rapidly truncating the gravity field as the orbits get higher. This may not be the best approach to accurately determine the orbit. Barker et al. [1] suggested a link in performance to the zonal truncation. Other studies have almost all examined the average behavior of the gravity field on the satellite orbit ephemeris. This may not tell the proper story for precise operations. Vallado [16] investigated the behavior of truncations for several satellites. One example is shown here for Japanese Earth Resources Satellite (JERS, about 500 km altitude circular orbit) (Figure 1.3).

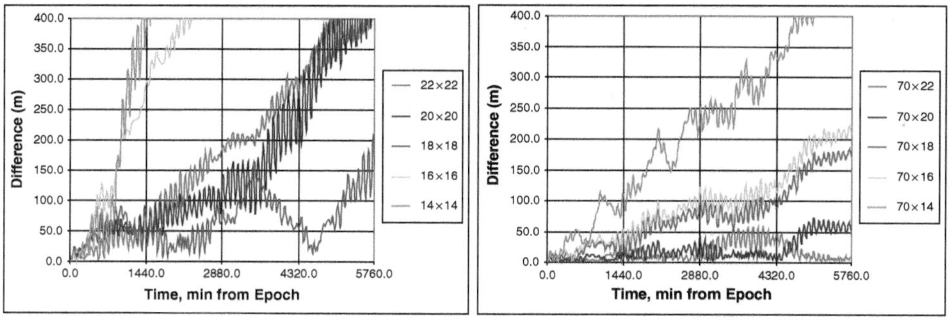

Fig. 1.3. Gravity field comparisons: Truncated gravity fields are compared to ephemeris runs for a complete EGM-96 70 × 70 field for a satellite at about 500 km altitude. The left plot is for a square gravity field. The right plot includes all the zonals in the truncations. The results do not always improve with a larger field (the differences for 22 × 22 are greater than 18 × 18 on the left, but the 70 × 22 is smaller than the 70 × 18 on right), but the accuracy generally improves as the non-square truncation is reduced.

Table 1.1
Fundamental Defining Parameters—EGM-96

Earth Semi-major Axis	$a = r_{\oplus}$	6378136.3 m
Flattening of the Earth	$1/f$	1.0/298.257
Angular Velocity of the Earth	$\omega_{\oplus}$	$7292115.8553 \times 10^{-11}$ rad/s
Earth's Gravitational Constant	GM, (μ)	$3.986004415 \times 10^{5}\,\text{km}^3/\text{s}^2$

Recognize that each time the gravity field is changed, the potential energy of the system changes, and an Orbit Determination (OD) process would produce a different state vector to reflect this change, based on the force models used during that evaluation. Although the most precise way to evaluate each force model would be to perform an OD on each individual case, the process would be unnecessarily long because we are only trying to establish the relative trends for each perturbation, not specific values for an individual case. As computers have become faster, the easiest approach is to simply use a complete gravity field (Table 1.1).

Models for gravitational perturbations are spherical harmonic expansions of the aspherical gravitational potential are in an Earth-centered, Earth-fixed reference frame.

1.4 Drag

The application of empirical atmospheric density models to astrodynamics in a real-world environment has been examined extensively since the launch of the first artificial satellites (Figure 1.4). Atmospheric density leads to significant drag effects for satellites below about 1000 km altitude, but its effects can be observed at altitudes well above this

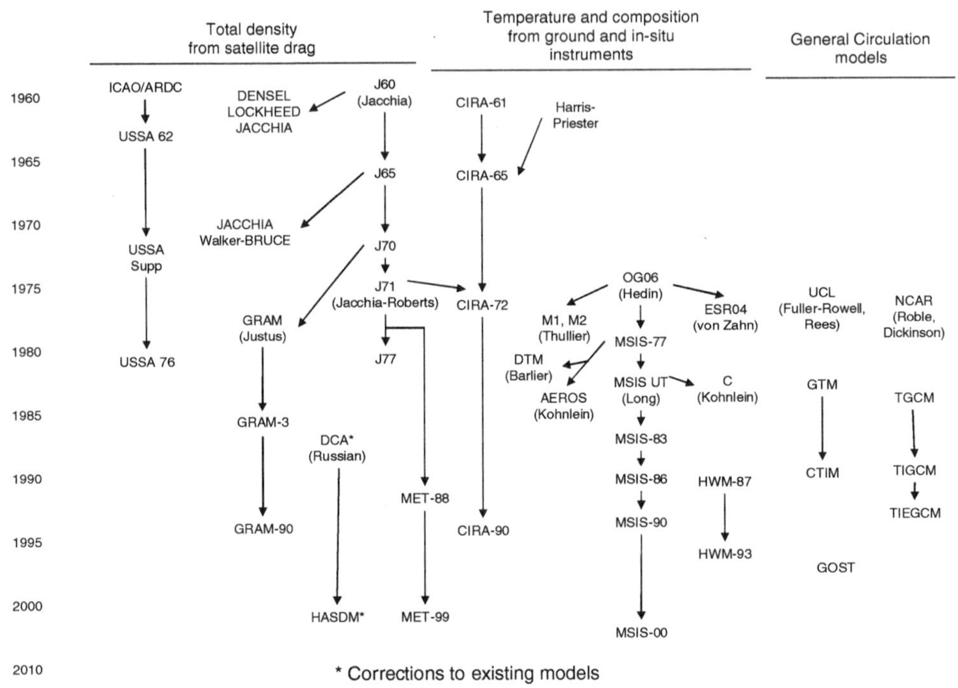

Fig. 1.4. Atmosphere Models: Notice the variety of models. Flow of information among the three overall categories is limited (Marcos, et al., 1993, 20). The main models in use today are the Standard Atmosphere, USSA76; variations of the Jacchia–Roberts, J71, J77, and GRAM90; COSPAR International Reference Atmosphere, CIRA90; Mass Spectrometer Incoherent Scatter, MSIS 00; Drag Temperature Model (DTM), Marshall Engineering Thermosphere (MET), the Russian GOST and general circulation models.

threshold. It is useful to review the basic acceleration equation. From Vallado [16], the following introduction and analysis is taken.

$$\vec{a}_{drag} = -\frac{1}{2}\rho\frac{c_D A}{m}v_{rel}^2\frac{\vec{v}_{rel}}{|\vec{v}_{rel}|} \tag{1.5}$$

ρ The density usually depends on the atmospheric model, Extreme Ultraviolet EUV, $F_{10.7}$, and geomagnetic indices a_p, prediction capability, atmospheric composition, etc. There is wide variability here, and many parameters that can cause significant changes. The popular parameters to examine today are the density and the exospheric temperatures. This single parameter represents the largest contribution to error in any orbit determination application.

c_D The coefficient of drag is related to the shape, but ultimately a difficult parameter to define. Gaposchkin [4] discusses that the c_D is affected by a complex interaction of reflection, molecular content, attitude, etc. It will vary, but typically not very much as the satellite materials usually remain constant.

- A — The cross-sectional area changes constantly (unless there is precise attitude control, or the satellite is a sphere). This variable can change by a factor of 10 or more depending on the specific satellite configuration. Macro models are often used for modeling solar pressure accelerations, but seldom if ever, for atmospheric drag.
- m — The mass is generally constant, but thrusting, ablation, etc., can change this quantity.
- $\vec{v}_{rel}$ — The velocity relative to the rotating atmosphere depends on the accuracy of the *a priori* estimate, and the results of any differential correction processes. Because it is generally large, and squared, it becomes a *very* important factor in the calculation of the acceleration.

The ballistic coefficient (BC $= m/c_D A$ – a variation is the inverse of this in some systems) is generally used to lump the mass, area, and coefficient of drag values together. It *will* vary, sometimes by a large factor. Several initiatives are examining the time-rate of change for this parameter, but not looking at the variable area, and its effect in this combined factor. It is probably best not to model this parameter because it includes several other time-varying parameters that are perhaps better modeled separately.

There are numerous atmospheric drag models. Figure 1.4 lists some of the more popular models.

The primary inputs in any program are the atmospheric density (handled via a specified model), and the *BC*. The mass and cross-sectional area are usually well known, and an estimate of the drag coefficient permits reasonable approximations. The atmospheric models also vary depending on several factors, including the satellite orbit, intensity of the solar activity, and the geomagnetic activity. Vallado and Kelso [17] discuss the files needed to compile a seamless file for operations. They are available at http://celestrak.com/SpaceData.

Unlike any other force model, atmospheric drag receives extensive analysis and near-continual updates. The bottom line for drag (and to a lesser extent solar radiation pressure, as we will see shortly) is to have as many options and choices as possible. While the programming and certification tasks becomes more complicated, this non-conservative force is often the most difficult to match in ephemeris comparisons and having these options provides the user with a much greater ability to minimize differences with other programs.

There are three general observations that are important—the difference between atmospheric models, the variability that can result from treating the input data differently, and the actual implementation of an approach. Vallado [16] conducted a series of tests to determine the variability of different atmospheric models for a given satellite using a single flight dynamics program, and the differences resulting from the diverse treatment of the input solar weather data. The state vectors, epoch, *BC*, and solar radiation pressure coefficient ($m/c_r\ A_{sun}$) were held constant for all runs. The baseline used the Jacchia–Roberts atmospheric model. The simulations were run during a time of "average"

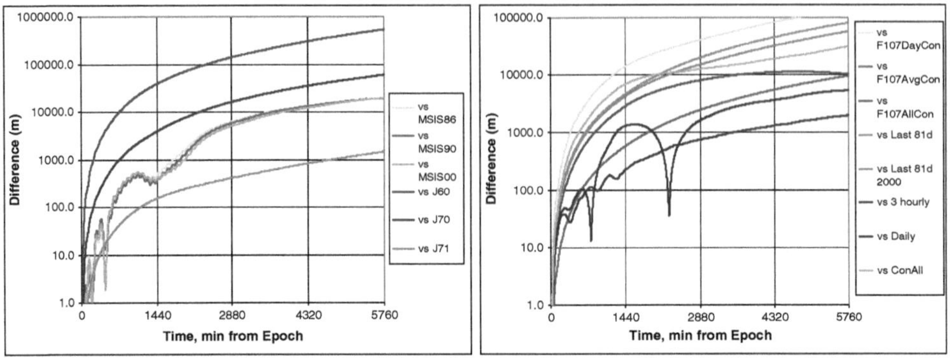

Fig. 1.5. Sample Atmospheric Drag Sensitivity: Positional differences are shown for JERS, about 500 km altitude and 97.6° inclination. Jacchia–Roberts drag is the baseline for all runs with 3-hourly interpolation. The left-hand graph shows the variations by simply selecting different atmospheric models. The right-hand graph shows the effect of various options for treating solar weather data. Specific options are discussed in the text. Note that the scales are the same, the relative effect of different models and solar data options are about the same, and any transient effects quickly disappear as the effect of drag overwhelms the contributions.

solar flux (January 4, 2003, $F_{10.7} \sim 140$). Minimum solar flux periods ($F_{10.7} \sim 70$) will show little difference. Maximum periods ($F_{10.7} \sim 220$) will show much larger excursions. Figure 1.5 shows the results for the JERS satellite, at about 500 km altitude and 97.6° inclination. Additional runs were performed with different satellites and as expected, the results were larger for lower and more eccentric orbits.

Most models as implemented in computer code, do not follow the exact technical derivation as defined in the literature. It is likely that none of the drag model implementations match the original technical definition. As a result, code contains numerous short cuts, and many additional features that may be the result of internal studies and information, but not the original work. This makes comparison of atmospheric models especially difficult.

Because atmospheric drag has perhaps the largest number of different models, defining an absolute standard is difficult to do, and would unnecessarily restrict research. There have been numerous studies to evaluate how well the atmospheric models perform, yet, no clear "winner" has ever emerged. Thus, we list models and present references that discuss the various merits of many of the models. An additional comment is necessary. Most models, as implemented in computer code, do not follow the exact technical derivation as defined in the literature. Numerous short cuts, and many additional features are included that may be the result of internal studies and information. This makes standardization of atmospheric models especially difficult.

For most of the simulations, the MSIS-86 and MSIS-90 models were quite close, as expected by the model descriptions. The Jacchia 1960 (J60) model appeared to be significantly different in all cases from the other models and J70 seemed to differ most from the J71 and JRob models. Because this chapter does not extensively examine comparisons with Precision Orbit Ephemerides (POEs), it is most important to come away with the overall level of variability within the different models. Essentially, if varying atmospheric models show differences that are significantly larger than differences between flight dynamics programs using the "same" models, which is right? After examining these data, we conclude that neither are right. Primarily, this is due to the results shown on the right-hand side of Fig. 1.5 which are discussed next. Although each atmospheric model is carefully designed, the treatment of solar weather data by each program adds so much variability, coupled with the lack of independent references and availability of observational data for comprehensive evaluation makes it highly unlikely that one approach is definitive for all cases.

The following recommendations are set forth.

1. There should be an option to use either the last $F_{10.7}$ 81-day average, or the centered 81-day average. Atmospheric model descriptions generally cite a centered average, but this is impractical for many operational systems, and a trailing 81-day average is often used.
2. Using a_p should be seamless, but there is the possibility of difficulties for certain conversions of average values. There are discrete values for which a_p and k_p exist in the daily data. Thus, a program needs to be careful not to input a derived value that does not exist in the other scale. Inside a program, however, conversions may proceed without restriction to value. Consistency should be maintained with the atmospheric model.
3. The cubic splines routine discussed in Vallado and Kelso [17] should be used to interpolate geomagnetic indices.
4. The codes should treat all $F_{10.7}$ measurements at the time the measurement is actually taken. The offset (2000 UTC after May 31, 1991, 1700 UTC before) should be used with all $F_{10.7}$ and average $F_{10.7}$ values. Any model specific "day before", "6.7 hours before", etc., should be done with this offset in mind. There is not an established approach, yet it is a big factor (sometimes km level) in the comparisons.
5. The options for using a_p should be
   a. daily—just the daily values are interpolated. All 3-hourly values are ignored.
   b. 3-hourly—just the 3-hourly values are used. The daily values are ignored and there is no interpolation. This will produce step function discontinuities, but that could be useful in some programs.
   c. 3-hourly interp.—this should use the cubic splines from Vallado and Kelso (2005). It should produce the smoothest transitions from one time to the next while preserving the discrete values. The measurements should reproduce exactly at the measurement times (0000, 0300, 0600, etc. UTC), and be smooth in between.
6. The lag time for a_p values is somewhat fixed to 6.7 hours, but others have been proposed. Since it is a variable option, it would be prudent to have a means to change it, without recompiling the entire program.
7. The drag coefficient, area, and mass need to be included in state vector transmissions to permit increased accuracy in subsequent calculations.

Many sources state that the current atmospheric models introduce about a 15% error in the determination of atmospheric drag effects on a satellite. In fact, this is a combination of the inaccuracy of the predictions of the solar flux and geomagnetic indices, the imperfect nature of the mathematical models, imprecise information about the molecular interaction of the satellite and the atmospheric particles, and several others.

1.5 3-Body

Third body effects include the perturbations induced by the gravitational influence of the Sun, Moon, and the planets. These are also called n-body perturbations acting on the satellite. The contributions are computed using a point-mass equation. However, the Sun and Moon also include an indirect effect as an interaction between a point-mass perturbing object and an oblate earth. Thus the third-body perturbation includes both direct and indirect terms of point mass third-body perturbations.

The general form of the acceleration due to third-body forces is

$$\vec{a}_{3\text{-body}} = -\frac{G(m_E + m_{sat})}{r_{Esat}^3}\vec{r}_{Esat} + Gm_3\left(\frac{\vec{r}_{sat3}}{\vec{r}_{sat3}^{\,3}} - \frac{\vec{r}_{E3}}{\vec{r}_{E3}^{\,3}}\right) \tag{1.6}$$

Analytical and numerically generated models dominate astrodynamic programs. Many applications use the analytical approaches because they provide adequate accuracy. However, numerical routines often require the additional accuracy of the JPL models.

1.6 Solar radiation pressure

The force due to solar radiation pressure (SRP) rises when photons from the Sun impinge on a satellite surface and are absorbed (or reflected-specular and diffuse) thus transferring photon impulse to the satellite. In contrast to drag, the SRP force does not vary with altitude and its main effect is a slight change in the eccentricity and longitude of perigee. The effect of SRP is most notable for satellites with large solar panels like communications satellites and GPS and depends on its mass and surface area. In cases of geodetic precision orbits, complex modeling of the exposed satellite surfaces have to modeled usually using finite-element computer codes. This is the case with GPS where SRP represents an important force.

Vallado [16] provides a background for SRP and is included herein. Although not studied as extensively in the literature, it poses many of the same challenges as atmospheric drag, but has a significantly smaller effect than the other forces. Consider the basic equation.

$$\vec{a}_{srp} = -p_{SR}\frac{c_R A_{Sun}}{m}\frac{\vec{r}_{sat-Sun}}{|\vec{r}_{sat-Sun}|} \tag{1.7}$$

p_{SR} The incoming solar pressure depends on the time of year, and the intensity of the solar output. It is derived from the incoming solar flux and values of about 1358–1373 W/m^2 are common.

c_R The coefficient of reflectivity indicates the absorptive and reflective properties of the material, and thus the susceptibility to incoming solar radiation.

A_{Sun} The cross-sectional area changes constantly (unless there is precise attitude control, or it is spherical). This variable can change iby a factor of 10 or more depending on the specific satellite configuration. Macro models are often used for geosynchronous satellites. This area is generally *not* the same as the cross-sectional area for drag.

m The mass is generally constant, but thrusting, ablation, etc., can change this quantity.

$r_{\text{sat}-\text{Sun}}$ The orientation of the force depends on the satellite–Sun vector—again a difference with atmospheric drag.

Despite the simple expression, accurate modeling of solar radiation pressure is challenging for several reasons. The major error sources are:

- Use of macro models/attitude—this is perhaps the largest difference between programs
- Use of differing shadow models (umbral/penumbral regions, cylindrical, none, etc.)
- Using a single value for the incoming solar luminosity, or equivalent flux at 1 AU
- Use of an effective Earth radius for shadow calculations (23 km additional altitude is common)—this approximates the effect of attenuation from the atmosphere
- Using different methods to account for seasonal variations in the solar pressure
- Not integrating to the exact points of arrival and departure at the shadow boundary
- Use of simplified treatment for the light-time travel from the Sun to the satellite (instantaneous (true), light delay to central body accounted for (app. to true), light delay to satellite (default))

A series of runs were made to determine the impact of each of these items on the results for a few selected satellites. Results are shown in Figure 1.6 for a nominal Global Positioning Satellite (GPS satellite).

1.7 Tides

Earth tidal effects on satellites are due to pole tides, ocean tides, and solid earth, tides. Most of the data that have resulted in a definitive model have come about within the last several years from satellites such as TOPEX and GRACE. The basis of the models for pole, solid earth, and ocean tide models can be found in IERS Conventions, with the latest update in McCarthy and Petit [12]. Tidal models do not enjoy the variety of the gravitational and atmospheric models yet, but there are several different approaches. These various models can be a factor if precise comparisons are desired. At this point in time, several models exist, and no clear "leader" has been recognized as the standard approach.

Pole tides define the rotational deformation of the pole due to an elastic earth. These are modeled by the C_{21} and S_{21} coefficients in the earth's potential.

Solid earth tidal contributions are computed as *corrections* to the spherical harmonics coefficients.

There are a wide variety of ocean tide models in existence that have been used since 1980 starting with the Swiderski hydrodynamic model. One current models is from the University of Texas, Center for Space Research (CSR) and are referred to by CSR4.0 which model the long wave-length characteristics. The early model was 1 degree by 1

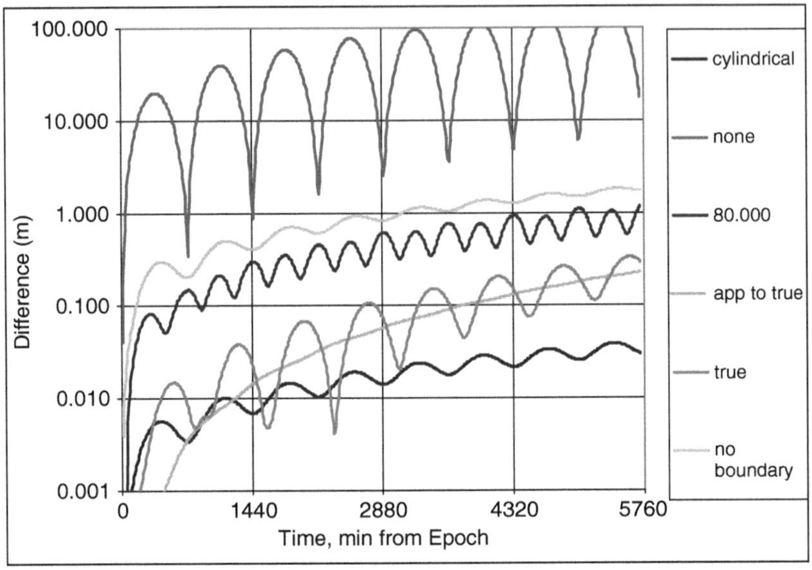

Fig. 1.6. Sample solar radiation pressure sensitivity. Positional differences are shown for a GPS satellite which is in eclipse. The baseline is a dual-cone (umbra/penumbra) shadow model. Using no shadow model (none) produces the largest differences. A simple cylindrical model introduces modest differences. Shadow boundary mitigation (no boundary) and the effective Earth size (23) contribute noticeable differences. The treatment of light travel time between the Sun and central body (app. to true) and instantaneous travel (true) produce smaller, but still detectable results.

degree model and the later model are 0.5×0.5 degree in extent. The IERS conventions describe another model. Most of these models have been developed since about 1995. The models use highly precise measurements from satellites such as JASON and TOPEX. Additional satellites such as GRACE and GOCE will contribute to a better future models.

1.8 Albedo

Albedo is the radiation pressure emitted from the Earth which causes a small perturbing force on a satellite. Although the effect of SRP is far larger, the effects of Earth's albedo can be comparable for certain configurations of orbits (e.g. sun-synchronous). The acceleration due to albedo is generally expressed in terms of a second degree zonal spherical harmonic model, and contributions from various Earth sectors are summed to determine the overall effect.

1.9 Other

As the accuracy of orbit determination and propagation increases, additional force models are included in analyses. In particular, applications using GPS data often must account for the [primarily] apsidal rotation caused by General Relativity. GPS signals msut also be corrected for General Relativity, as well as atomic clock corrections. The effects of General Relativity are very small and only become important where high orbit precision at the cm level is needed. Satellites, such as the Gravity Probe B, will try to measure and quantify this effect in its verification and validation of Einstein's theory of General Relativity. Satellite thrusting can also be a significant perturbing force. Many satellites use maneuvers for mission operation and for orbit maintenance. The forces induced by these motor firings can be large and small. We do not describe these in any detail, but introduce the forces as something needed to be considered in mission planning and precision orbit determination modeling.

1.10 Propagating the orbit

There are several techniques to propagate an orbit. Generally, analytical, numerical, and semianalytical techniques encompass the potential choices. However, a primary technique to find analytical solutions, the variation of parameters, may be used in either analytical or numerical applications. The fundamental distinction is the use of position and velocity state vectors, or orbital elements as the elements of the state. Recalling the previous force model discussion, one can find benefits with each approach.

Direct integration where possible. Analytical methods are accurate and yield a quick solution; however, the series truncations may be difficult depending on the equations of motion. The numerical method is very accurate with the correct step size, but this determination may be tricky, and long propagations can still be time-consuming. The semianalytical technique combines the analytical and numerical approaches.

1.11 Analytical

General perturbation techniques replace the original equations of motion with an analytical approximation that captures the essential character of the motion over some limited time interval and which also permits analytical integration. Such methods rely on series expansions of the perturbing accelerations and are usually derived from variation of parameter equations which will be addressed shortly. In practice, we truncate the resulting expressions to allow simpler expressions in the theory. This trade-off speeds up computation but decreases accuracy. Unlike numerical techniques, analytical methods produce approximate, or "general" results that hold for some limited time interval and accept any initial input conditions. The quality of the solution degrades over time, but remember that the numerical solution also degrades—at different rates and for different reasons. Analytical techniques are generally more difficult to develop than numerical techniques, but they often lead to a better understanding of the perturbation source.

1.12 Numerical

Special perturbation techniques numerically integrate the equations of motion including all necessary perturbing accelerations. Because numerical integration is involved, we can think of numerical formulations as producing a *specific*, or *special*, answer that is valid only for the given data (initial conditions and force-model parameters). To numerically integrate Cowell's formulation, we must have mathematical models for each perturbing force. The general form is usually taken as the following Albedo.

$$\vec{a} = \frac{\mu r}{r^3}\vec{r} + \vec{a}_{non\text{-}spherical} + \vec{a}_{drag} + \vec{a}_{3\text{-}body} + \vec{a}_{srp} + \vec{a}_{tides} + \vec{a}_{other} \tag{1.8}$$

Numerical integration may also be applied to the variation of parameter (VOP) equations, in which case a set of orbital elements is numerically integrated. To form an ephemeris, one then needs to convert the osculating orbital elements into the appropriate state vectors.

Although numerical methods can give very accurate results and often establish the "truth" in analyses, they suffer from their specificity, which keeps us from using them in a different problem. Thus, new data means new integration, which can add lengthy computing times. NASA began the first complex numerical integrations during the late 1960s and early 1970s. Personal computers now compute sufficiently fast enough to perform complex perturbation analyses using numerical techniques. However, numerical methods suffer from errors that build up with truncation and round-off due to fixed computer word-length. These errors can cause numerical solutions to degrade as the propagation interval lengthens.

1.13 Semianalytical

Semianalytical techniques combine the best features of numerical and analytical methods to get the best mix of accuracy and efficiency. The result can be a very accurate, relatively fast algorithm which applies to most situations. But semianalytical techniques vary widely. We choose a semianalytical technique mainly for its ability to handle varying orbital applications, its documentation, and the fidelity and the number of force models it includes. Most semianalytical techniques have improved accuracy and computational efficiency, but the availability of documentation (including very structured computer code) and flexibility are often important discriminators. We consider a technique semianalytical if it is not *entirely* analytical or numerical.

1.14 Variation of parameters

Most analytical, and some numerical solutions rely on the *variation of parameters* (VOP) form of the equations of motion originally developed by Euler and improved by Lagrange [8]. The overall process is called *variation of parameters* (VOP) because the orbital elements (the constant parameters in the two-body equations) are changing. Lagrange and Gauss both developed VOP methods to analyze perturbations—Lagrange's technique applies to conservative accelerations, whereas Gauss's technique also works for non-conservative accelerations. Depending on the orbital elements chosen, the form will differ. I will show a form for the classical orbital elements.

Using the VOP technique, we can analyze the effects of perturbations on specific orbital elements. This is very useful in mission planning and analysis. We want any theory to model as many perturbing forces as possible. Most operational analytical theories are limited to central body and drag. Analytical expressions for third-body and solar-radiation forces are far less common, mainly because their effects are much smaller for many orbits. Also, whenever accuracy requires us to use effects of third bodies and solar-radiation pressure, numerical integration is usually just as easy for all the perturbing forces.

1.15 Lagrangian VOP—conservative forces

The VOP method is a formulation of the equations of motion that are well-suited to perturbed, dynamical systems. The concept is based on the premise that we can use the solution for the unperturbed system to represent the solution of the perturbed system, provided that we can generalize the constants in the solution to be time-varying parameters. The unperturbed system is the two-body system, and it represents a collection of formulas that provide the position and velocity vectors at a desired time. Remember, these formulas depend only on the six orbital elements and time. In principle, however, we could use any set of constants of the unperturbed motion, including the initial position and velocity vectors. Time is related to the equations of motion through the conversions of mean, eccentric, and true anomaly.

The general theory for finding the rates of change of the osculating elements is known as the *Lagrange planetary equations of motion*, or simply the *Lagrangian VOP*, and is attributed to Lagrange because he was the first person to obtain these equations for all six orbital elements. He was concerned with the small disturbances on planetary motion about the Sun due to the gravitational attraction of the planets. He chose to model the disturbing acceleration due to this conservative perturbation as the gradient of a potential function. From Vallado [15],

$$\frac{\mathrm{d}a}{\mathrm{d}t} = \frac{2}{na}\frac{\partial R}{\partial M_o}$$

$$\frac{\mathrm{d}e}{\mathrm{d}t} = \frac{1-e^2}{na^2e}\frac{\partial R}{\partial M_o} - \frac{\sqrt{1-e^2}}{na^2e}\frac{\partial R}{\partial \omega}$$

$$\frac{\mathrm{d}i}{\mathrm{d}t} = \frac{1}{na^2\sqrt{1-e^2}\sin(i)}\left\{\cos(i)\frac{\partial R}{\partial \omega} - \frac{\partial R}{\partial \Omega}\right\}$$

$$\frac{\mathrm{d}\omega}{\mathrm{d}t} = \frac{\sqrt{1-e^2}}{na^2e}\frac{\partial R}{\partial e} - \frac{\cot(i)}{na^2\sqrt{1-e^2}}\frac{\partial R}{\partial i}$$

$$\frac{\mathrm{d}\Omega}{\mathrm{d}t} = \frac{1}{na^2\sqrt{1-e^2}\sin(i)}\frac{\partial R}{\partial i}$$

$$\frac{\mathrm{d}M_o}{\mathrm{d}t} = \frac{1-e^2}{na^2e}\frac{\partial R}{\partial e} - \frac{2}{na}\frac{\partial R}{\partial a}$$

1.16 Gaussian VOP—nonconservative forces

For many applications, it is convenient to express the rates of change of the elements explicitly in terms of the disturbing forces—actually acceleration (*specific* forces) to match the units in the equations. Gauss's form of VOP is advantageous for non-conservative forces because it is expressed directly from the disturbing acceleration. But it works equally well for conservative forces because the forces are simply gradients of the potential functions. It is also easy to visualize this representation because we're familiar with the concept of a force. Gauss's form of the VOP requires the partial derivatives of the elements with respect to the velocity. We must determine these for particular element sets. Gauss chose to develop the equations in the RSW system.[1] Let the components of the disturbing force (per unit mass) be along the radius vector, perpendicular to the *R*axis in the orbit plane in the *direction* of satellite motion, and normal to the orbit plane. The disturbing (specific) force become

$$
\begin{aligned}
\frac{\mathrm{d}a}{\mathrm{d}t} &= \frac{2}{n\sqrt{1-e^2}}\left(e\sin(v)F_R + \frac{p}{r}rF_S\right)\\
\frac{\mathrm{d}e}{\mathrm{d}t} &= \frac{\sqrt{1-e^2}}{na}\left(\sin(v)F_R + \left(\cos(v) + \frac{e+\cos(v)}{1+e\cos(v)}r\right)F_S\right)\\
\frac{\mathrm{d}i}{\mathrm{d}t} &= \frac{r\cos(u)F_W}{na^2\sqrt{1-e^2}}\\
\frac{\mathrm{d}\Omega}{\mathrm{d}t} &= \frac{r\sin(u)F_W}{na^2\sqrt{1-e^2}\sin(i)}\\
\frac{\mathrm{d}\omega}{\mathrm{d}t} &= \frac{\sqrt{1-e^2}}{na^2\sin(i)}\left\{-\cos(v)F_R + \sin(v)\left(1+\frac{r}{p}\right)F_S\right\} - \frac{r\cot(i)\sin(u)F_W}{h}\\
\frac{\mathrm{d}M_o}{\mathrm{d}t} &= \frac{1}{na^2e}\{(p\cos(v) - -2er)F_R - (p+r)\sin(v)F_S\}
\end{aligned}
$$

These VOP equations in classical orbital elements have some limitations. First, they are limited to eccentricities less than 1.0 because of the presence of the eccentricity in the denominator and in square roots. Also note that they suffer from the same singularities as Lagrange's form of the VOP equations because the singularities are due to the particular element set, not how the disturbing forces are characterized. The rate of change of Ω has $\sin(i)$ in the denominator. This causes the equation to be indeterminate for small inclinations. A similar problem exists for ω with small values of eccentricity. Thus, this particular set of equations is not recommended for small values of eccentricity or inclination.

---

[1] In the RSW system, the *R* axis is parallel to the position vector. Along-track displacements are normal to the position vector (along the *S* axis), but not necessarily aligned with the velocity vector. The *W* axis points in the instantaneous direction of the angular momentum vector.

1.17 Effect on orbits

Given the numerous techniques to analyze and account for the effects of perturbations, there are several different trends that can be noted for different satellite orbits. However, be aware that the results are specific to individual orbits. Also note that the increasing popularity of numerical methods has positive and negative effects. These techniques include all effects from the perturbations, but they do not indicate the source of dominant errors, and can thus be problematic for the satellite mission designer. We examine two major areas here—a J_2 secular perturbation, and the effects of combined forces resulting from numerical simulations.

1.18 J_2 Only

Due to its simplicity, studies are often conducted with only the secular effects of J_2 included. While this is true for many systems, the modern computer renders many of these analyses obsolete for precise studies. Still, the effects are illustrative of the effect on satellite orbits. Consider the nodal regression (Figure 1.7) and the apsidal rotation (Figure 1.8). These two effects are common for satellite mission planners, and they result from the secular effect of J_2.

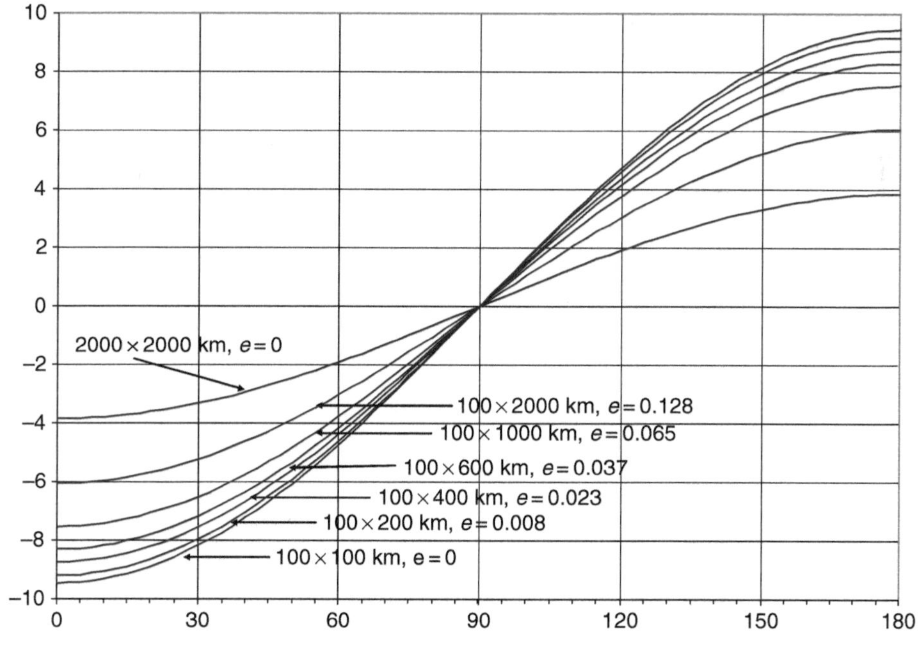

Fig. 1.7. Daily nodal regression. For the eccentric orbits, I used a perigee altitude of 100 km with the apogee values as indicated. The 100 × 2000 km orbit shows that the perturbing effect for each of the *eccentric* orbits would be smaller if the orbit were circular at that apogee altitude (2000 × 2000 km) [15].

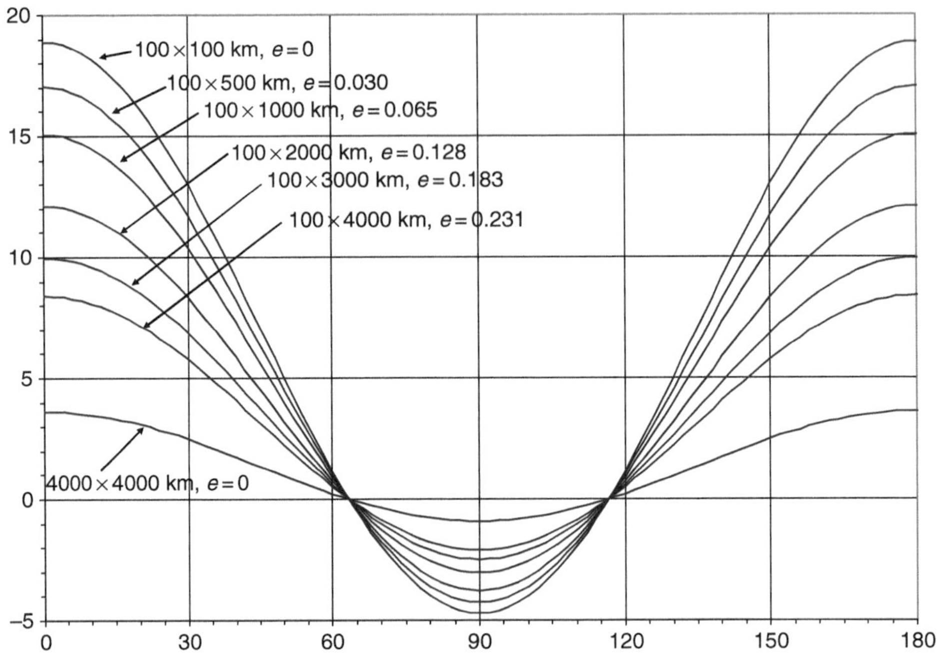

Fig. 1.8. Daily Apsidal Regression. As with nodal regression, circular orbits at an altitude (say, 4000 × 4000) would have a smaller daily change than an eccentric orbit with apogee at the same altitude (100 × 4000 km) [15].

1.19 Comparative force model effects

The effects of all physical forces affecting the motion of the earth satellite or spacecraft can be complex, and is often best handled by numerical integration. By far the largest effect is due to gravitation, usually followed by atmospheric drag, solar radiation pressure (SRP) effects and several other effects such as tides, third body perturbations and others. Vallado (2005) shows the relative effect of various forces on several satellites. Figure 1.9 reproduces these quantitative effects of all physical forces in terms of positional differences for two satellites—one is in an 800 by 110 km altitude, 50° inclined orbit, while the other is a geosynchronous satellite. Note that most of the effects like tides and third body forces and relativity are very small, but need to be taken into account when precision is of importance. The satellite parameters were chosen to illustrate force model effects. Each spacecraft parameter was held constant (coefficient of drag $c_D = 2.2$, coefficient of solar radiation pressure $c_R = 1.2$, area to mass ratio $A/m = 0.04\,\mathrm{m}^2/\mathrm{kg}$). The simulation time, January 4, 2003, was chosen as the epoch to propagate as this was a moderate period of solar activity (solar flux $F_{10.7} \sim 140$). The baseline for comparison in all cases was a 12 × 12 EGM-96 gravity field (degree and order, 12 zonal harmonics plus 12 sectoral terms). Except for the two-body (0 × 0) and 70 × 70 cases, all the other force model comparisons included a 12 × 12

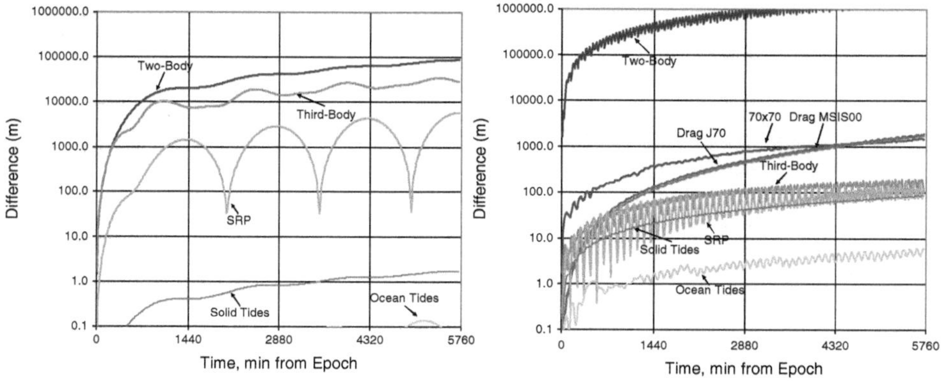

Fig. 1.9. Force model comparisons: This figure shows the positional difference over time (four days) when using various force models on the same initial state for a geosynchronous satellite. Each comparison is made with respect to a two-body ephemeris except for the gravity runs which compare to the nearest gravity case. Thus, "12 × 12" is a comparison of a 12 × 12 WGS84/EGM96 gravity field to a WGS84/EGM96 2 × 10 gravity field ephemeris, etc.

EGM-96 gravity field. Thus, for example, the "vs Drag JRob" case indicates a comparison of 12 × 12 EGM-96 gravity and a 12 × 12 EGM-96 gravity with Jacchia–Roberts drag.

In general, gravity was the largest single perturbation source, so additional tests were conducted to determine the sensitivity of this perturbation force. Atmospheric drag was generally second for lower orbits, but third-body effects were much higher for higher altitude satellites. Because the study results indicated the conservative forces could be matched to cm-level, no additional studies were performed on third-body forces. Drag was considered separately. It is important to note that these are prediction differences are based on the propagation of identical state vectors with differing acceleration models. A study of orbit determination accuracy using differing acceleration models would produce a very different set of results.

1.20 Conclusions

There are numerous forces that affect a satellite in orbit. Proper treatment of these forces is essential to proper mission planning and satellite operations. The selection of the type of propagation scheme and consistency with the chosen technique are important.

References

1. Barker, William N., S. J. Casali, and R. N. Wallner. (1995). The Accuracy of General Perturbations and Semianalytic Satellite Ephemeris Theories. Paper AAS-95-432 presented at the AAS/ AIAA Astrodynamics Specialist Conference. Halifax, Nova Scotia, Canada.
2. Battin, Richard H. (1987). *An Introduction to the Mathematics and Methods of Astrodynamics.* AIAA Education Series, New York.

3. Brouwer, Dirk, and G. M. Clemence. (1961). *Methods of Celestial Mechanics.* New York: Academic Press, Inc.
4. Gaposchkin, E. M. (1994). Calculation of Satellite Drag Coefficients. Technical Report 998. MIT Lincoln Laboratory, MA.
5. Kaplan, Marshall H. (1976). *Modern Spacecraft Dynamics and Control.* New York: John Wiley & Sons.
6. Kaula, William M. (1966). *Theory of Satellite Geodesy.* Waltham MA: Blaisdell Publishing Co.
7. Kreyzig, Erwin. (1983). *Advanced Engineering Mathematics.* 5th edn. New York: John Wiley Publishing.
8. Lagrange, J. L. (1873). *Collected Works.* Vol. 6. Paris: Gauthier-Villars.
9. Lambeck, Kurt. (1988). *Geophysical Geodesy.* Oxford: Clarendon Press.
10. Long, Anne C. et al. (1989). *Goddard Trajectory Determination System (GTDS) Mathematical Theory (Revision 1).* FDD/552-89/001 and CSC/TR-89/6001. Goddard Space Flight Center: National Aeronautics and Space Administration.
11. Marcos, Frank A. et al. (1993). Satellite Drag Models: Current Status and Prospects. Paper AAS-93-621 presented at the AAS/AIAA Astrodynamics Specialist Conference. Victoria BC, Canada.
12. McCarthy, Dennis, and Gerard Petit. (2003). *IERS Technical Note #32.* U.S. Naval Observatory.
13. Moritz, Helmut, and I. Mueller. (1987). *Earth Rotation—Theory and Observation.* New York: Ungar Publishing Company.
14. Roy, Archie E. (1988). *Orbital Motion.* New York: John Wiley & Sons.
15. Vallado, David A. (2004). *Fundamentals of Astrodynamics and Applications.* 2nd edn, second printing. Microcosm, El Segundo, CA.
16. Vallado, David A. (2005). An Analysis of State Vector Propagation using Differing Flight Dynamics Programs. Paper AAS 05-199 presented at the AAS/AIAA Space Flight Mechanics Conference. Copper Mountain, CO.
17. Vallado, David A., and T. S. Kelso. (2005). Using EOP and Solar Weather Data for Real-time Operations. Paper USR 05-S7.3 presented at the US/Russian Space Surveillance Workshop, August 22–26, 2005. St Petersburg, Russia.
18. Vetter, Jerome R. (1994). The Evolution of Earth Gravity Models used in Astrodynamics. *APL Technical Digest*, John Hopkins. **15**(4), pp. 319–335.
19. Vetter, Jerome R. (2001). Private communication.

2 Gauge Freedom in Astrodynamics

MICHAEL EFROIMSKY
US Naval Observatory

Contents

2.1 Introduction

2.1.1 What this chapter is about

Both orbital and attitude dynamics employ the method of variation of parameters. In a non-perturbed setting, the coordinates (or the Euler angles) get expressed as functions of the time and six adjustable constants called elements. Under disturbance, each such expression becomes ansatz, the "constants" being endowed with time dependence. The perturbed velocity (linear or angular) consists of a partial time derivative and a convective term containing time derivatives of the "constants." It can be shown that this construction leaves one with a freedom to impose three arbitrary conditions upon the "constants" and/or their derivatives. Out of convenience, the Lagrange constraint is often imposed. It nullifies the convective term and thereby guarantees that under perturbation the functional dependence of the velocity upon the time and "constants" stays the same as in the undisturbed case. "Constants" obeying this condition are called osculating elements.

The "constants" chosen to be canonical are called Delaunay elements, in the orbital case, or Andoyer elements, in the spin case. (As some of the Andoyer elements are time-dependent even in the free-spin case, the role of "constants" is played by these elements' initial values.) The Andoyer and Delaunay sets of elements share a feature not readily apparent: in certain cases the standard equations render these elements non-osculating.

In orbital mechanics, elements calculated via the standard planetary equations come out non-osculating when perturbations depend on velocities. To keep elements osculating under such perturbations, the equations must be amended with extra terms that are not parts of the disturbing function [1, 2]. For the Kepler elements, this merely complicates the equations. In the case of Delaunay parameterisation, these extra terms not only complicate the equations, but also destroy their canonicity. So under velocity-dependent disturbances, osculation and canonicity are incompatible.

Similarly, in spin dynamics the Andoyer elements come out non-osculating under angular-velocity-dependent perturbation (a switch to a non-inertial frame being one such case). Amendment of the dynamical equations only with extra terms in the Hamiltonian makes the equations render non-osculating Andoyer elements. To make them osculating, more terms must enter the equations (and the equations will no longer be canonical).

It is often convenient to deliberately deviate from osculation by substituting the Lagrange constraint with an arbitrary condition that gives birth to a family of non-osculating elements. The freedom in choosing this condition is analogous to the gauge freedom. Calculations in non-osculating variables are mathematically valid and sometimes highly advantageous, but their physical interpretation is non-trivial. For example, non-osculating orbital elements parameterise instantaneous conics not tangent to the orbit, so the non-osculating inclination will be different from the real inclination of the physical orbit.

We present examples of situations in which ignoring of the gauge freedom (and of the unwanted loss of osculation) leads to oversights.

2.1.2 Historical prelude

The orbital dynamics is based on the variation-of-parameters method, invention whereof is attributed to Euler [3, 4] and Lagrange [5–9]. Though both greatly contributed to this approach, its initial sketch was offered circa 1687 by Newton in his unpublished *Portsmouth Papers*. Very succinctly, Newton brought up this issue also in Cor. 3 and 4 of Prop. 17 in the first book of his *Principia*.

Geometrically, the part and parcel of this method is representation of an orbit as a set of points, each of which is contributed by a member of some chosen family of curves $C(\kappa)$, where κ stands for a set of constants that number a particular curve within the family. (For example, a set of three constants $\kappa = \{a, b, c\}$ defines one particular hyperbola $y = ax^2 + bx + c$ out of many). This situation is depicted on Fig. 2.1. Point A of the orbit coincides with some point λ_1 on a curve $C(\kappa_1)$. Point B of the orbit coincides with point λ_2 on some other curve $C(\kappa_2)$ of the same family, etc. This way, orbital motion from A to B becomes a superposition of motion along C_κ from λ_1 to λ_2 and a gradual distortion of the curve C_κ from the shape $C(\kappa_1)$ to the shape $C(\kappa_2)$. In a loose language, the motion along the orbit consists of steps along an instantaneous curve $C(\kappa)$ which itself is evolving while those steps are being made.

Normally, the family of curves C_κ is chosen to be that of ellipses or that of hyperbolae, κ being six orbital elements, and λ being the time. However, if we disembody this idea of its customary implementation, we shall see that it is of a far more general nature and contains three aspects:

1. A trajectory may be assembled of points contributed by a family of curves of an essentially arbitrary type, not just conics.
2. It is not necessary to choose the family of curves tangent to the orbit. As we shall see below, it is often beneficial to choose those non-tangent. We shall also see examples

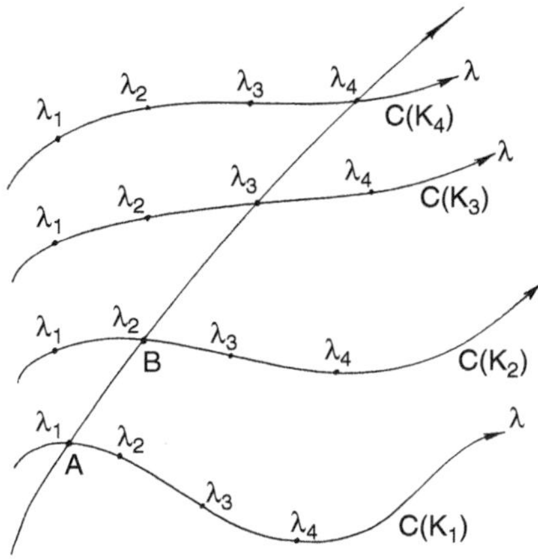

Fig. 2.1. Each point of the orbit is contributed by a member of some family of curves $C(\kappa)$ of a certain type, κ standing for a set of constants that number a particular curve within the family. Motion from A to B is, first, due to the motion along the curve $C(\kappa)$ from λ_1 to λ_2 and, second, due to the fact that during this motion the curve itself was evolving from $C(\kappa_1)$ to $C(\kappa_2)$.

when in orbital calculations this loss of tangentiality (loss of osculation) takes place and goes unnoticed.

3. The approach is general and can be applied, for example, to Euler's angles. A disturbed rotation can be thought of as a series of steps (small turns) along different Eulerian cones. An Eulerian cone is an orbit (on the Euler angles' manifold) corresponding to an unperturbed spin state. Just as a transition from one instantaneous Keplerian conic to another is caused by disturbing forces, so a transition from one instantaneous Eulerian cone to another is dictated by external torques or other perturbations. Thus, in the attitude mechanics, the Eulerian cones play the same role as the Keplerian conics do in the orbital dynamics. Most importantly, a perturbed rotation may be "assembled" of the Eulerian cones in an osculating or in a non-osculating manner. An unwanted loss of osculation in attitude mechanics happens in the same way as in the theory of orbits, but is much harder to notice. On the other hand, a deliberate choice of non-osculating rotational elements in attitude mechanics may sometimes be beneficial.

From the viewpoint of calculus, the concept of variation of parameters looks as follows. We have a system of differential equations to solve ("system in question") and a system of differential equations ("fiducial system") whose solution is known and contains arbitrary constants. We then use the known solution to the fiducial system as an ansatz for solving the system in question. The constants entering this ansatz are endowed with time dependence of their own, and the subsequent substitution of this known solution into the system in question yields equations for the "constants." The number of "constants" often exceeds that of equations in the system to solve. In this case we impose, by hand, arbitrary constraints upon the "constants." For example, in the case of a reduced N-body problem, we begin with $3(N-1)$ unconstrained second-order equations for $3(N-1)$ Cartesian coordinates. After a change of variables from the Cartesian coordinates to the orbital parameters, we end up with $3(N-1)$ differential equations for the $6(N-1)$ orbital variables. Evidently, $3(N-1)$ constraints are necessary.[1] To this end, the so-called Lagrange constraint (the condition of the instantaneous conics being tangent to the physical orbit) is introduced almost by default, because it is regarded natural. Two things should be mentioned in this regard:

First, what seems natural is not always optimal. The freedom of choice of the supplementary condition (the gauge freedom) gives birth to an internal symmetry (the gauge symmetry) of the problem. Most importantly, it can be exploited for simplifying the equations of motion for the "constants." On this issue we shall dwell in the current paper.

Second, the entire scheme may, in principle, be reversed and used to solve systems of differential equations with constraints. Suppose we have $N+M$ variables $C_j(t)$ obeying a system of N differential equations of the second order and M constraints expressed with first-order differential equations or with algebraic expressions. One possible approach to solving this system will be to assume that the variables C_j come about as constants emerging in a solution to some fiducial system of differential equations. Then our N second-order differential equations for $C_j(t)$ will be interpreted as a result of substitution of such an ansatz into the fiducial system with some perturbation, while our M constraints will be interpreted as weeding out of the redundant degrees of freedom. This subject is out of the scope of our paper and will not be developed here.

2.1.3 The simplest example of gauge freedom

Variation of constants first emerged in the non-linear context of celestial mechanics and later became a universal tool. We begin with a simple example offered in Newman and Efroimsky [10].

A harmonic oscillator disturbed by a force $\Delta F(t)$ gives birth to the initial-condition problem

$$\ddot{x}+x=\Delta F(t), \quad \text{with } x(0) \text{ and } \dot{x}(0) \text{ known,} \tag{2.1}$$

[1] In a fixed Cartesian frame, any solution to the unperturbed reduced 2-body problem can be written as

$$x_j = f_j(t, C_1, \ldots, C_6), \quad j=1,2,3,$$

$$\dot{x}_j = g_j(t, C_1, \ldots, C_6), \quad g_j \equiv \left(\frac{\partial f_j}{\partial t}\right)_C$$

the adjustable constants C standing for orbital elements. Under disturbance, the solution is sought as

$$x_j = f_j(t, C_1(t), \ldots, C_6(t)), \quad j=1,2,3,$$

$$\dot{x}_j = g_j(t, C_1(t), \ldots, C_6(t)) + \Phi_j(t, C_1(t), \ldots, C_6(t)), \quad g_j \equiv \left(\frac{\partial f_j}{\partial t}\right)_C, \quad \Phi_j \equiv \sum_r \frac{\partial f_j}{\partial C_r}\dot{C}_r.$$

Insertion of $x_j = f_j(t, C)$ into the perturbed gravity law yields three scalar equations for six functions $C_r(t)$. This necessitates imposition of three conditions upon C_r and $\dot{C}_r$. Under the simplest choice $\Phi_j = 0$, $j=1,2,3$, the perturbed physical velocity $\dot{x}_j(t, C)$ has the same functional form as the unperturbed $g_j(t, C)$. Therefore, the instantaneous conics become tangent to the orbit (and the orbital elements $C_r(t)$ are called osculating).

whose solution may be sought using ansatz

$$x = C_1(t)\sin t + C_2(t)\cos t. \tag{2.2}$$

This will lead us to

$$\dot{x} = \left[\dot{C}_1(t)\sin t + \dot{C}_2(t)\cos t\right] + C_1(t)\cos t - C_2(t)\sin t. \tag{2.3}$$

It is common, at this point, to put the sum $\left[\dot{C}_1(t)\sin t + \dot{C}_2(t)\cos t\right]$ equal to zero, in order to remove the ambiguity stemming from the fact that we have only one equation for two variables. Imposition of this constraint is convenient but not obligatory. A more general way of fixing the ambiguity may be expressed as

$$\dot{C}_1(t)\sin t + \dot{C}_2(t)\cos t = \phi(t), \tag{2.4}$$

$\phi(t)$ being an arbitrary function of time. This entails:

$$\ddot{x} = \dot{\phi} + \dot{C}_1(t)\cos t - \dot{C}_2(t)\sin t - C_1(t)\sin t - C_2(t)\cos t, \tag{2.5}$$

summation whereof with Eq. (2.2) gives:

$$\ddot{x} + x = \dot{\phi} + \dot{C}_1(t)\cos t - \dot{C}_2(t)\sin t. \tag{2.6}$$

Substitution thereof into Eq. (2.1) yields the dynamical equation re-written in terms of the "constants" C_1, C_2. This equation, together with identity (2.4), will constitute the following system:

$$\begin{aligned} \dot{\phi} + \dot{C}_1(t)\cos t - \dot{C}_2(t)\sin t &= \Delta F(t), \\ \dot{C}_1(t)\sin t + \dot{C}_2(t)\cos t &= \phi(t), \end{aligned} \tag{2.7}$$

This leads to

$$\begin{aligned} \dot{C}_1 &= \Delta F \cos t - \frac{\mathrm{d}}{\mathrm{d}t}(\phi\cos t) \\ \dot{C}_2 &= -\Delta F \sin t + \frac{\mathrm{d}}{\mathrm{d}t}(\phi\sin t), \end{aligned} \tag{2.8}$$

the function $\phi(t)$ still remaining arbitrary.[2] Integration of Eq. (2.8) entails:

$$\begin{aligned} C_1 &= \int^t \Delta F \cos t'\mathrm{d}t' - \phi\cos t + a_1 \\ C_2 &= -\int^t \Delta F \sin t'\mathrm{d}t' + \phi\sin t + a_2. \end{aligned} \tag{2.9}$$

[2] Function $\phi(t)$ can afford being arbitrary, no matter what the initial conditions are to be. Indeed, for fixed $x(0)$ and $\dot{x}(0)$, the system $C_2(0) = x(0)$, $\phi(0) + C_1(0) = \dot{x}(0)$ solves for $C_1(0)$ and $C_2(0)$ for an arbitrary choice of $\phi(0)$.

Substitution of Eq. (2.9) into Eq. (2.2) leads to complete cancellation of the ϕ terms:

$$x = C_1 \sin t + C_2 \cos t = -\cos t \int^t \Delta F \sin t' \mathrm{d}t' + \sin t \int^t \Delta F \cos t' \mathrm{d}t' + a_1 \sin t + a_2 \cos t \tag{2.10}$$

Naturally, the physical trajectory $x(t)$ remains invariant under the choice of gauge function $\phi(t)$, even though the mathematical description (2.9) of this motion in terms of the parameters C is gauge dependent. It is, however, crucial that a numerical solution of the system (2.8) will come out ϕ-dependent, because the numerical error will be sensitive to the choice of $\phi(t)$. This issue is now being studied by P. Gurfil and I. Klein [11], and the results are to be published soon.

It remains to notice that (2.8) is a simple analogue to the Lagrange-type system of planetary equations, system that, too, admits gauge freedom. (See subsection 2.2.2 below.)

2.1.4 *Gauge freedom under a variation of the Lagrangian*

The above example permits an evident extension [12, 13]. Suppose some mechanical system obeys the equation

$$\ddot{\boldsymbol{r}} = F(t, \boldsymbol{r}, \dot{\boldsymbol{r}}), \tag{2.11}$$

whose solution is known and has a functional form

$$\boldsymbol{r} = \boldsymbol{f}(t, C_1, \ldots, C_6), \tag{2.12}$$

C_j being adjustable constants to vary only under disturbance.

When a perturbation $\Delta \boldsymbol{F}$ gets switched on, the system becomes:

$$\ddot{\boldsymbol{r}} = \boldsymbol{F}(t, \boldsymbol{r}, \dot{\boldsymbol{r}}) + \Delta \boldsymbol{F}(t, \boldsymbol{r}, \dot{\boldsymbol{r}}), \tag{2.13}$$

and its solution will be sought in the form of

$$\boldsymbol{r} = \boldsymbol{f}(t, C_1(t), \ldots, C_6(t)). \tag{2.14}$$

Evidently,

$$\dot{\boldsymbol{r}} = \frac{\partial \boldsymbol{f}}{\partial t} + \boldsymbol{\Phi}, \quad \boldsymbol{\Phi} \equiv \sum_{j=1}^{6} \frac{\partial \boldsymbol{f}}{\partial C_j} \dot{C}_j. \tag{2.15}$$

In defiance of what the textbooks advise, we do *not* put $\boldsymbol{\Phi}$ nil. Instead, we proceed further to

$$\ddot{\boldsymbol{r}} = \frac{\partial^2 \boldsymbol{f}}{\partial t^2} + \sum_{j=1}^{6} \frac{\partial^2 \boldsymbol{f}}{\partial t\, \partial C_j} \dot{C}_j + \dot{\boldsymbol{\Phi}}, \tag{2.16}$$

dot standing for a *full* time derivative. If we now insert the latter into the perturbed equation of motion (2.13) and if we recall that, according to (2.11),[3] $\partial^2 \boldsymbol{f}/\partial t^2 = \boldsymbol{F}$, then we shall obtain the equation of motion for the new variables $C_j(t)$:

$$\sum_{j=1}^{6} \frac{\partial^2 \boldsymbol{f}}{\partial t\, \partial C_j} \dot{C}_j + \dot{\boldsymbol{\Phi}} = \Delta \boldsymbol{F} \tag{2.17}$$

where

$$\boldsymbol{\Phi} \equiv \sum_{j=1}^{6} \frac{\partial \boldsymbol{f}}{\partial C_j} \dot{C}_j \tag{2.18}$$

so far is merely an identity. It will become a constraint after we choose a particular functional form $\boldsymbol{\Phi}(t; C_1, \ldots, C_6)$ for the gauge function $\boldsymbol{\Phi}$, i.e., if we choose that the sum $\sum \frac{\partial \boldsymbol{f}}{\partial C_j} \dot{C}_j$ be equal to some arbitrarily fixed function $\boldsymbol{\Phi}(t; C_1, \ldots, C_6)$ of the time and of the variable "constants." This arbitrariness exactly parallels the gauge invariance in electrodynamics: on the one hand, the choice of the functional form of $\boldsymbol{\Phi}(t; C_1, \ldots, C_6)$ will never[4] influence the eventual solution for the physical variable $\boldsymbol{r}$; on the other hand, though, a qualified choice may considerably simplify the process of finding the solution. To illustrate this, let us denote by $\mathbf{g}(t, C_1, \ldots, C_6)$ the functional dependence of the unperturbed velocity on the time and adjustable constants:

$$\mathbf{g}(t, C_1, \ldots, C_6) \equiv \frac{\partial}{\partial t} \boldsymbol{f}(t, C_1, \ldots, C_6), \tag{2.19}$$

and rewrite the above system as

$$\sum_j \frac{\partial \mathbf{g}}{\partial C_j} \dot{C}_j = -\dot{\boldsymbol{\Phi}} + \Delta \boldsymbol{F} \tag{2.20}$$

$$\sum_j \frac{\partial \boldsymbol{f}}{\partial C_j} \dot{C}_j = \boldsymbol{\Phi}. \tag{2.21}$$

If we now dot-multiply the first equation with $\partial \boldsymbol{f}/\partial C_i$ and the second one with $\partial \mathbf{g}/\partial C_i$, and then take the difference of the outcomes, we shall arrive at

$$\sum_j [C_n C_j] \dot{C}_j = (\Delta \boldsymbol{F} - \dot{\boldsymbol{\Phi}}) \cdot \frac{\partial \boldsymbol{f}}{\partial C_n} - \boldsymbol{\Phi} \cdot \frac{\partial \mathbf{g}}{\partial C_n}, \tag{2.22}$$

[3] We remind that in Eq. (2.11) there was no difference between a partial and a full time derivative, because at that point the integration "constants" C_i were indeed constant. Later, they acquired time dependence, and therefore the full time derivative implied in Eqs. (2.15–2.16) became different from the partial one implied in Eq. (2.11).

[4] Our usage of words "arbitrary" and "never" should be limited to the situations where the chosen gauge (2.21) does not contradict the equations of motion (2.20). This restriction, too, parallels a similar one present in field theories. Below we shall encounter a situation where this restriction becomes crucial.

the Lagrange brackets being defined in a gauge-invariant (i.e., $\boldsymbol{\Phi}$-independent) fashion.[5] If we agree that $\boldsymbol{\Phi}$ is a function of both the time and the parameters C_n, but not of their derivatives,[6] then the right-hand side of Eq. (2.22) will implicitly contain the first time derivatives of C_n. It will then be reasonable to move these to the left-hand side. Hence, Eq. (2.22) will be reshaped into

$$\sum_j \left([C_n C_j] + \frac{\partial \boldsymbol{f}}{\partial C_n} \cdot \frac{\partial \boldsymbol{\Phi}}{\partial C_j} \right) \frac{\mathrm{d}C_j}{\mathrm{d}t} = \frac{\partial \mathbf{f}}{\partial C_n} \cdot \Delta \boldsymbol{F} - \frac{\partial \mathbf{f}}{\partial C_n} \cdot \frac{\partial \boldsymbol{\Phi}}{\partial t} - \frac{\partial \mathbf{g}}{\partial C_n} \cdot \boldsymbol{\Phi}. \tag{2.23}$$

This is the general form of the gauge-invariant perturbation equations, that follows from the variation-of-parameters method applied to problem (2.13), for an arbitrary perturbation $\boldsymbol{F}(\boldsymbol{r}, \dot{\boldsymbol{r}}, t)$ and under the simplifying assumption that the arbitrary gauge function $\boldsymbol{\Phi}$ is chosen to depend on the time and the parameters C_n, but not on their derivatives.[7] Assume that our problem (2.13) is not simply mathematical but is an equation of motion for some physical setting, so that $\boldsymbol{F}$ is a physical force corresponding to some undisturbed Lagrangian $\mathcal{L}_o$, and $\Delta \boldsymbol{F}$ is a force perturbation generated by a Lagrangian variation $\Delta \mathcal{L}$. If, for example, we begin with $\mathcal{L}_o(\boldsymbol{r}, \dot{\boldsymbol{r}}, t) = \dot{\boldsymbol{r}}^2/2 - U(\boldsymbol{r}, t)$, momentum $\mathbf{p} = \dot{\boldsymbol{r}}$, and Hamiltonian $\mathcal{H}_o(\boldsymbol{r}, \mathbf{p}, t) = \mathbf{p}^2/2 + U(\boldsymbol{r}, t)$, then their disturbed counterparts will read:

$$\mathcal{L}(\boldsymbol{r}, \dot{\boldsymbol{r}}, t) = \frac{\dot{\boldsymbol{r}}^2}{2} - U(\boldsymbol{r}) + \Delta \mathcal{L}(\boldsymbol{r}, \dot{\boldsymbol{r}}, t), \tag{2.24}$$

$$\boldsymbol{p} = \dot{\boldsymbol{r}} + \frac{\partial \Delta \mathcal{L}}{\partial \dot{\boldsymbol{r}}}, \tag{2.25}$$

$$\mathcal{H} = \mathbf{p}\dot{\boldsymbol{r}} - \mathcal{L} = \frac{\mathbf{p}^2}{2} + U + \Delta \mathcal{H}, \tag{2.26}$$

$$\Delta \mathcal{H} \equiv -\Delta \mathcal{L} - \frac{1}{2} \left(\frac{\partial \Delta \mathcal{L}}{\partial \dot{\boldsymbol{r}}} \right)^2. \tag{2.27}$$

---

[5] The Lagrange-bracket matrix is defined in a gauge-invariant way:

$$\sum_j [C_n C_j] \equiv \frac{\partial \boldsymbol{f}}{\partial C_n} \cdot \frac{\partial \mathbf{g}}{\partial C_j} - \frac{\partial \vec{\mathbf{g}}}{\partial C_n} \cdot \frac{\partial \boldsymbol{f}}{\partial C_j}.$$

and so is its inverse, the matrix composed of the Poisson brackets

$$\{C_n C_j\} \equiv \frac{\partial C_n}{\partial \boldsymbol{f}} \cdot \frac{\partial C_j}{\partial \mathbf{g}} - \frac{\partial C_n}{\partial \mathbf{g}} \cdot \frac{\partial C_j}{\partial \boldsymbol{f}}.$$

Evidently, Eq. (2.22) yields

$$\dot{C}_n = \sum_j \{C_n C_j\} \left[\frac{\partial \boldsymbol{f}}{\partial C_j} \cdot (\Delta \boldsymbol{F} - \dot{\boldsymbol{\Phi}}) - \boldsymbol{\Phi} \cdot \frac{\partial \mathbf{g}}{\partial C_j} \right].$$

[6] The necessity to fix a functional form of $\boldsymbol{\Phi}(t; C_1, \ldots, C_6)$, i.e., to impose three arbitrary conditions upon the "constants" C_j, evidently follows from the fact that, on the one hand, in the ansatz (2.14) we have six variables $C_n(t)$ and, on the other hand, the number of scalar equations of motion (i.e., Cartesian projections of the perturbed vector equation (2.13)) is only three. This necessity will become even more mathematically transparent after we cast the perturbed equation (2.13) into the normal form of Cauchy. (see Appendix1)

[7] We may also impart the gauge function with dependence upon the parameters' time derivatives of all orders. This will yield higher-than-first-order derivatives in Eq. (2.23). In order to close this system, one will then have to impose additional initial conditions, beyond those on $\boldsymbol{r}$ and $\dot{\boldsymbol{r}}$.

The Euler–Lagrange equation written for the perturbed Lagrangian (2.24) is:

$$\ddot{\mathbf{r}} = -\frac{\partial U}{\partial \boldsymbol{r}} + \Delta \boldsymbol{F}, \tag{2.28}$$

where the disturbing force is given by

$$\Delta \boldsymbol{F} \equiv \frac{\partial \Delta \mathcal{L}}{\partial \boldsymbol{r}} - \frac{\mathrm{d}}{\mathrm{d}t}\left(\frac{\partial \Delta \mathcal{L}}{\partial \dot{\boldsymbol{r}}}\right). \tag{2.29}$$

Its substitution in Eq. (2.23) yields the generic form of the equations in terms of the Lagrangian disturbance [2]:

$$\sum_j \left([C_n C_j] + \frac{\partial f}{\partial C_n} \cdot \frac{\partial}{\partial C_j}\left(\frac{\partial \Delta \mathcal{L}}{\partial \dot{\boldsymbol{r}}} + \boldsymbol{\Phi}\right)\right) \frac{\mathrm{d}C_j}{\mathrm{d}t} = \frac{\partial}{\partial C_n}\left[\Delta \mathcal{L} + \frac{1}{2}\left(\frac{\partial \Delta \mathcal{L}}{\partial \dot{\boldsymbol{r}}}\right)^2\right]$$
$$- \left(\frac{\partial \mathbf{g}}{\partial C_n} + \frac{\partial f}{\partial C_n}\frac{\partial}{\partial t} + \frac{\partial \Delta \mathcal{L}}{\partial \dot{\boldsymbol{r}}}\frac{\partial}{\partial C_n}\right) \cdot \left(\boldsymbol{\Phi} + \frac{\partial \Delta \mathcal{L}}{\partial \dot{\boldsymbol{r}}}\right). \tag{2.30}$$

This equation not only reveals the convenience of the special gauge

$$\boldsymbol{\Phi} = -\frac{\partial \Delta \mathcal{L}}{\partial \dot{\boldsymbol{r}}}, \tag{2.31}$$

(which reduces to $\boldsymbol{\Phi} = 0$ in the case of velocity-independent perturbations), but also explicitly demonstrates how the Hamiltonian variation comes into play: it is easy to notice that, according to Eq. (2.27), the sum in square brackets on the right-hand side of Eq. (2.30) is equal to $-\Delta \mathcal{H}$, so the above equation takes the form $\sum_j [C_n\, C_j] \dot{C}_j = -\partial \Delta \mathcal{H} / \partial C_n$. All in all, it becomes clear that the trivial gauge, $\boldsymbol{\Phi} = 0$, leads to the maximal simplification of the variation-of-parameters equations expressed through the disturbing force: it follows from Eq. (2.22) that

$$\sum_j \left[C_n\, C_j\right] \dot{C}_j = \Delta \boldsymbol{F} \cdot \frac{\partial f}{\partial C_n}, \quad \text{provided we have chosen } \boldsymbol{\Phi} = 0. \tag{2.32}$$

However, the choice of the special gauge (2.31) entails the maximal simplification of the variation-of-parameters equations when they are formulated via a variation of the Hamiltonian:

$$\sum_j [C_n\, C_j] \frac{\mathrm{d}C_j}{\mathrm{d}t} = -\frac{\partial \Delta \mathcal{H}}{\partial C_n}, \quad \text{provided we have chosen } \boldsymbol{\Phi} = -\frac{\partial \Delta \mathcal{L}}{\partial \dot{\boldsymbol{r}}}. \tag{2.33}$$

It remains to spell out the already obvious fact that, in case the unperturbed force $\boldsymbol{F}$ is given by the Newton gravity law (i.e., when the undisturbed setting is the reduced two-body problem), then the variable "constants" C_n are merely the orbital elements parameterising a sequence of instantaneous conics out of which we "assemble" the perturbed trajectory through Eq. (2.14). When the conics' parameterisation is chosen to be via the Kepler or the Delaunay variables, then Eq. (2.30) yields the gauge-invariant version of the Lagrange-type or the Delaunay-type planetary equations, accordingly. Similarly, Eq. (2.22) implements the gauge-invariant generalisation of the planetary equations in the Euler–Gauss form.

From Eq. (2.22) we see that the Euler–Gauss-type planetary equations will always assume their simplest form (2.32) under the gauge choice $\boldsymbol{\Phi} = 0$. In astronomy this choice is called "the Lagrange constraint." It makes the orbital elements osculating, i.e., guarantees that the instantaneous conics, parameterised by these elements, are tangent to the perturbed orbit.

From Eq. (2.33) one can easily notice that the Lagrange- and Delaunay-type planetary equations simplify maximally under the condition (2.31). This condition coincides with the Lagrange constraint $\boldsymbol{\Phi} = 0$ when the perturbation depends only upon positions (not upon velocities or momenta). Otherwise, condition (2.31) deviates from that of Lagrange, and the orbital elements rendered by Eq. (2.33) are no longer osculating (so that the corresponding instantaneous conics are no longer tangent to the physical trajectory).

Of an even greater importance will be the following observation. If we have a velocity-dependent perturbing force, we can always find the appropriate Lagrangian variation and, therefrom, the corresponding variation of the Hamiltonian. If now we simply add the negative of this Hamiltonian variation to the disturbing function, then the resulting Eq. (2.33) will render not the osculating elements but orbital elements of a different type, ones satisfying the non-Lagrange constraint (2.31). Since the instantaneous conics, parameterised by such non-osculating elements, will not be tangent to the orbit, then physical interpretation of such elements may be non-trivial. Besides, they will return a velocity different from the physical one.[8] This pitfall is well-camouflaged and is easy to fall in.

These and other celestial-mechanics applications of the gauge freedom will be considered in detail in Section 2.2 below.

2.1.5 Canonicity versus osculation

One more relevant development will come from the theory of canonical perturbations. Suppose that in the absence of disturbances we start out with a system

$$\dot{q} = \frac{\partial \mathcal{H}^{(o)}}{\partial p}, \quad \dot{p} = -\frac{\partial \mathcal{H}^{(o)}}{\partial q}. \tag{2.34}$$

q and p being the Cartesian or polar coordinates and their conjugated momenta, in the orbital case, or the Euler angles and their momenta, in the rotation case. Then we switch, via a canonical transformation

$$q = f(Q, P, t), \quad p = \chi(Q, P, t) \tag{2.35}$$

to

$$\dot{Q} = \frac{\partial \mathcal{H}^{*}}{\partial P} = 0, \quad \dot{P} = -\frac{\partial \mathcal{H}^{*}}{\partial Q} = 0, \quad \mathcal{H}^{*} = 0, \tag{2.36}$$

[8] We mean that substitution of the values of these elements in $\mathbf{g}(t; C_1(t), \ldots, C_6(t))$ will not give the right velocity. The correct physical velocity will be given by $\dot{\boldsymbol{r}} = \mathbf{g} + \boldsymbol{\Phi}$.

where Q and P denote the set of Delaunay elements, in the orbital case, or the initial values of the Andoyer variables, in the case of rigid-body rotation.

This scheme relies on the fact that, for an unperturbed motion (i.e., for an unperturbed Keplerian conic, in an orbital case; or for an undisturbed Eulerian cone, in the spin case) a six-constant parameterisation may be chosen so that:

1. the parameters are constants and, at the same time, are canonical variables $\{Q, P\}$ with a zero Hamiltonian $\mathcal{H}^*(Q, P) = 0$;
2. for constant Q and P, the transformation equations (2.35) are mathematically equivalent to the dynamical equations (2.34).

Under perturbation, the "constants" Q, P begin to evolve so that, after their substitution into

$$q = f\,(Q(t), P(t), t)\,, \quad p = \chi(Q(t), P(t), t), \tag{2.37}$$

(f, χ being the same functions as in (2.35)), the resulting motion obeys the disturbed equations

$$\dot{q} = \frac{\partial(\mathcal{H}^{(o)} + \Delta\mathcal{H})}{\partial p}, \quad \dot{p} = -\frac{\partial(\mathcal{H}^{(o)} + \Delta\mathcal{H})}{\partial q}. \tag{2.38}$$

We also want our "constants" Q and P to remain canonical and to obey

$$\dot{Q} = \frac{\partial\,(\mathcal{H}^* + \Delta\mathcal{H}^*)}{\partial P}, \quad \dot{P} = -\frac{\partial\,(\mathcal{H}^* + \Delta\mathcal{H}^*)}{\partial Q} \tag{2.39}$$

where

$$\mathcal{H}^* = 0 \quad \text{and} \quad \Delta\mathcal{H}^*(Q, P, t) = \Delta\mathcal{H}\,(q(Q, P, t), p(Q, P, t), t)\,. \tag{2.40}$$

Above all, we wish the perturbed "constants" $C = Q, P$ (the Delaunay elements, in the orbital case; or the initial values of the Andoyer elements, in the spin case) to osculate. This means that we want the perturbed velocity to be expressed by the same function of $C_j(t)$ and t as the unperturbed velocity. Let us check when this is possible. The perturbed velocity is

$$\dot{q} = g + \Phi, \tag{2.41}$$

where

$$g(C(t), t) \equiv \frac{\partial q(C(t), t)}{\partial t} \tag{2.42}$$

is the functional expression for the unperturbed velocity, while

$$\Phi(C(t), t) \equiv \sum_{j=1}^{6} \frac{\partial q(C(t), t)}{\partial C_j} \dot{C}_j(t) \tag{2.43}$$

is the convective term. Since we chose the "constants" C_j to make canonical pairs (Q, P) obeying Eq. (2.39–2.40), then insertion of Eq. (2.39) into Eq. (2.43) will result in

$$\Phi = \sum_{n=1}^{3} \frac{\partial q}{\partial Q_n} \dot{Q}_n(t) + \sum_{n=1}^{3} \frac{\partial q}{\partial P_n} \dot{P}_n(t) = \frac{\partial \Delta\mathcal{H}(q, p)}{\partial p}. \tag{2.44}$$

So canonicity is incompatible with osculation when $\Delta\mathcal{H}$ depends on p. Our desire to keep the perturbed equations (2.39) canonical makes the orbital elements Q, P non-osculating in a particular manner prescribed by Eq. (2.44). This breaking of gauge invariance reveals that the canonical description is marked with "gauge stiffness" (term suggested by Peter Goldreich).

We see that, under a momentum-dependent perturbation, we still can use the ansatz (2.37) for calculation of the coordinates and momenta, but can no longer use $\dot{q} = \partial q/\partial t$ for calculating the velocities. Instead, we must use $\dot{q} = \partial q/\partial t + \partial\Delta\mathcal{H}/\partial p$, and the elements C_j will no longer be osculating. In the case of orbital motion (when C_j are the non-osculating Delaunay elements), this will mean that the instantaneous ellipses or hyperbolae parameterised by these elements will not be tangent to the orbit [1]. In the case of spin, the situation will be similar, except that, instead of instantaneous Keplerian conics, one will be dealing with instantaneous Eulerian cones—a set of trajectories on the Euler-angles manifold, each of which corresponds to some non-perturbed spin state [14].

The main conclusion to be derived from this example is the following: whenever we encounter a disturbance that depends not only upon positions but also upon velocities or momenta, implementation of the afore described canonical-perturbation method necessarily yields equations that render non-osculating canonical elements. It is possible to keep the elements osculating, but only at the cost of sacrificing canonicity. For example, under velocity-dependent orbital perturbations (like inertial forces, or atmospheric drag, or relativistic correction) the equations for osculating Delaunay elements will no longer be Hamiltonian [12, 13].

Above in this subsection we discussed the disturbed velocity $\dot{q}$. How about the disturbed momentum? For sufficiently simple unperturbed Hamiltonians, it can be written down very easily. For example, for $\mathcal{H} = \mathcal{H}_o + \Delta\mathcal{H} = p^2/2m + U(q) + \Delta\mathcal{H}$ we get:

$$p = \dot{q} + \frac{\partial\Delta\mathcal{L}}{\partial\dot{q}} = g + \Phi + \frac{\partial\Delta\mathcal{L}}{\partial\dot{q}} = g + \left(\Phi - \frac{\partial\Delta\mathcal{H}}{\partial\dot{q}}\right) = g. \tag{2.45}$$

In this case, the perturbed momentum p coincides with the unperturbed one, g. In application to the orbital motion, this means that contact elements (i.e., the non-osculating orbital elements obeying Eq. (2.31)), when substituted in $g(t; C_1, \ldots, C_6)$, furnish not the correct perturbed velocity but the correct perturbed momentum, i.e., they osculate the orbit *in phase space*. That such elements must exist was pointed out long ago by Goldreich [15] and Brumberg et al. [16], though these authors did not study their properties in detail.

2.2 Gauge freedom in the theory of orbits

2.2.1 Geometrical meaning of the arbitrary gauge function $\mathbf{\Phi}$

As explained above, the content of subsection 2.1.4 becomes merely a formulation of the Lagrange theory of orbits, provided $\boldsymbol{F}$ stands for the Newton gravity force, so that the undisturbed setting is the two-body problem. Then Eq. (2.22) expresses the gauge-invariant (i.e., taken with an arbitrary gauge $\mathbf{\Phi}(t; C_1, \ldots, C_6)$) planetary equations in the Euler–Gauss form. These equations render orbital elements that are, generally, not osculating. Equation (2.32) stands for the customary Euler–Gauss-type system for osculating (i.e., obeying $\mathbf{\Phi} = 0$) orbital elements.

Similarly, Eq. (2.30) stands for the gauge-invariant Lagrange-type or Delaunay-type (dependent upon whether C_i stand for the Kepler or Delaunay variables) equations. Such equations yield elements, which, generally, are not osculating. In those equations, one could fix the gauge by putting $\mathbf{\Phi} = 0$, thus making the resulting orbital elements osculating. However, this would be advantageous only in the case of velocity-independent $\Delta\mathcal{L}$. Otherwise, a maximal simplification is achieved through a deliberate refusal from osculation: by choosing the gauge as in Eq. (2.31) one ends up with simple equations (2.33). Thus, gauge (2.31) simplifies the planetary equations. (See Eqs. (2.46–2.57) below.) Besides, in the case when the Delaunay parameterisation is employed, this gauge makes the equations for the Delaunay variables canonical for reasons explained above in subsection 2.1.4.

The geometrical meaning of the convective term $\mathbf{\Phi}$ becomes evident if we recall that a perturbed orbit is assembled of points, each of which is donated by one representative of a sequence of conics, as on Fig. 2.2 and Fig. 2.3 where the "walk" over the instantaneous conics may be undertaken either in a non-osculating manner or in the osculating manner. The physical velocity $\dot{\boldsymbol{r}}$ is always tangent to the perturbed orbit, while the unperturbed Keplerian velocity $\mathbf{g} \equiv \partial \boldsymbol{f}/\partial t$ is tangent to the instantaneous conic. Their difference is

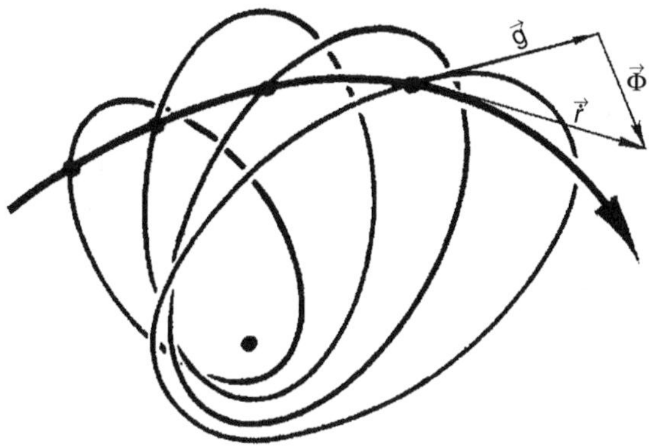

Fig. 2.2. The orbit is a set of points, each of which is donated by one of the confocal instantaneous ellipses that are not supposed to be tangent or even coplanar to the orbit. As a result, the physical velocity $\dot{\boldsymbol{r}}$ (tangent to the orbit) differs from the unperturbed Keplerian velocity $\mathbf{g}$ (tangent to the ellipse). To parameterise the depicted sequence of non-osculating ellipses, and to single it out of the other sequences, it is suitable to employ the difference between $\dot{\boldsymbol{r}}$ and $\mathbf{g}$, expressed as a function of time and six (non-osculating) orbital elements: $\mathbf{\Phi}(t, C_1, \ldots, C_6) = \dot{\boldsymbol{r}}(t, C_1, \ldots, C_6) - \mathbf{g}(t, C_1, \ldots, C_6)$. Evidently,

$$\dot{\boldsymbol{r}} = \frac{\partial \boldsymbol{r}}{\partial t} + \sum_{j=1}^{6} \frac{\partial C_j}{\partial t} \dot{C}_j = \mathbf{g} + \mathbf{\Phi},$$

where the unperturbed Keplerian velocity is $\vec{\mathbf{g}} \equiv \partial \boldsymbol{r}/\partial t$. The convective term, which emerges under perturbation, is $\mathbf{\Phi} \equiv \sum(\partial \boldsymbol{r}/\partial C_j)\dot{C}_j$. When a particular functional dependence of $\mathbf{\Phi}$ on time and the elements is fixed, this function, $\mathbf{\Phi}(t, C_1, \ldots, C_6)$, is called gauge function or gauge velocity or, simply, gauge.

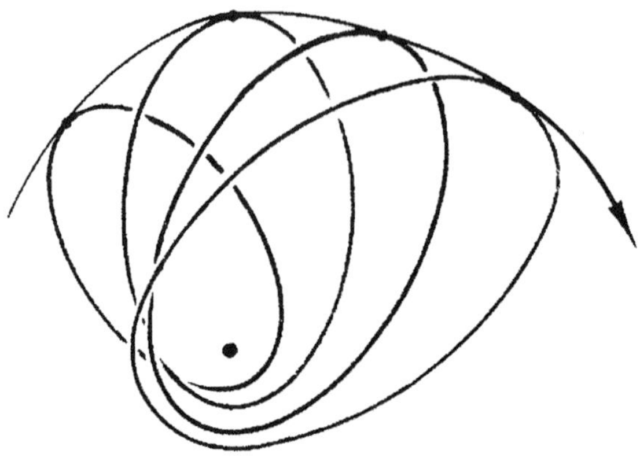

Fig. 2.3. The orbit is represented by a sequence of confocal instantaneous ellipses that are tangent to the orbit, i.e., osculating. Now, the physical velocity $\dot{\boldsymbol{r}}$ (tangent to the orbit) coincides with the unperturbed Keplerian velocity $\vec{\mathbf{g}}$ (tangent to the ellipse), so that their difference $\boldsymbol{\Phi}$ vanishes everywhere:

$$\boldsymbol{\Phi}(t, C_1, \ldots, C_6) \equiv \dot{\boldsymbol{r}}(t, C_1, \ldots, C_6) - \mathbf{g}(t, C_1, \ldots, C_6) = \sum_{j=1}^{6} \frac{\partial C_j}{\partial t} \dot{C}_j = 0.$$

This equality, called Lagrange constraint or Lagrange gauge, is the necessary and sufficient condition of osculation.

the convective term $\boldsymbol{\Phi}$. So if we use non-osculating orbital elements, then insertion of their values in $\boldsymbol{f}(t; C_1, \ldots, C_6)$ will yield a correct position of the body. However, their insertion in $\mathbf{g}(t; C_1, \ldots, C_6)$ will *not* give the right velocity. To get the correct velocity, one will have to add $\boldsymbol{\Phi}$. (See Appendix 1 for a more formal mathematical treatment in the normal form of Cauchy.)

When using non-osculating orbital elements, we must always be careful about their physical interpretation. On Fig. 2.2, the instantaneous conics are not supposed to be tangent to the orbit, nor are they supposed to be even coplanar thereto. (They may be even perpendicular to the orbit!—why not?) This means that, for example, the non-osculating element i may considerably differ from the real, physical inclination of the orbit.

We would add that the arbitrariness of choice of the function $\boldsymbol{\Phi}(t, C_1(t), \ldots, C_6(t))$ had been long known but never used in astronomy until a recent effort undertaken by several authors [1, 2, 10, 12, 13, 17, 18, 24] (Efroimsky 2005c). Substitution of the Lagrange constraint $\boldsymbol{\Phi} = 0$ with alternative choices does not influence the physical motion, but alters its mathematical description (i.e., renders different values of the orbital parameters $C_i(t)$). Such invariance of a physical picture under a change of parameterisation goes under the name of gauge freedom. It is a part and parcel of electrodynamics and other field theories. In mathematics, it is described in terms of fiber bundles. A clever choice of gauge often simplifies solution of the equations of motion. On the other hand, the gauge invariance may have implications upon numerical procedures. We mean the so-called "gauge drift," i.e., unwanted displacement in the gauge function $\boldsymbol{\Phi}$, caused by accumulation of numerical errors in the constants.

2.2.2 Gauge-invariant planetary equations of the Lagrange and Delaunay types

We present the gauge-invariant Lagrange-type equations, following Efroimsky and Goldreich [1]. These equations follow from (2.30) if we take into account the gauge-invariance (i.e., the $\mathbf{\Phi}$-independence) of the Lagrange-bracket matrix $[C_i\, C_j]$.

$$\frac{\mathrm{d}a}{\mathrm{d}t} = \frac{2}{na}\left[\frac{\partial(-\Delta\mathcal{H})}{\partial M_o} - \frac{\partial \Delta\mathcal{L}}{\partial \dot{\boldsymbol{r}}}\frac{\partial}{\partial M_o}\left(\mathbf{\Phi}+\frac{\partial\Delta\mathcal{L}}{\partial\dot{\boldsymbol{r}}}\right) - \left(\mathbf{\Phi}+\frac{\partial\Delta\mathcal{L}}{\partial\dot{\boldsymbol{r}}}\right)\frac{\partial \mathbf{g}}{\partial M_o} \right.$$
$$\left. - \frac{\partial \boldsymbol{f}}{\partial M_o}\frac{\mathrm{d}}{\mathrm{d}t}\left(\mathbf{\Phi}+\frac{\partial\Delta\mathcal{L}}{\partial\dot{\boldsymbol{r}}}\right)\right], \tag{2.46}$$

$$\frac{\mathrm{d}e}{\mathrm{d}t} = \frac{1-e^2}{na^2e}\left[\frac{\partial(-\Delta\mathcal{H})}{\partial M_o} - \frac{\partial \Delta\mathcal{L}}{\partial \dot{\boldsymbol{r}}}\frac{\partial}{\partial a}\left(\mathbf{\Phi}+\frac{\partial\Delta\mathcal{L}}{\partial\dot{\boldsymbol{r}}}\right) - \left(\mathbf{\Phi}+\frac{\partial\Delta\mathcal{L}}{\partial\dot{\boldsymbol{r}}}\right)\frac{\partial \mathbf{g}}{\partial M_o} \right.$$
$$\left. - \frac{\partial \boldsymbol{f}}{\partial M_o}\frac{\mathrm{d}}{\mathrm{d}t}\left(\mathbf{\Phi}+\frac{\partial\Delta\mathcal{L}}{\partial\dot{\boldsymbol{r}}}\right)\right] - \frac{(1-e^2)^{1/2}}{na^2e}\left[\frac{\partial(-\Delta\mathcal{H})}{\partial \omega} - \frac{\partial \Delta\mathcal{L}}{\partial \dot{\boldsymbol{r}}}\frac{\partial}{\partial \omega}\left(\mathbf{\Phi}+\frac{\partial\Delta\mathcal{L}}{\partial\dot{\boldsymbol{r}}}\right)\right.$$
$$\left. - \left(\mathbf{\Phi}+\frac{\partial\Delta\mathcal{L}}{\partial\dot{\boldsymbol{r}}}\right)\frac{\partial \mathbf{g}}{\partial \omega} - \frac{\partial \boldsymbol{f}}{\partial \omega}\frac{\mathrm{d}}{\mathrm{d}t}\left(\mathbf{\Phi}+\frac{\partial\Delta\mathcal{L}}{\partial\dot{\boldsymbol{r}}}\right)\right], \tag{2.47}$$

$$\frac{\mathrm{d}\omega}{\mathrm{d}t} = \frac{-\cos i}{na^2(1-e^2)^{1/2}\sin i}\left[\frac{\partial(-\Delta\mathcal{H})}{\partial i} - \frac{\partial \Delta\mathcal{L}}{\partial \dot{\boldsymbol{r}}}\frac{\partial}{\partial i}\left(\mathbf{\Phi}+\frac{\partial\Delta\mathcal{L}}{\partial\dot{\boldsymbol{r}}}\right) - \left(\mathbf{\Phi}+\frac{\partial\Delta\mathcal{L}}{\partial\dot{\boldsymbol{r}}}\right)\frac{\partial \mathbf{g}}{\partial i} \right.$$
$$\left. - \frac{\partial \boldsymbol{f}}{\partial i}\frac{\mathrm{d}}{\mathrm{d}t}\left(\mathbf{\Phi}+\frac{\partial\Delta\mathcal{L}}{\partial\dot{\boldsymbol{r}}}\right)\right] + \frac{(1-e^2)^{1/2}}{na^2e}\left[\frac{\partial(-\Delta\mathcal{H})}{\partial e} - \frac{\partial \Delta\mathcal{L}}{\partial \dot{\boldsymbol{r}}}\frac{\partial}{\partial e}\left(\mathbf{\Phi}+\frac{\partial\Delta\mathcal{L}}{\partial\dot{\boldsymbol{r}}}\right)\right.$$
$$\left. - \left(\mathbf{\Phi}+\frac{\partial\Delta\mathcal{L}}{\partial\dot{\boldsymbol{r}}}\right)\frac{\partial \mathbf{g}}{\partial e} - \frac{\partial \boldsymbol{f}}{\partial e}\frac{\mathrm{d}}{\mathrm{d}t}\left(\mathbf{\Phi}+\frac{\partial\Delta\mathcal{L}}{\partial\dot{\boldsymbol{r}}}\right)\right], \tag{2.48}$$

$$\frac{\mathrm{d}i}{\mathrm{d}t} = \frac{\cos i}{na^2(1-e^2)^{1/2}\sin i}\left[\frac{\partial(-\Delta\mathcal{H})}{\partial \omega} - \frac{\partial \Delta\mathcal{L}}{\partial \dot{\boldsymbol{r}}}\frac{\partial}{\partial \omega}\left(\mathbf{\Phi}+\frac{\partial\Delta\mathcal{L}}{\partial\dot{\boldsymbol{r}}}\right) - \left(\mathbf{\Phi}+\frac{\partial\Delta\mathcal{L}}{\partial\dot{\boldsymbol{r}}}\right)\frac{\partial \mathbf{g}}{\partial \omega} \right.$$
$$\left. - \frac{\partial \boldsymbol{f}}{\partial \omega}\frac{\mathrm{d}}{\mathrm{d}t}\left(\mathbf{\Phi}+\frac{\partial\Delta\mathcal{L}}{\partial\dot{\boldsymbol{r}}}\right)\right] - \frac{1}{na^2(1-e^2)^{1/2}\sin i}\left[\frac{\partial(-\Delta\mathcal{H})}{\partial \Omega} - \frac{\partial \Delta\mathcal{L}}{\partial \dot{\boldsymbol{r}}}\frac{\partial}{\partial \Omega}\left(\Phi+\frac{\partial\Delta\mathcal{L}}{\partial\dot{\boldsymbol{r}}}\right)\right.$$
$$\left. - \left(\mathbf{\Phi}+\frac{\partial\Delta\mathcal{L}}{\partial\dot{\boldsymbol{r}}}\right)\frac{\partial \mathbf{g}}{\partial \Omega} - \frac{\partial \boldsymbol{f}}{\partial \Omega}\frac{\mathrm{d}}{\mathrm{d}t}\left(\mathbf{\Phi}+\frac{\partial\Delta\mathcal{L}}{\partial\dot{\boldsymbol{r}}}\right)\right], \tag{2.49}$$

$$\frac{\mathrm{d}\Omega}{\mathrm{d}t} = \frac{1}{na^2(1-e^2)^{1/2}\sin i}\left[\frac{\partial(-\Delta\mathcal{H})}{\partial i} - \frac{\partial \Delta\mathcal{L}}{\partial \dot{\boldsymbol{r}}}\frac{\partial}{\partial i}\left(\Phi+\frac{\partial\Delta\mathcal{L}}{\partial\dot{\boldsymbol{r}}}\right) - \left(\mathbf{\Phi}+\frac{\partial\Delta\mathcal{L}}{\partial\dot{\boldsymbol{r}}}\right)\frac{\partial \mathbf{g}}{\partial i} \right.$$
$$\left. - \frac{\partial \boldsymbol{f}}{\partial i}\frac{\mathrm{d}}{\mathrm{d}t}\left(\mathbf{\Phi}+\frac{\partial\Delta\mathcal{L}}{\partial\dot{\boldsymbol{r}}}\right)\right], \tag{2.50}$$

$$\frac{\mathrm{d}M_o}{\mathrm{d}t} = -\frac{1-e^2}{na^2 e}\left[\frac{\partial(-\Delta\mathcal{H})}{\partial e} - \frac{\partial\Delta\mathcal{L}}{\partial\dot{\boldsymbol{r}}}\frac{\partial}{\partial e}\left(\boldsymbol{\Phi}+\frac{\partial\Delta\mathcal{L}}{\partial\dot{\boldsymbol{r}}}\right) - \left(\boldsymbol{\Phi}+\frac{\partial\Delta\mathcal{L}}{\partial\dot{\boldsymbol{r}}}\right)\frac{\partial\mathbf{g}}{\partial e}\right.$$
$$\left. -\frac{\partial\boldsymbol{f}}{\partial e}\frac{\mathrm{d}}{\mathrm{d}t}\left(\boldsymbol{\Phi}+\frac{\partial\Delta\mathcal{L}}{\partial\dot{\boldsymbol{r}}}\right)\right] - \frac{2}{na}\left[\frac{\partial(-\Delta\mathcal{H})}{\partial a} - \frac{\partial\Delta\mathcal{L}}{\partial\dot{\boldsymbol{r}}}\frac{\partial}{\partial a}\left(\boldsymbol{\Phi}+\frac{\partial\Delta\mathcal{L}}{\partial\dot{\boldsymbol{r}}}\right)\right.$$
$$\left. -\left(\boldsymbol{\Phi}+\frac{\partial\Delta\mathcal{L}}{\partial\dot{\boldsymbol{r}}}\right)\frac{\partial\mathbf{g}}{\partial a} - \frac{\partial\boldsymbol{f}}{\partial a}\frac{\mathrm{d}}{\mathrm{d}t}\left(\boldsymbol{\Phi}+\frac{\partial\Delta\mathcal{L}}{\partial\dot{\boldsymbol{r}}}\right)\right]. \tag{2.51}$$

Similarly, the gauge-invariant Delaunay-type system can be written down as:

$$\frac{\mathrm{d}L}{\mathrm{d}t} = \frac{\partial(-\Delta\mathcal{H})}{\partial M_o} - \frac{\partial\Delta L}{\partial\dot{\boldsymbol{r}}}\frac{\partial}{\partial M_o}\left(\boldsymbol{\Phi}+\frac{\partial\Delta\mathcal{L}}{\partial\dot{\boldsymbol{r}}}\right) - \left(\boldsymbol{\Phi}+\frac{\partial\Delta\mathcal{L}}{\partial\dot{\boldsymbol{r}}}\right)\frac{\partial\mathbf{g}}{\partial M_o}$$
$$-\frac{\partial\boldsymbol{r}}{\partial M_o}\frac{\mathrm{d}}{\mathrm{d}t}\left(\boldsymbol{\Phi}+\frac{\partial\Delta\mathcal{L}}{\partial\dot{\boldsymbol{r}}}\right), \tag{2.52}$$

$$\frac{\mathrm{d}M_o}{\mathrm{d}t} = -\frac{\partial(-\Delta\mathcal{H})}{\partial L} + \frac{\partial\Delta\mathcal{L}}{\partial\dot{\boldsymbol{r}}}\frac{\partial}{\partial L}\left(\boldsymbol{\Phi}+\frac{\partial\Delta\mathcal{L}}{\partial\dot{\boldsymbol{r}}}\right) + \left(\vec{\boldsymbol{\Phi}}+\frac{\partial\Delta\mathcal{L}}{\partial\dot{\boldsymbol{r}}}\right)\frac{\partial\mathbf{g}}{\partial L}$$
$$+\frac{\partial\boldsymbol{r}}{\partial L}\frac{\mathrm{d}}{\mathrm{d}t}\left(\boldsymbol{\Phi}+\frac{\partial\Delta\mathcal{L}}{\partial\dot{\boldsymbol{r}}}\right), \tag{2.53}$$

$$\frac{\mathrm{d}G}{\mathrm{d}t} = \frac{\partial(-\Delta\mathcal{H})}{\partial\omega} - \frac{\partial\Delta\mathcal{L}}{\partial\dot{\boldsymbol{r}}}\frac{\partial}{\partial\omega}\left(\boldsymbol{\Phi}+\frac{\partial\Delta\mathcal{L}}{\partial\dot{\boldsymbol{r}}}\right) - \left(\boldsymbol{\Phi}+\frac{\partial\Delta\mathcal{L}}{\partial\dot{\boldsymbol{r}}}\right)\frac{\partial\mathbf{g}}{\partial\omega}$$
$$-\frac{\partial\boldsymbol{r}}{\partial\omega}\frac{\mathrm{d}}{\mathrm{d}t}\left(\boldsymbol{\Phi}+\frac{\partial\Delta\mathcal{L}}{\partial\dot{\boldsymbol{r}}}\right), \tag{2.54}$$

$$\frac{\mathrm{d}\omega}{\mathrm{d}t} = -\frac{\partial(-\Delta\mathcal{H})}{\partial G} + \frac{\partial\Delta\mathcal{L}}{\partial\dot{\boldsymbol{r}}}\frac{\partial}{\partial G}\left(\boldsymbol{\Phi}+\frac{\partial\Delta\mathcal{L}}{\partial\dot{\boldsymbol{r}}}\right) + \left(\boldsymbol{\Phi}+\frac{\partial\Delta\mathcal{L}}{\partial\dot{\boldsymbol{r}}}\right)\frac{\partial\mathbf{g}}{\partial G}$$
$$+\frac{\partial\boldsymbol{r}}{\partial G}\frac{\mathrm{d}}{\mathrm{d}t}\left(\boldsymbol{\Phi}+\frac{\partial\Delta\mathcal{L}}{\partial\dot{\boldsymbol{r}}}\right), \tag{2.55}$$

$$\frac{\mathrm{d}H}{\mathrm{d}t} = \frac{\partial(-\Delta\mathcal{H})}{\partial\Omega} - \frac{\partial\Delta\mathcal{L}}{\partial\dot{\boldsymbol{r}}}\frac{\partial}{\partial\Omega}\left(\boldsymbol{\Phi}+\frac{\partial\Delta\mathcal{L}}{\partial\dot{\boldsymbol{r}}}\right) - \left(\boldsymbol{\Phi}+\frac{\partial\Delta\mathcal{L}}{\partial\dot{\boldsymbol{r}}}\right)\frac{\partial\mathbf{g}}{\partial\Omega}$$
$$-\frac{\partial\boldsymbol{f}}{\partial\Omega}\frac{\mathrm{d}}{\mathrm{d}t}\left(\boldsymbol{\Phi}+\frac{\partial\Delta\mathcal{L}}{\partial\dot{\boldsymbol{r}}}\right), \tag{2.56}$$

$$\frac{\mathrm{d}\Omega}{\mathrm{d}t} = -\frac{\partial(-\Delta\mathcal{H})}{\partial H} + \frac{\partial\Delta\mathcal{L}}{\partial\dot{\boldsymbol{r}}}\frac{\partial}{\partial H}\left(\boldsymbol{\Phi}+\frac{\partial\Delta\mathcal{L}}{\partial\dot{\boldsymbol{r}}}\right) + \left(\boldsymbol{\Phi}+\frac{\partial\Delta\mathcal{L}}{\partial\dot{\boldsymbol{r}}}\right)\frac{\partial\mathbf{g}}{\partial H}$$
$$+\frac{\partial\boldsymbol{r}}{\partial H}\frac{\mathrm{d}}{\mathrm{d}t}\left(\boldsymbol{\Phi}+\frac{\partial\Delta\mathcal{L}}{\partial\dot{\boldsymbol{r}}}\right). \tag{2.57}$$

where μ stands for the reduced mass, while

$$L \equiv \mu^{1/2} a^{1/2}, \quad G \equiv \mu^{1/2} a^{1/2} \left(1 - e^2\right)^{1/2}, \quad H \equiv \mu^{1/2} a^{1/2} \left(1 - e^2\right)^{1/2} \cos i. \tag{2.58}$$

The symbols $\mathbf{\Phi}, \boldsymbol{f}, \mathbf{g}$ now denote the functional dependencies of the gauge, position, and velocity upon the Delaunay, not Keplerian elements, and therefore these are functions different from $\mathbf{\Phi}, \boldsymbol{f}, \mathbf{g}$ used in Eqs. (2.46–2.51) where they stood for the dependencies upon the Kepler elements. (In Efroimsky [12, 13] the dependencies $\mathbf{\Phi}, \boldsymbol{f}, \mathbf{g}$ upon the Delaunay variables were equipped with tilde, to distinguish them from the dependencies upon the Kepler coordinates.)

To employ the gauge-invariant equations in analytical calculations is a delicate task: one should always keep in mind that, in case $\mathbf{\Phi}$ is chosen to depend not only upon time but also upon the "constants" (but not upon their derivatives), the right-hand sides of these equations will implicitly contain the first derivatives $\mathrm{d}C_i/\mathrm{d}t$, and one will have to move them to the left-hand sides (like in the transition from Eq. (2.22) to (2.23)). The choices $\mathbf{\Phi} = 0$ and $\mathbf{\Phi} = -\partial \Delta \mathcal{L}/\partial \dot{\boldsymbol{r}}$ are exceptions. (The most general exceptional gauge reads as $\mathbf{\Phi} = -\partial \Delta \mathcal{L}/\partial \dot{\boldsymbol{r}} + \eta(t)$, where $\eta(t)$ is an arbitrary function of time.)

As was expected from (2.30), both the Lagrange and Delaunay systems simplify in the gauge (2.31). Since for orbital motions we have $\partial \mathcal{H}/\partial \boldsymbol{p} = -\partial \Delta \mathcal{L}/\partial \dot{\boldsymbol{r}}$, then (2.31) coincides with Eq. (2.44). Hence, the Hamiltonian analysis (2.34–2.44) explains why it is exactly in the gauge (2.31) that the Delaunay system becomes symplectic. In physicists' parlance, the canonicity condition breaks the gauge symmetry by stiffly fixing the gauge (2.44), gauge that is equivalent, in the orbital case, to (2.31)—phenomenon called "gauge stiffness." The phenomenon may be looked upon also from a different angle. Above we emphasized that the gauge freedom implies essential arbitrariness in our choice of the functional form of $\mathbf{\Phi}(t; C_1, \ldots, C_6)$, provided the choice does not come into a contradiction with the equations of motion—an important clause that shows its relevance in the Delaunay-type Eqs. (2.52–2.57): we see that, for example, the Lagrange choice $\mathbf{\Phi} = 0$ (as well as any other choice different from Eq. (2.31)) is incompatible with the canonical structure of the equations of motion for the elements.

2.3 A practical example on gauges: a satellite orbiting a precessing oblate planet

Above we presented the Lagrange- and Delaunay-type planetary equations in the gauge-invariant form (i.e., for an arbitrary choice of the gauge function $\mathbf{\Phi}(t; C_1, \ldots, C_6)$) and for a generic perturbation $\Delta \mathcal{L}$ that may depend not only upon positions but also upon velocities and the time. We saw that the disturbing function is the negative Hamiltonian variation (which differs from the Lagrangian variation when the perturbation depends on velocities). Below, we shall also see that the functional dependence of $\Delta \mathcal{H}$ upon the orbital elements is gauge-dependent.

2.3.1 The gauge freedom and the freedom of frame choice

In the most compressed form, implementation of the variation-of-constants method in orbital mechanics looks like this. A generic solution to the two-body problem is expressed with

$$\boldsymbol{r} = \boldsymbol{f}\,(C, t)\,, \tag{2.59}$$

$$\left(\frac{\partial \boldsymbol{f}}{\partial t}\right)_C = \mathbf{g}\,(C, t)\,, \tag{2.60}$$

$$\left(\frac{\partial \mathbf{g}}{\partial t}\right)_C = -\frac{\mu}{f^2}\frac{\boldsymbol{f}}{f} \tag{2.61}$$

and is used as an ansatz to describe the perturbed motion:

$$\boldsymbol{r} = \boldsymbol{f}(C(t), t), \tag{2.62}$$

$$\dot{\boldsymbol{r}} = \frac{\partial \boldsymbol{f}}{\partial t} + \frac{\partial \boldsymbol{f}}{\partial C_i}\frac{\mathrm{d}C_i}{\mathrm{d}t} = \mathbf{g} + \boldsymbol{\Phi}, \tag{2.63}$$

$$\ddot{\mathbf{r}} = \frac{\partial \mathbf{g}}{\partial t} + \frac{\partial \mathbf{g}}{\partial C_i}\frac{\mathrm{d}C_i}{\mathrm{d}t} + \frac{\mathrm{d}\boldsymbol{\Phi}}{\mathrm{d}t} = -\frac{\mu}{f^2}\frac{\boldsymbol{f}}{f} + \frac{\partial \mathbf{g}}{\partial C_i}\frac{\mathrm{d}C_i}{\mathrm{d}t} + \frac{\mathrm{d}\boldsymbol{\Phi}}{\mathrm{d}t}. \tag{2.64}$$

As can be seen from Eq. (2.63), our choice of a particular gauge is equivalent to a particular way of decomposition of the physical motion into a movement with velocity $\mathbf{g}$ along the instantaneous conic, and a movement caused by the conic's deformation at the rate $\boldsymbol{\Phi}$. Beside the fact that we decouple the physical velocity $\dot{\boldsymbol{r}}$ in a certain proportion between these two movements, $\mathbf{g}$ and $\boldsymbol{\Phi}$, it also matters *what* physical velocity (i.e., velocity relative to what frame) is decoupled in this proportion. Thus, the choice of gauge does not exhaust all freedom: one can still choose *in what frame* to write ansatz (2.62)—one can write it in inertial axes or in some accelerated or/and rotating ones. For example, in the case of a satellite orbiting a precessing oblate primary it is most *convenient* to write the ansatz in a frame co-precessing (but not corotating with the planet's equator.

The kinematic formulae (2.62–2.64) do not yet contain information about our choice of the reference system wherein to employ variation of constants. This information shows up only when (2.62) and (2.64) get inserted into the equation of motion $\ddot{\boldsymbol{r}} + (\mu \boldsymbol{r}/r^3) = \Delta \boldsymbol{F}$ to render

$$\frac{\partial \mathbf{g}}{\partial C_i}\frac{\mathrm{d}C_i}{\mathrm{d}t} + \frac{\mathrm{d}\boldsymbol{\Phi}}{\mathrm{d}t} = \Delta \boldsymbol{f} = \frac{\partial \Delta \mathcal{L}}{\partial \boldsymbol{r}} - \frac{\mathrm{d}}{\mathrm{d}t}\left(\frac{\partial \Delta \mathcal{L}}{\partial \dot{\boldsymbol{r}}}\right). \tag{2.65}$$

Information about the reference frame, where we employ the method and define the elements C_i, is contained in the expression for the perturbing force $\Delta \boldsymbol{F}$. If the operation is carried out in an inertial system, $\Delta \boldsymbol{F}$ contains only physical forces. If we work in a frame moving with a linear acceleration $\vec{\mathbf{a}}$, then $\Delta \boldsymbol{F}$ also contains the inertial force $-\vec{\mathbf{a}}$. In case this coordinate frame also rotates relative to inertial ones at a rate $\boldsymbol{\mu}$, then $\Delta \boldsymbol{F}$ also includes the inertial contributions $-2\boldsymbol{\mu} \times \dot{\boldsymbol{r}} - \dot{\boldsymbol{\mu}} \times \boldsymbol{r} - \boldsymbol{\mu} \times (\boldsymbol{\mu} \times \dot{\boldsymbol{r}})$. When studying orbits about an oblate precessing planet, it is most convenient (though not obligatory) to apply the variation-of-parameters method in axes coprecessing with the planet's equator of date: it is in this coordinate system that one should write ansatz (2.62) and decompose $\dot{\boldsymbol{r}}$ into $\mathbf{g}$ and $\boldsymbol{\Phi}$. This convenient choice of coordinate system will still leave one with the freedom of gauge nomination: in the said coordinate system, one will still have to decide what function $\boldsymbol{\Phi}$ to insert in (2.63).

2.3.2 *The disturbing function in a frame co-precessing with the equator of date*

The equation of motion in the inertial frame is

$$\boldsymbol{r}'' = -\frac{\partial U}{\partial \boldsymbol{r}}, \tag{2.66}$$

where U is the total gravitational potential, and time derivatives in the inertial axes are denoted by primes. In a coordinate system precessing at angular rate $\boldsymbol{\mu}(t)$, Eq. (2.66) becomes:

$$\begin{aligned}\ddot{\mathbf{r}} &= -\frac{\partial U}{\partial \boldsymbol{r}} - 2\boldsymbol{\mu}\times\dot{\boldsymbol{r}} - \dot{\boldsymbol{\mu}}\times\boldsymbol{r} - \boldsymbol{\mu}\times(\boldsymbol{\mu}\times\boldsymbol{r}) \\ &= -\frac{\partial U_o}{\partial \boldsymbol{r}} - \frac{\partial \Delta U}{\partial \boldsymbol{r}} - 2\boldsymbol{\mu}\times\dot{\boldsymbol{r}} - \dot{\boldsymbol{\mu}}\times\boldsymbol{r} - \boldsymbol{\mu}\times(\boldsymbol{\mu}\times\boldsymbol{r}),\end{aligned} \tag{2.67}$$

dots standing for time derivatives in the co-precessing frame, and $\boldsymbol{\mu}$ being the coordinate system's angular velocity relative to an inertial frame. Formula (2.125) in the Appendix gives the expression for $\boldsymbol{\mu}$ in terms of the longitude of the node and the inclination of the equator of date relative to that of epoch. The physical (i.e., not associated with inertial forces) potential $U(\boldsymbol{r})$ consists of the (reduced) two-body part $U_o(\boldsymbol{r}) \equiv -GM\boldsymbol{r}/r^3$ and a term $\Delta U(\boldsymbol{r})$ caused by the planet's oblateness (or, generally, by its triaxiality).

To implement variation of the orbital elements defined in this frame, we note that the disturbing force on the right-hand side of Eq. (2.67) is generated, according to Eq. (2.65), by

$$\Delta\mathcal{L}(\boldsymbol{r},\dot{\boldsymbol{r}},t) = -\Delta U(\boldsymbol{r}) + \dot{\boldsymbol{r}}\cdot(\boldsymbol{\mu}\times\boldsymbol{r}) + \frac{1}{2}(\boldsymbol{\mu}\times\boldsymbol{r})\cdot(\boldsymbol{\mu}\times\boldsymbol{r}). \tag{2.68}$$

Since

$$\frac{\partial \Delta\mathcal{L}}{\partial \dot{\mathbf{r}}} = \boldsymbol{\mu}\times\boldsymbol{r}, \tag{2.69}$$

then

$$\boldsymbol{p} = \dot{\boldsymbol{r}} + \frac{\partial \Delta\mathcal{L}}{\partial \dot{\mathbf{r}}} = \dot{\boldsymbol{r}} + \boldsymbol{\mu}\times\boldsymbol{r} \tag{2.70}$$

and, therefore, the corresponding Hamiltonian perturbation reads:

$$\begin{aligned}\Delta\mathcal{H} &= -\left[\Delta\mathcal{L} + \frac{1}{2}\left(\frac{\partial \Delta\mathcal{L}}{\partial \dot{\mathbf{r}}}\right)^2\right] = -[-\Delta U + \boldsymbol{p}\cdot(\boldsymbol{\mu}\times\boldsymbol{r})] \\ &= -[-\Delta U + (\boldsymbol{r}\times\boldsymbol{p})\cdot\boldsymbol{\mu}] = \Delta U - \mathbf{J}\cdot\boldsymbol{\mu},\end{aligned} \tag{2.71}$$

with vector $\mathbf{J} \equiv \boldsymbol{r} \times \boldsymbol{p}$ being the satellite's orbital angular momentum in the inertial frame.

According to (2.63) and (2.70), the momentum can be written as

$$\boldsymbol{p} = \mathbf{g} + \boldsymbol{\Phi} + \boldsymbol{\mu} \times \boldsymbol{f}, \tag{2.72}$$

whence the Hamiltonian perturbation becomes

$$\Delta \mathcal{H} = -\left[\Delta \mathcal{L} + \frac{1}{2}\left(\frac{\partial \Delta \mathcal{L}}{\partial \dot{\boldsymbol{r}}}\right)^2\right] = -\left[-\Delta U + (\boldsymbol{f} \times \mathbf{g}) \cdot \boldsymbol{\mu} + (\boldsymbol{\Phi} + \boldsymbol{\mu} \times \boldsymbol{f}) \cdot (\boldsymbol{\mu} \times \boldsymbol{f})\right]. \tag{2.73}$$

This is what one is supposed to plug in (2.30) or, the same, in (2.46–2.57).

2.3.3 *Planetary equations in a precessing frame, written in terms of contact elements*

In the subsection 2.3.2 we fixed our choice of the frame wherein to describe the orbit. By writing the Lagrangian and Hamiltonian variations as (2.68) and (2.73), we stated that our elements would be defined in the frame coprecessing with the equator. The frame being fixed, we are still left with the freedom of gauge choice. As evident from (2.33) or (2.46–2.57), the special gauge (2.31) ideally simplifies the planetary equations. Indeed, (2.31) and (2.69) together yield

$$\boldsymbol{\Phi} = -\frac{\partial \Delta \mathcal{L}}{\partial \dot{\boldsymbol{r}}} = -\boldsymbol{\mu} \times \boldsymbol{r} \equiv -\boldsymbol{\mu} \times \boldsymbol{f}, \tag{2.74}$$

wherefrom the Hamiltonian (2.73) becomes

$$\Delta \mathcal{H}^{(cont)} = -\left[-\Delta U(\boldsymbol{f}) + \boldsymbol{\mu} \cdot (\boldsymbol{f} \times \mathbf{g})\right], \tag{2.75}$$

while the planetary equations (2.30) get the shape

$$[C_r C_i]\frac{\mathrm{d}C_i}{\mathrm{d}t} = \frac{\partial\left(-\Delta \mathcal{H}^{(cont)}\right)}{\partial C_r}, \tag{2.76}$$

or, the same,

$$[C_r\, C_i]\frac{\mathrm{d}C_i}{\mathrm{d}t} = \frac{\partial}{\partial C_r}\left[-\Delta U(\boldsymbol{f}) + \boldsymbol{\mu} \cdot (\boldsymbol{f} \times \mathbf{g})\right], \tag{2.77}$$

where $\boldsymbol{f}$ and $\mathbf{g}$ stand for the undisturbed (two-body) functional expressions (2.59) and (2.60) of the position and velocity via the time and the chosen set of orbital elements. Planetary equations (2.76) were obtained with aid of (2.74), and therefore they render non-osculating orbital elements that are called contact elements. This is why we equipped the Hamiltonian (2.75) with superscript "(*cont*)." In distinction from the osculating elements, the contact ones osculate *in phase space*: (2.72) and (2.74) entail that $\boldsymbol{p} = \mathbf{g}$. As already mentioned in the end of section 2.1, existence of such elements was pointed out by Goldreich [15] and Brumberg et al. [16] long before the concept of gauge freedom was introduced in celestial mechanics. Brumberg et al. [16] simply *defined* these elements by the condition that their insertion in $\mathbf{g}(t; C_1, \ldots, C_6)$ returns not the perturbed velocity, but the perturbed momentum. Goldreich [15] defined these elements (without calling them "contact") differently. Having in mind inertial forces (2.67), he wrote down the corresponding Hamiltonian equation (2.71) and added its negative to the disturbing function of the standard planetary equations (without enriching the equations with any other terms). Then he noticed that those equations furnished non-osculating elements. Now we can easily see that both Goldreich's and Brumberg's definitions correspond to the gauge choice (2.31).

When one chooses the Keplerian parameterisation, then Eq. (2.77) becomes:

$$\frac{\mathrm{d}a}{\mathrm{d}t} = \frac{2}{na}\frac{\partial\left(-\Delta\mathcal{H}^{(cont)}\right)}{\partial M_o}, \tag{2.78}$$

$$\frac{\mathrm{d}e}{\mathrm{d}t} = \frac{1-e^2}{na^2e}\frac{\partial\left(-\Delta\mathcal{H}^{(cont)}\right)}{\partial M_o} - \frac{(1-e^2)^{1/2}}{na^2e}\frac{\partial\left(-\Delta\mathcal{H}^{(cont)}\right)}{\partial\omega}, \tag{2.79}$$

$$\frac{\mathrm{d}\omega}{\mathrm{d}t} = \frac{-\cos i}{na^2(1-e^2)^{1/2}\sin i}\frac{\partial\left(-\Delta\mathcal{H}^{(cont)}\right)}{\partial i} + \frac{(1-e^2)^{1/2}}{na^2e}\frac{\partial\left(-\Delta\mathcal{H}^{(cont)}\right)}{\partial e} \tag{2.80}$$

$$\frac{\mathrm{d}i}{\mathrm{d}t} = \frac{\cos i}{na^2(1-e^2)^{1/2}\sin i}\frac{\partial\left(-\Delta\mathcal{H}^{(cont)}\right)}{\partial\omega} - \frac{1}{na^2(1-e^2)^{1/2}\sin i}\frac{\partial\left(-\Delta\mathcal{H}^{(cont)}\right)}{\partial\Omega}, \tag{2.81}$$

$$\frac{\mathrm{d}\Omega}{\mathrm{d}t} = \frac{1}{na^2(1-e^2)^{1/2}\sin i}\frac{\partial\left(-\Delta\mathcal{H}^{(cont)}\right)}{\partial i}, \tag{2.82}$$

$$\frac{\mathrm{d}M_o}{\mathrm{d}t} = -\frac{1-e^2}{na^2e}\frac{\partial\left(-\Delta\mathcal{H}^{(cont)}\right)}{\partial e} - \frac{2}{na}\frac{\partial\left(-\Delta\mathcal{H}^{(cont)}\right)}{\partial a}. \tag{2.83}$$

The above equations implement an interesting pitfall. When describing orbital motion relative to a frame coprecessing with the equator of date, it is tempting to derive the Hamiltonian variation caused by the inertial forces, and to simply plug it, with a negative sign, into the disturbing function. This would entail equations (2.76–2.83) which, as demonstrated above, belong to the non-Lagrange gauge (2.31). The elements furnished by these equations are non-osculating, so that the conics parameterised by these elements are not tangent to the perturbed trajectory. For example, i gives the inclination of the instantaneous non-tangent conic, but differs from the real, physical physical (i.e., osculating), inclination of the orbit. This approach—when an inertial term is simply added to the disturbing function—was employed by Goldreich [15], Brumberg et al. [16], and Kinoshita [19], and many others. Goldreich and Brumberg noticed that this destroyed the osculation.

Goldreich [15] studied how the equinoctial precession of Mars influences the long-term evolution of Phobos' and Deimos' orbit inclinations. Goldreich assumed that the elements a and e stay constant; he also substituted the Hamiltonian variation (2.75) with its orbital average, which made his planetary equations render the secular parts of the elements. He assumed that the averaged physical term $\langle \Delta U \rangle$ is only due to the primary's oblateness:

$$\langle \Delta U \rangle = -\frac{n^2 J_2}{4} \rho^2 \frac{3\cos^2 i - 1}{(1-e^2)^{3/2}}, \tag{2.84}$$

ρ being the mean radius of the planet,[9] and n being the satellite's mean motion. To simplify the inertial term, Goldreich employed the well known formula

$$\boldsymbol{r} \times \mathbf{g} = \sqrt{Gma(1-e^2)}\,\boldsymbol{w}, \tag{2.85}$$

where

$$\boldsymbol{w} = \hat{\mathbf{x}}_1 \sin i \sin \Omega - \hat{\mathbf{x}}_2 \sin i \cos \Omega + \hat{\mathbf{x}}_3 \cos i \tag{2.86}$$

is a unit vector normal to the instantaneous ellipse, expressed through unit vectors $\hat{\mathbf{x}}_1, \hat{\mathbf{x}}_2, \hat{\mathbf{x}}_3$ associated with the co-precessing frame x_1, x_2, x_3 (axes x_1 and x_2 lying in the planet's equatorial plane of date, and x_1 pointing along the fiducial line wherefrom the longitude of the ascending node of the satellite orbit, Ω, is measured). This resulted in

$$\langle \Delta \mathcal{H}^{(cont)} \rangle = -[-\langle \Delta U \rangle + \langle \boldsymbol{\mu} \cdot (\boldsymbol{f} \times \mathbf{g}) \rangle] = -\frac{GmJ_2}{4} \frac{\rho_e^2}{a^3} \frac{3\cos^2 i - 1}{(1-e^2)^{3/2}} - \sqrt{Gma(1-e^2)}$$
$$(\mu_1 \sin i \sin \Omega - \mu_2 \sin i \cos \Omega + \mu_3 \cos i), \tag{2.87}$$

all letters now standing not for the appropriate variables but for their orbital averages. Substitution of this averaged Hamiltonian in (2.81–2.82) lead Goldreich, in assumption that both $|\dot{\boldsymbol{\mu}}|/(n^2 J_2 \sin i)$ and $|\boldsymbol{\mu}|/(nJ_2 \sin i)$ are much less than unity, to the following system:

$$\frac{\mathrm{d}\Omega}{\mathrm{d}t} \approx -\frac{3}{2} n J_2 \left(\frac{\rho_e}{a}\right)^2 \frac{\cos i}{(1-e^2)^2}, \tag{2.88}$$

$$\frac{\mathrm{d}i}{\mathrm{d}t} \approx -\mu_1 \cos \Omega - \mu_2 \sin \Omega, \tag{2.89}$$

whose solution,

$$i = -\frac{\mu_1}{\chi} \cos[-\chi(t-t_o) + \Omega_o] + \frac{\mu_2}{\chi} \sin[-\chi(t-t_o) + \Omega_o] + i_o,$$

$$\Omega = -\chi(t-t_o) + \Omega_o \quad \text{where} \quad \chi \equiv \frac{3}{2} n J_2 \left(\frac{\rho_e}{a}\right)^2 \frac{\cos i}{(1-e^2)^2}, \tag{2.90}$$

---

[9] Goldreich used the non-sphericity parameter $J \equiv (3/2)(\rho_e/\rho)^2 J_2$, where ρ_e is the mean *equatorial* radius.

tells us that in the course of equinoctial precession the satellite inclination oscillates about i_o.

Goldreich [15] noticed that his i and the other elements were not osculating, but he assumed that their secular parts would differ from those of the osculating ones only in the orders higher than $O(|\boldsymbol{\mu}|)$. Below we shall probe the applicability limits for this assumption. (See the end of subsection 2.3.5.)

2.3.4 *Planetary equations in a precessing frame, in terms of osculating elements*

When one introduces elements in the precessing frame and also demands that they osculate in this frame (i.e., obey the Lagrange constraint $\boldsymbol{\Phi}=0$), the Hamiltonian variation reads:[10]

$$\Delta\mathcal{H}^{(osc)} = -\left[-\Delta U + \boldsymbol{\mu}\cdot(\boldsymbol{f}\times\mathbf{g}) + (\boldsymbol{\mu}\times\boldsymbol{f})\cdot(\boldsymbol{\mu}\times\boldsymbol{f})\right], \tag{2.91}$$

while Eq. (2.30) becomes:

$$[C_n C_i]\frac{\mathrm{d}C_i}{\mathrm{d}t} = -\frac{\partial\Delta\mathcal{H}^{(osc)}}{\partial C_n} + \boldsymbol{\mu}\cdot\left(\frac{\partial\boldsymbol{f}}{\partial C_n}\times\mathbf{g} - \boldsymbol{f}\times\frac{\partial\mathbf{g}}{\partial C_n}\right) - \dot{\boldsymbol{\mu}}\cdot\left(\boldsymbol{f}\times\frac{\partial\boldsymbol{f}}{\partial C_n}\right)$$
$$- (\boldsymbol{\mu}\times\boldsymbol{f})\frac{\partial}{\partial C_n}(\boldsymbol{\mu}\times\boldsymbol{f}). \tag{2.92}$$

To ease the comparison of this equation with (2.77), it is convenient to split the expression (2.91) for $\Delta\mathcal{H}^{(osc)}$ into two parts:

$$\Delta\mathcal{H}^{(cont)} = -\left[R_{oblate}(\boldsymbol{f},t) + \boldsymbol{\mu}\cdot(\boldsymbol{f}\times\mathbf{g})\right] \tag{2.93}$$

and

$$-(\boldsymbol{\mu}\times\boldsymbol{f})\cdot(\boldsymbol{\mu}\times\boldsymbol{f}), \tag{2.94}$$

and then to group the latter part with the last term on the right-hand side of (2.35):

$$[C_n C_i]\frac{\mathrm{d}C_i}{\mathrm{d}t} = -\frac{\partial\Delta\mathcal{H}^{(cont)}}{\partial C_n} + \boldsymbol{\mu}\cdot\left(\frac{\partial\boldsymbol{f}}{\partial C_n}\times\mathbf{g} - \boldsymbol{f}\times\frac{\partial\mathbf{g}}{\partial C_n}\right) - \dot{\boldsymbol{\mu}}\cdot\left(\boldsymbol{f}\times\frac{\partial\boldsymbol{f}}{\partial C_n}\right)$$
$$+ (\boldsymbol{\mu}\times\boldsymbol{f})\frac{\partial}{\partial C_n}(\boldsymbol{\mu}\times\boldsymbol{f}). \tag{2.95}$$

Comparison of this analytical theory with a straightforward numerical integration[11] has confirmed that the $O(|\boldsymbol{\mu}|^2)$ term in (2.95) may be neglected over time scales of, at

[10] Both $\Delta\mathcal{H}^{(cont)}$ and $\Delta\mathcal{H}^{(osc)}$ are equal to $-[-\Delta U(\boldsymbol{f},t)+\boldsymbol{\mu}\cdot\mathbf{J}] = -[-\Delta U(\boldsymbol{f},t)+\boldsymbol{\mu}\cdot(\boldsymbol{f}\times\boldsymbol{p})]$. However, the canonical momentum now is different from $\mathbf{g}$ and reads as: $\boldsymbol{p}=\mathbf{g}+(\boldsymbol{\mu}\times\boldsymbol{f})$. Hence, the functional forms of $\Delta\mathcal{H}^{(osc)}(\boldsymbol{f},\boldsymbol{p})$ and $\Delta\mathcal{H}^{(can)}(\boldsymbol{f},\boldsymbol{p})$ are different, though their values coincide.

[11] Credit for this comparison goes to Pini Gurfil and Valery Lainey.

least, hundreds of millions of years. In this approximation there is no difference between $\Delta\mathcal{H}^{(cont)}$ and $\Delta\mathcal{H}^{(osc)}$, so we shall write down the equations as:

$$[C_n\, C_i]\frac{\mathrm{d}C_i}{\mathrm{d}t} = -\frac{\partial \Delta\mathcal{H}^{(cont)}}{\partial C_n} + \boldsymbol{\mu}\cdot\left(\frac{\partial \boldsymbol{f}}{\partial C_n}\times \mathbf{g} - \boldsymbol{f}\times\frac{\partial \vec{\mathbf{g}}}{\partial C_n}\right) - \dot{\boldsymbol{\mu}}\cdot\left(\boldsymbol{f}\times\frac{\partial \boldsymbol{f}}{\partial C_n}\right). \tag{2.96}$$

For C_i chosen as the Kepler elements, inversion of the Lagrange brackets in (2.90) will yield the following Lagrange-type system:

$$\frac{\mathrm{d}a}{\mathrm{d}t} = \frac{2}{na}\left[\frac{\partial\left(-\Delta\mathcal{H}^{(cont)}\right)}{\partial M_o} - \dot{\boldsymbol{\mu}}\cdot\left(\boldsymbol{f}\times\frac{\partial \boldsymbol{f}}{\partial M_o}\right)\right], \tag{2.97}$$

$$\begin{aligned}\frac{\mathrm{d}e}{\mathrm{d}t} = {} & \frac{1-e^2}{na^2e}\left[\frac{\partial\left(-\Delta\mathcal{H}^{(cont)}\right)}{\partial M_o} - \dot{\boldsymbol{\mu}}\cdot\left(\boldsymbol{f}\times\frac{\partial \boldsymbol{f}}{\partial M_o}\right)\right] - \frac{(1-e^2)^{1/2}}{na^2e} \\ & \times\left[\frac{\partial\left(-\Delta\mathcal{H}^{(cont)}\right)}{\partial \omega} + \boldsymbol{\mu}\cdot\left(\frac{\partial \boldsymbol{f}}{\partial \omega}\times\mathbf{g} - \boldsymbol{f}\times\frac{\partial \mathbf{g}}{\partial \omega}\right) - \dot{\boldsymbol{\mu}}\cdot\left(\boldsymbol{f}\times\frac{\partial \boldsymbol{f}}{\partial \omega}\right)\right],\end{aligned} \tag{2.98}$$

$$\begin{aligned}\frac{\mathrm{d}\omega}{\mathrm{d}t} = {} & \frac{-\cos i}{na^2(1-e^2)^{1/2}\sin i}\left[\frac{\partial\left(-\Delta\mathcal{H}^{(cont)}\right)}{\partial i} + \boldsymbol{\mu}\cdot\left(\frac{\partial \boldsymbol{f}}{\partial i}\times\mathbf{g} - \boldsymbol{f}\times\frac{\partial \mathbf{g}}{\partial i}\right) - \dot{\boldsymbol{\mu}}\cdot\left(\boldsymbol{f}\times\frac{\partial \boldsymbol{f}}{\partial i}\right)\right] \\ & + \frac{(1-e^2)^{1/2}}{na^2e}\left[\frac{\partial\left(-\Delta\mathcal{H}^{(cont)}\right)}{\partial e} + \boldsymbol{\mu}\cdot\left(\frac{\partial \boldsymbol{f}}{\partial e}\times\mathbf{g} - \boldsymbol{f}\times\frac{\partial \mathbf{g}}{\partial e}\right) - \dot{\boldsymbol{\mu}}\cdot\left(\boldsymbol{f}\times\frac{\partial \boldsymbol{f}}{\partial e}\right)\right],\end{aligned} \tag{2.99}$$

$$\begin{aligned}\frac{\mathrm{d}i}{\mathrm{d}t} = {} & \frac{\cos i}{na^2(1-e^2)^{1/2}\sin i}\left[\frac{\partial\left(-\Delta\mathcal{H}^{(cont)}\right)}{\partial \omega} + \boldsymbol{\mu}\cdot\left(\frac{\partial \boldsymbol{f}}{\partial \omega}\times\mathbf{g} - \boldsymbol{f}\times\frac{\partial \mathbf{g}}{\partial \omega}\right) - \dot{\boldsymbol{\mu}}\cdot\left(\boldsymbol{f}\times\frac{\partial \boldsymbol{f}}{\partial \omega}\right)\right] \\ & - \frac{1}{na^2(1-e^2)^{1/2}\sin i}\left[\frac{\partial\left(-\Delta\mathcal{H}^{(cont)}\right)}{\partial \Omega} + \boldsymbol{\mu}\cdot\left(\frac{\partial \boldsymbol{f}}{\partial \Omega}\times\mathbf{g} - \boldsymbol{f}\times\frac{\partial \mathbf{g}}{\partial \Omega}\right) - \dot{\boldsymbol{\mu}}\cdot\left(\boldsymbol{f}\times\frac{\partial \boldsymbol{f}}{\partial \Omega}\right)\right],\end{aligned} \tag{2.100}$$

$$\begin{aligned}\frac{\mathrm{d}\Omega}{\mathrm{d}t} = {} & \frac{1}{na^2(1-e^2)^{1/2}\sin i}\left[\frac{\partial\left(-\Delta\mathcal{H}^{(cont)}\right)}{\partial i} + \boldsymbol{\mu}\cdot\left(\frac{\partial \boldsymbol{f}}{\partial i}\times\mathbf{g} - \boldsymbol{f}\times\frac{\partial \mathbf{g}}{\partial i}\right)\right. \\ & \left. - \dot{\boldsymbol{\mu}}\cdot\left(\boldsymbol{f}\times\frac{\partial \boldsymbol{f}}{\partial i}\right)\right],\end{aligned} \tag{2.101}$$

$$\frac{\mathrm{d}M_o}{\mathrm{d}t} = -\frac{1-e^2}{na^2e}\left[\frac{\partial\left(-\Delta\mathcal{H}^{(cont)}\right)}{\partial e} + \boldsymbol{\mu}\cdot\left(\frac{\partial \boldsymbol{f}}{\partial e}\times\mathbf{g} - \boldsymbol{f}\times\frac{\partial \mathbf{g}}{\partial e}\right) - \dot{\boldsymbol{\mu}}\cdot\left(\boldsymbol{f}\times\frac{\partial \boldsymbol{f}}{\partial e}\right)\right]$$
$$-\frac{2}{na}\left[\frac{\partial\left(-\Delta\mathcal{H}^{(cont)}\right)}{\partial a} + \boldsymbol{\mu}\cdot\left(\frac{\partial \boldsymbol{f}}{\partial a}\times\mathbf{g} - \boldsymbol{f}\times\frac{\partial \mathbf{g}}{\partial a}\right) - \dot{\boldsymbol{\mu}}\cdot\left(\boldsymbol{f}\times\frac{\partial \boldsymbol{f}}{\partial a}\right)\right], \tag{2.102}$$

terms $\boldsymbol{\mu}\cdot((\partial \boldsymbol{f}/\partial M_o)\times\mathbf{g} - (\partial\mathbf{g}/\partial M_o)\times\boldsymbol{f})$ being omitted in (2.97–2.98), because these terms vanish identically (see the Appendix to Efroimsky [14]).

2.3.5 Comparison of calculations performed in the two above gauges

Simply from looking at Eqs. (2.76–2.83) and (2.96–2.102) we notice that the difference in orbit descriptions performed in the two gauges emerges already in the first order of the precession rate $\boldsymbol{\mu}$ and in the first order of $\dot{\boldsymbol{\mu}}$.

Calculation of the $\boldsymbol{\mu}$- and $\dot{\boldsymbol{\mu}}$-dependent terms emerging in Eqs. (2.97–2.102) takes more than 20 pages of algebra. The resulting expressions are published in Efroimsky [20], their detailed derivation being available in web-archive preprint Efroimsky [20]. As an illustration, we present a couple of expressions:

$$-\dot{\boldsymbol{\mu}}\cdot\left(\boldsymbol{f}\times\frac{\partial \boldsymbol{f}}{\partial i}\right) = a^2\frac{\left(1-e^2\right)^2}{(1+e\cos\nu)^2}\{\dot{\mu}_1[-\cos\Omega\sin(\omega+\nu)$$
$$-\sin\Omega\cos(\omega+\nu)\cos i]\sin(\omega+\nu) + \dot{\mu}_2[-\sin\Omega\sin(\omega+\nu)$$
$$+\cos\Omega\cos(\omega+\nu)\cos i]\sin(\omega+\nu)$$
$$+\dot{\mu}_3\sin(\omega+\nu)\cos(\omega+\nu)\sin i\}, \tag{2.103}$$

$$\boldsymbol{\mu}\cdot\left(\frac{\partial \boldsymbol{f}}{\partial e}\times\mathbf{g} - \boldsymbol{f}\times\frac{\partial \vec{\mathbf{g}}}{\partial e}\right) = -\mu_\perp\frac{na^2\left(3e+2\cos\nu+e^2\cos\nu\right)}{(1+e\cos\nu)\sqrt{1-e^2}}, \tag{2.104}$$

ν denoting the true anomaly. The fact that almost none of these terms vanish reveals that Eqs. (2.76–2.83) and (2.96–2.102) may yield very different results, i.e., that the contact elements may differ from their osculating counterparts already in the first order of $\boldsymbol{\mu}$.

Luckily, in the practical situations we need not the elements *per se* but their secular parts. To calculate these, one can substitute both the Hamiltonian variation and the $\boldsymbol{\mu}$- and $\dot{\boldsymbol{\mu}}$-dependent terms with their orbital averages[12] calculated through

$$\langle\ldots\rangle \equiv \frac{\left(1-e^2\right)^{3/2}}{2\pi}\int_0^{2\pi}\ldots\frac{\mathrm{d}\nu}{(1+e\cos\nu)^2}. \tag{2.105}$$

[12] Mathematically, this procedure is, to say the least, not rigorous. In practical calculations it works well, at least over not too long time scales.

The situation might simplify very considerably if we could also assume that the precession rate $\boldsymbol{\mu}$ stays constant. Then in equations (2.97–2.102), we would assume $\boldsymbol{\mu} = constant$ and proceed with averaging the expressions $((\partial \boldsymbol{f}/\partial C_j) \times \mathbf{g} - \boldsymbol{f} \times (\partial \mathbf{g}/\partial C_j))$ only (while all the terms with $\dot{\boldsymbol{\mu}}$ will now vanish).

Averaging of the said terms is lengthy and is presented in the Appendix to Efroimsky [14]. All in all, we get, for constant $\boldsymbol{\mu}$:

$$\boldsymbol{\mu} \cdot \left\langle \left(\frac{\partial \boldsymbol{f}}{\partial a} \times \vec{\mathbf{g}} - \boldsymbol{f} \times \frac{\partial \mathbf{g}}{\partial a} \right) \right\rangle = \boldsymbol{\mu} \cdot \left(\frac{\partial \boldsymbol{f}}{\partial a} \times \vec{\mathbf{g}} - \boldsymbol{f} \times \frac{\partial \mathbf{g}}{\partial a} \right) = \frac{3}{2} \mu_{\perp} \sqrt{\frac{Gm\,(1-e^2)}{a}}, \tag{2.106}$$

$$\boldsymbol{\mu} \cdot \left\langle \left(\frac{\partial \boldsymbol{f}}{\partial C_j} \times \mathbf{g} - \boldsymbol{f} \times \frac{\partial \mathbf{g}}{\partial C_j} \right) \right\rangle = 0, \quad C_j = e, \Omega, \omega, i, M_o. \tag{2.107}$$

Since the orbital averages (2.107) vanish, then e will, along with a, stay constant for as long as our approximation remains valid. Besides, no trace of $\boldsymbol{\mu}$ will be left in the equations for Ω and i. This means that, in the assumed approximation and under the extra assumption of constant $\boldsymbol{\mu}$, the afore quoted analysis (2.84–2.90), offered by Goldreich [15], will remain valid at time scales which are not tool long.

In the realistic case of time-dependent precession, the averages of terms containing $\boldsymbol{\mu}$ and $\dot{\boldsymbol{\mu}}$ do not vanish (except for $\boldsymbol{\mu} \cdot ((\partial \boldsymbol{f}/\partial M_o) \times \mathbf{g} - \boldsymbol{f} \times (\partial \mathbf{g}/\partial M_o))$, which is identically nil). These terms show up in all equations (except in that for a) and influence the motion. Integration that includes these terms gives results very close to the Goldreich approximation (approximation (2.90) that neglects the said terms and approximates the secular parts of the non-osculating elements with those of their osculating counterparts). However, this agreement takes place only at time scales of order millions to dozens of millions of years. At larger time scales, differences begin to accumulate [21].

In real life, the equinoctial-precession rate of the planet, $\boldsymbol{\mu}$, is not constant. Since the equinoctial precession is caused by the solar torque acting on the oblate planet, this precession is regulated by the relative location and orientation of the Sun and the planetary equator. This is why $\boldsymbol{\mu}$ of a planet depends upon this planet's orbit precession caused by the pull from the other planets. This dependence is described by a simple model developed by Colombo [22].

2.4 Conclusions: how we benefit from the gauge freedom

In this chapter we gave a review of the gauge concept in orbital and attitude dynamics. Essentially, this is the freedom of choosing non-osculating orbital (or rotational) elements, i.e., the freedom of making them deviate from osculation in a known, prescribed, manner.

The advantage of elements introduced in a non-trivial gauge is that in certain situations the choice of such elements considerably simplifies the mathematical description of orbital and attitude problems. One example of such simplification is the Goldreich [15] approximation (2.90) for satellite orbiting a precessing oblate planet. Although performed in terms of non-osculating elements, Goldreich's calculation has the advantage of mathematical simplicity. Most importantly, later studies [20, 23] have confirmed that Goldreich's results, obtained for non-osculating elements, serves as a very good approximation for the osculating elements. To be more exact, the secular parts of these non-osculating elements coincide, in the first order over the precession-caused perturbation, with those of their osculating counterparts, the difference accumulating only at very long time scales—see the end of Section 2.3 above. A comprehensive investigation into this topic, with the relevant numerics, will be presented in Lainey et al. [21].

On the other hand, *neglect of the gauge freedom may sometimes produce camouflaged pitfalls caused by the fact that non-osculating elements lack evident physical meaning.* For example, the non-osculating "inclination" does not coincide with the real, physical inclination of the orbit. This happens because non-osculating elements parameterise instantaneous conics non-tangent to the orbit. Similar difficulties emerge in the theory of rigid-body rotation, when non-osculating Andoyer variables are employed.

Appendix 1. Mathematical formalities: Orbital dynamics in the normal form of Cauchy

Let us cast the perturbed equation

$$\ddot{\mathbf{r}} = \boldsymbol{F} + \Delta \boldsymbol{f} = -\frac{\mu}{r^2}\frac{\boldsymbol{r}}{r} + \Delta \boldsymbol{f} \tag{2.108}$$

into the normal form of Cauchy:

$$\dot{\boldsymbol{r}} = \mathbf{v}, \tag{2.109}$$

$$\dot{\mathbf{v}} = -\frac{\mu}{r^2}\frac{\boldsymbol{r}}{r} + \Delta \boldsymbol{f}\,(\boldsymbol{r}(t, C_1, \ldots, C_6), \mathbf{v}(t, C_1, \ldots, C_6), t)\,. \tag{2.110}$$

Insertion of our ansatz

$$\boldsymbol{r} = \boldsymbol{f}\,(t, C_1(t), \ldots, C_6(t))\,, \tag{2.111}$$

will make (2.109) equivalent to

$$\mathbf{v} = \frac{\partial \boldsymbol{f}}{\partial t} + \sum_i \frac{\partial \boldsymbol{f}}{\partial C_i}\dot{C}_i. \tag{2.112}$$

The function $\boldsymbol{f}$ is, by definition, the generic solution to the unperturbed equation

$$\ddot{\mathbf{r}} = \boldsymbol{F} = -\frac{\mu}{r^2}\frac{\boldsymbol{r}}{r}. \tag{2.113}$$

This circumstance, along with (2.112), will transform (2.109) into

$$\sum_i \frac{\partial \mathbf{g}}{\partial C_i}\dot{C}_i + \dot{\boldsymbol{\Phi}} = \Delta \mathbf{F}\,(\boldsymbol{f}(t, C_1, \ldots, C_6),\ \mathbf{g}(t, C_1, \ldots, C_6) + \boldsymbol{\Phi}) \tag{2.114}$$

where

$$\boldsymbol{\Phi} \equiv \sum_i \frac{\partial \boldsymbol{f}}{\partial C_j}\dot{C}_j \tag{2.115}$$

is an identity, $f(t, C_1, \ldots, C_6)$ and $\mathbf{g}(t, C_1, \ldots, C_6) \equiv \partial f/\partial t$ being known functions. Now (2.114–2.115) make an incomplete system of six first-order equations for nine variables $(C_1, \ldots, C_6, \Phi_1, \ldots, \Phi_3)$. So one has to impose three arbitrary conditions on C, Φ, for example as

$$\boldsymbol{\Phi} = \boldsymbol{\Phi}(t, C_1, \ldots, C_6). \tag{2.116}$$

This will result in a closed system of six equations for six variables C_j:

$$\sum_i \frac{\partial \mathbf{g}}{\partial C_i} \dot{C}_i = \Delta \mathbf{F}\left(\mathbf{f}(t, C_1, \ldots, C_6), \mathbf{g}(t, C_1, \ldots, C_6) + \boldsymbol{\Phi}\right) - \dot{\boldsymbol{\Phi}} \tag{2.117}$$

$$\sum_i \frac{\partial \mathbf{f}}{\partial C_i} \frac{\mathrm{d}C_i}{\mathrm{d}t} = \boldsymbol{\Phi}, \tag{2.118}$$

$\boldsymbol{\Phi} = \boldsymbol{\Phi}(t, C_1, \ldots, C_6)$ now being some fixed function (gauge).[13] A trivial choice is $\boldsymbol{\Phi}(t, C_1, \ldots, C_6) = 0$, and this is what is normally taken by default. This choice is only one out of infinitely many, and often is not optimal. Under an arbitrary, non-zero, choice of the function $\boldsymbol{\Phi}(t, C_1, \ldots, C_6)$, the system (2.117–2.118) will have some different solution $C_j(t)$. To get the appropriate solution for the Cartesian components of the position and velocity, one will have to use formulae

$$\boldsymbol{r} = \boldsymbol{f}(t, C_1, \ldots, C_6), \tag{2.119}$$

$$\dot{\boldsymbol{r}} \equiv \mathbf{v} = \mathbf{g}(t, C_1, \ldots, C_6) + \boldsymbol{\Phi}(t, C_1, \ldots, C_6), \tag{2.120}$$

Appendix 2. Precession of the equator of date relative to the equator of epoch

The afore introduced vector $\boldsymbol{\mu}$ is the precession rate of the equator of date relative to the equator of epoch. Let the inertial axes (X, Y, Z) and the corresponding unit vectors $(\hat{\mathbf{X}}, \hat{\mathbf{Y}}, \hat{\mathbf{Z}})$ be fixed in space so that X and Y belong to the equator of epoch. A rotation within the equator-of-epoch plane by longitude h_p, from axis X, will define the line of nodes, x. A rotation about this line by an inclination angle I_p will give us the planetary equator of date. The line of nodes x, along with axis y naturally chosen within the equator-of-date plane, and with axis z orthogonal to this plane, will constitute the precessing coordinate system, with the appropriate basis denoted by $(\hat{\mathbf{x}}, \hat{\mathbf{y}}, \hat{\mathbf{z}})$.

In the inertial basis $(\hat{\mathbf{X}}, \hat{\mathbf{Y}}, \hat{\mathbf{Z}})$, the direction to the North Pole of date is given by

$$\hat{\mathbf{z}} = \left(\sin I_p \sin h_p, -\sin I_p \cos h_p, \cos I_p\right)^T \tag{2.121}$$

while the total angular velocity reads:

$$\boldsymbol{\omega}_{\text{total}}^{(\text{inertial})} = \hat{\mathbf{z}}\,\Omega_z + \boldsymbol{\mu}^{(\text{inertial})}, \tag{2.122}$$

[13] Generally, $\boldsymbol{\Phi}$ may depend also upon the variables' time derivatives of all orders: $\boldsymbol{\Phi}(t;\, C_i, \dot{C}_i, \ddot{C}_i, \ldots)$. This will give birth to higher time derivatives of C in subsequent developments and will require additional initial conditions, beyond those on $\boldsymbol{r}$ and $\dot{\boldsymbol{r}}$, to be fixed to close the system. So it is practical to accept (2.116).

the first term denoting the rotation about the precessing axis $\hat{\mathbf{z}}$, and the second term being the precession rate of $\hat{\mathbf{z}}$ relative to the inertial frame $(\hat{\mathbf{X}}, \hat{\mathbf{Y}}, \hat{\mathbf{Z}})$. This precession rate is given by

$$\boldsymbol{\mu}^{(\text{inertial})} = \left(\dot{I}_p \cos h_p, \dot{I}_p \sin h_p, \dot{h}_p\right)^T, \tag{2.123}$$

because this expression satisfies $\boldsymbol{\mu}^{(\text{inertial})} \times \hat{\mathbf{z}} = \dot{\hat{\mathbf{z}}}$.

In a frame co-precessing with the equator of date, the precession rate will be represented by vector

$$\boldsymbol{\mu} = \hat{\mathbf{R}}_{i\to p}\, \boldsymbol{\mu}^{(\text{inertial})}, \tag{2.124}$$

where the matrix of rotation from the equator of epoch to that of date (i.e., from the inertial frame to the precessing one) is given by

$$\hat{\mathbf{R}}_{i\to p} = \begin{bmatrix} \cos h_p & \sin h_p & 0 \\ -\cos I_p \sin h_p & \cos I_p \sin h_p & \sin I_p \\ \sin I_p \sin h_p & -\sin I_p \sin h_p & \cos I_p \end{bmatrix}$$

From here we get the components of the precession rate, as seen in the co-precessing coordinate frame (x, y, z):

$$\boldsymbol{\mu} = (\mu_1, \mu_2, \mu_3)^T = \left(\dot{I}_p, \dot{h}_p \sin I_p, \dot{h}_p \cos I_p\right)^T. \tag{2.125}$$

References

1. Efroimsky, Michael, and Peter Goldreich. (2003). Gauge symmetry of the N-body problem in the Hamilton–Jacobi approach. *Journal of Mathematical Physics*, **44**, pp. 5958–5977 astro-ph/0305344.
2. Efroimsky, Michael, and Peter Goldreich (2004). Gauge freedom in the N-body problem of celestial mechanics. *Astronomy & Astrophysics*, **415**, pp. 1187–1199 astro-ph/0307130.
3. Euler, L. (1748). *Recherches sur la question des inegalites du mouvement de Saturne et de Jupiter, sujet propose pour le prix de l'annee*. Berlin. Modern edition: L. Euler *Opera mechanica et astronomica*. Birkhauser-Verlag, Switzerland, 1999.
4. Euler, L. (1753). *Theoria motus Lunae exhibens omnes ejus inaequalitates etc.* Impensis Academiae Imperialis Scientarum Petropolitanae. St. Petersburg, Russia 1753. Modern edition: L. Euler *Opera mechanica et astronomica*. Birkhauser-Verlag, Switzerland 1999.
5. Lagrange, J.-L. (1778). *Sur le Problème de la détermination des orbites des comètes d'après trois observations, 1-er et 2-ième mémoires*. Nouveaux Mémoires de l'Académie de Berlin, 1778. Later edition: *Œuvres de Lagrange*. **IV**, Gauthier-Villars, Paris 1869.
6. Lagrange, J.-L. (1783). *Sur le Problème de la détermination des orbites des comètes d'après trois observations, 3-ième mémoire*. Ibidem, 1783. Later edition: *Œuvres de Lagrange*. **IV**, Gauthier-Villars, Paris 1869.
7. Lagrange, J.-L. (1808). Sur la théorie des variations des éléments des planètes et en particulier des variations des grands axes de leurs orbites. Lu, le 22 août 1808 à l'Institut de France. Later edition: *Œuvres de Lagrange*. **VI**, pp. 713–768, Gauthier-Villars, Paris 1877.
8. Lagrange, J.-L. (1809). Sur la théorie générale de la variation des constantes arbitraires dans tous les problèmes de la mécanique. Lu, le 13 mars 1809 à l'Institut de France. Later edition: *Œuvres de Lagrange*. **VI**, pp. 771–805, Gauthier-Villars, Paris 1877.

9. Lagrange, J.-L. (1810). Second mémoire sur la théorie générale de la variation des constantes arbitraires dans tous les problèmes de la mécanique. Lu, le 19 février 1810 à l'Institut de France. Later edition: *Œuvres de Lagrange*. **VI**, pp. 809–816, Gauthier-Villars, Paris 1877.
10. Newman, W., and M. Efroimsky. (2003). The Method of Variation of Constants and Multiple Time Scales in Orbital Mechanics. *Chaos*, **13**, pp. 476–485.
11. Gurfil, P., and Klein, I. (2006). Mitigating the Integration Error in Numerical Simulations of Newtonian Systems. Submitted to *The International Journal for Numerical Methods in Engineering*.
12. Efroimsky, Michael (2002a). Equations for the orbital elements. Hidden symmetry. Preprint 1844 of the Institute of Mathematics and its Applications, University of Minnesota http://www.ima.umn.edu/preprints/ feb02/feb02.html.
13. Efroimsky, Michael (2002b). The implicit gauge symmetry emerging in the N-body problem of celestial mechanics. astro-ph/0212245.
14. Efroimsky, M. (2004). Long-term evolution of orbits about a precessing oblate planet. The case of uniform precession. astro-ph/0408168 (This preprint is a very extended version of the published paper Efroimsky (2005). It contains all technical calculations omitted in the said publication.)
15. Goldreich, P. (1965). Inclination of satellite orbits about an oblate precessing planet. *The Astronomical Journal*, **70**, pp. 5–9.
16. Brumberg, V. A., L. S. Evdokimova, and N. G. Kochina. (1971). Analytical methods for the orbits of artificial satellites of the moon. *Celestial Mechanics*, **3**, pp. 197–221.
17. Slabinski, V. (2003). Satellite orbit plane perturbations using an Efroimsky gauge velocity. Talk at the 34th Meeting of the AAS Division on Dynamical Astronomy, Cornell University, May 2003.
18. Gurfil, P. (2004). Analysis of J_2–perturbed Motion using Mean Non-Osculating Orbital Elements. *Celestial Mechanics & Dynamical Astronomy*, **90**, pp. 289–306.
19. Kinoshita, T. (1993). Motion of the Orbital Plane of a Satellite due to a Secular Change of the Obliquity of its Mother Planet. *Celestial Mechanics and Dynamical Astronomy*, **57**, pp. 359–368.
20. Efroimsky, M. (2005a). Long-term evolution of orbits about a precessing oblate planet. 1. The case of uniform precession. *Celestial Mechanics and Dynamical Astronomy*, **91**, pp. 75–108.
21. Lainey, V., Gurfil, P., and Efroimsky, M. (2005). Long-term evolution of orbits about a precessing oblate planet. 3. A semianalytical and a purely numerical approaches. *Celestial Mechanics and Dynamical Astronomy* (submitted).
22. Colombo, G. (1966). Cassini's second and third laws. *The Astronomical Journal.*, **71**, pp. 891–896.
23. Efroimsky, M. (2005b). Long-term evolution of orbits about a precessing oblate planet. 2. The case of variable precession. *Celestial Mechanics & Dynamical Astronomy* (submitted).
24. Efroimsky, M. (2005c). The theory of canonical perturbations applied to attitude dynamics and to the Earth rotation. astro-ph/0506427.

3 Solving Two-Point Boundary Value Problems Using Generating Functions: Theory and Applications to Astrodynamics

VINCENT M. GUIBOUT* AND DANIEL J. SCHEERES†

* *MBDA France*

† *University of Michigan, Ann Arbor, Michigan*

Contents

3.1 Introduction

Two-point boundary value problems have a central place in the field of astrodynamics. In general, most of the hard problems in this field revolve around solving such problems. Examples include the targeting problem for mission design, the computation of periodic orbits for the analysis of systems, and the solution of optimal control problems.

In this chapter, a new methodology for solving two-point boundary value problems in phase space for Hamiltonian systems is presented. Using the Hamilton–Jacobi theory in conjunction with the canonical transformation induced by the phase flow, we show that the generating functions for this transformation solve any two-point boundary value problem in phase space. Properties of the generating functions are exposed, we especially emphasize multiple solutions, singularities, relations with the state transition matrix and symmetries. Next, we show that using Hamilton's principal function we are also able to solve two-point boundary value problems, nevertheless both methodologies have fundamental differences that we explore. Then we present and study an algorithm to compute the generating functions specialized to such two-point boundary value problems. This algorithm is able to compute the generating functions for a large class of practical two-point boundary value problems. Specifically, the algorithm naturally avoids singularities and allows one to specify the initial conditions as a function of a parameter. Finally, we present applications of this method to two difficult problems of astrodynamics to show its generality, computation of periodic orbits and solution of an optimal control problem.

One of the most famous two-point boundary value problems in astrodynamics is Lambert's problem, which consists of finding a trajectory in the two-body problem which goes through two given points in a given lapse of time. Even though the two-body problem is integrable, no explicit solution to this problem exists. Many other two-point boundary value problems in astrodynamics can also be couched within a Hamiltonian formalism. These include all problems of orbital motion, excepting the effect of atmospheric drag on an orbiter, and also include all instances of optimal control problems. It is important to note that optimal control problems can all be recast into Hamiltonian systems via the necessary conditions. Thus, even non-conservative dynamical systems can be treated using the formalism we develop here when dealing with their optimal control. In the following, we do not make a distinction between whether a Hamiltonian dynamical system arises out of mechanics or out of optimal control, as the basic results of the Hamilton–Jacobi theory that we use apply to both.

For a general Hamiltonian dynamical system, a two-point boundary value problem is generally solved using iterative techniques such as shooting and relaxation methods. The shooting method [5, 29] consists of choosing values for all of the dependent variables at one boundary. These values must be consistent with any boundary conditions for that boundary, but otherwise are initially guessed "randomly". After integration of the differential equations, we in general find discrepancies between the desired boundary values at the other boundary. Then, we adjust the initial guess to reduce these discrepancies and reiterate this procedure again. The method provides a systematic approach to solving boundary value problems, but suffers several inherent limitations. As summarized by Bryson and Ho ([7] p. 214),

> The main difficulty with this method is getting started; i.e., finding a first estimate of the unspecified conditions at one end that produces a solution reasonably close to the specified conditions at the other end. The reason for this peculiar difficulty is that the extremal solutions are often very sensitive to small changes in the unspecified boundary conditions.

To get rid of the sensitivity to small changes in initial guesses, techniques such as the multiple shooting method [21] were developed. They consist of breaking the time domain into segments and solving a boundary value problem on each of these segments. In this manner, non-linear effects are limited over each segment, but on the other hand the size of the problem is increased considerably. However, the choice of the initial conditions still remains as the main hurdle to successfully apply shooting methods to general problems.

Relaxation methods [30] use a different approach. The differential equations are replaced by finite-difference equations on a mesh of points that covers the range of the integration. A trial solution consists of values for the dependent variables at each mesh point, not satisfying the desired finite-difference equations, nor necessarily even satisfying the required boundary conditions. The iteration, now called relaxation, consists of adjusting all the values on the mesh so as to bring them into successively closer agreement with the finite-difference equations and simultaneously with the boundary conditions. In general, relaxation works better than shooting when the boundary conditions are especially delicate or subtle. However, if the solution is highly oscillatory then many grid points are required for accurate representation. Also, the number and positions of the required mesh points are not known a priori and must be adjusted manually for each problem. In addition, if solutions to the differential equations develop singularities, attempts to refine the mesh to improve accuracy may fail.

With the advent of computers, these two methods are able to solve most of the two-point boundary value problems. They may require substantial time to find an appropriate initial guess and/or computer memory to refine the mesh, but they generally succeed. However, there are problems for which these methods reach their limits. For instance, the design of space missions involving several spacecraft in formation requires one to solve a large number of boundary value problems for which the boundary conditions may in turn depend on parameters. Most research in this area to date has focused on the solution of such boundary value problems for linearized motion. Extension of these techniques to non-linear dynamics is much more difficult, with most progress being limited to non-linear expansions of the two-body problem with minimal perturbations added, if any [2, 24]. However, for precise control of formations over long periods of time or over large distances, it is crucial that non-linear solutions to these problems be available. For example, to reconfigure a formation of N spacecraft, there are $N!$ possibilities in general, i.e., $N!$ boundary value problems need to be solved [31]. Similarly, suppose that we plan to reconfigure a spacecraft formation to achieve a specific goal, such as for an interferometry mission where they may be required to be equally spaced on a circle perpendicular to the line of sight they observe. In that case, the final positions are specified in terms of the angle that indicates the position of the spacecraft on the circle. In order to find the value of the angle that minimizes fuel expenditure, infinitely many boundary value problems may need to be solved, if evaluated in a formal sense. As a result, the algorithms mentioned above are no longer appropriate as they require excessive computation and time for modeling non-linear situations. To address these complex problems arising in spacecraft formation design, Guibout and Scheeres [16–18] developed a novel approach for solving boundary value problems which outperforms traditional methods for spacecraft formation design. In the present contribution we generalize their method and study its properties. We first prove that it allows us to formally solve a non-linear two-point boundary value problem at a cost of a single function evaluation once generating functions are known.

In addition, properties of the generating functions are studied. In particular, for linear systems we show that generating functions and state transition matrices are closely related. The state transition matrix allows one to predict singularities of the generating functions whereas the generating functions provide information on the structure of the state transition matrix. This relationship also allows us to recover and extend some results on the perturbation matrices introduced by Battin in Ref. [4]. For non-linear systems, generating functions may also develop singularities (called caustics). Using the Legendre transformation, we propose a technique to study the geometry of these caustics. We illustrate our method with the study of the singularities of the F_1 generating function in the Hill three-body problem and relate the existence of singularities to the presence of multiple solutions to boundary value problems. Furthermore, we discuss Hamilton's principal function, a function *similar* to the generating functions that also solves two-point boundary value problems. We highlight the differences between Hamilton's function and generating functions and justify our choice of focusing on generating functions. Next we outline and evaluate a method for constructing solutions for the generating functions. Finally, we present direct applications of this theory that have been identified in previous papers. These applications are presented to illustrate the application of our method to different problems in astrodynamics. We only present the main ideas and refer to previous papers for details.

3.2 Solving two-point boundary value problems

In this section, we review the principle of least action for Hamiltonian systems and derive the Hamilton–Jacobi equation. Local existence of generating functions is proved, but we underline that we do not study global properties. In general, we do not know a priori if the generating functions will be defined for all time, and in most of the cases we found that they develop singularities. We refer the reader to Refs. [1, 3, 9, 11, 14, 22, 23] for more details on local Hamilton–Jacobi theory, Refs. [1, 3, 23] for global theory and Refs. [1, 3, 8] and Section 3.2.4 of this chapter for the study of singularities.

3.2.1 The Hamilton–Jacobi theory

Let $(\mathcal{P}, \omega, X_H)$ be a Hamiltonian system with n degrees of freedom, and $H: \mathcal{P} \times \mathbb{R} \to \mathbb{R}$ the Hamiltonian function. We consider the symplectic charts whose existence is guaranteed by Darboux's theorem [6]. We denote the component functions (also called canonical coordinates) by (q_i, p_i) so that, in the symplectic chart, ω is locally written as:

$$\omega = \sum_{i=1}^{n} \mathrm{d}q_i \wedge \mathrm{d}p_i .$$

In the extended phase space $\mathcal{P} \times \mathbb{R}$, we consider an integral curve of the vector field X_H connecting the points (q_0, p_0, t_0) and (q_1, p_1, t_1). The principle of least action [23] reads:

Theorem 3.2.1. ***(The principle of least action in phase space)*** *Critical points of $\int_0^1 p\mathrm{d}q - H\mathrm{d}t$ in the class of curves γ whose ends lie in the n-dimensional subspaces $(t = t_0, q = q_0)$ and $(t = t_1, q = q_1)$ correspond to trajectories of the Hamiltonian system whose ends are q_0 at t_0 and q_1 at t_1.*

Proof. We proceed to the computation of the variation.

$$\begin{aligned} \delta \int_\gamma (p\dot{q} - H)\mathrm{d}t &= \int_\gamma \left(\dot{q}\delta p + p\delta\dot{q} - \frac{\partial H}{\partial q}\delta q - \frac{\partial H}{\partial p}\delta p \right) \mathrm{d}t \\ &= [p\delta q]_0^1 + \int_\gamma \left[\left(\dot{q} - \frac{\partial H}{\partial p} \right) \delta p - \left(\dot{p} + \frac{\partial H}{\partial q} \right) \delta q \right] \mathrm{d}t \end{aligned} \tag{3.1}$$

Therefore, since the variation vanishes at the end points, the integral curves of the Hamiltonian vector field are the only extrema. □

Now let $(\mathcal{P}_1, \omega_1)$ and $(\mathcal{P}_2, \omega_2)$ be symplectic manifolds,

Definition 3.2.1. A smooth map $f: \mathcal{P}_1 \times \mathbb{R} \to \mathcal{P}_2 \times \mathbb{R}$ is a canonical transformation from (q, p, t) to (Q, P, T) if and only if:
1. f is a diffeomorphism,
2. f preserves the time, i.e., there exists a function g_t such that $f(x, t) = (g_t(x), t)$ (from here we assume that $t = T$),
3. Critical points of $\int_{t_0}^{t_1}(\langle P, \dot{Q}\rangle - K(Q, P, t))\mathrm{d}t$ correspond to trajectories of the Hamiltonian system, where $K(Q, P, t) = H(q(Q, P, t), p(Q, P, t), t)$ is the Hamiltonian function expressed in the new set of coordinates.

Consider a canonical transformation between two sets of coordinates in the phase space $f: (q_i, p_i, t) \mapsto (Q_i, P_i, t)$ and let $H(q, p, t)$ and $K(Q, P, t)$ be the Hamiltonian functions of the same system expressed in different sets of coordinates. From Def. 3.2.1, trajectories correspond to critical points of $\int_{t_0}^{t_1}(\langle P, \dot{Q}\rangle - K(Q, P, t))\mathrm{d}t$. Therefore, they are integral of:

$$\begin{cases} \dot{Q}_i = \dfrac{\partial K}{\partial P_i}, \\ \dot{P}_i = -\dfrac{\partial K}{\partial Q_i}, \end{cases} \tag{3.2}$$

i.e., f preserves the canonical form of Hamilton's equations.

Conversely, suppose that f is a coordinate transformation that preserves the canonical form of Hamilton's equations and leaves the time invariant. Let $K(Q, P, t)$ be the Hamiltonian in the new set of coordinates, then from the modified Hamilton principle (Thm. 3.2.1), critical points of

$$\int_{t_0}^{t_1} \big(\langle P, \dot{Q}\rangle - K(Q, P, t)\big)\,\mathrm{d}t$$

correspond to trajectories of the system. Thus, f is a canonical map. These last two remarks are summarized in the following lemma:

Lemma 3.2.2. *The third item in Def. 3.2.1 is equivalent to:*

(4)—f preserves the canonical form of Hamilton's equations and the new Hamiltonian function is $K(Q, P, t)$.

Remark 3.2.1. The definition we give is different from the one given in many textbooks but in agreement with Arnold [3], Abraham and Marsden [1], and Marsden and Ratiu [23]. Often the third item reduces to:

(5)—f preserves the canonical form of Hamilton's equations.

We consider again a canonical transformation $f: (q_i, p_i, t) \mapsto (Q_i, P_i, t)$ and a Hamiltonian system defined by H. Along trajectories, we have by definition:

$$\delta \int_{t_0}^{t_1} \left(\sum_{i=1}^{n} p_i \dot{q}_i - H(q, p, t) \right) \mathrm{d}t = 0, \tag{3.3}$$

$$\delta \int_{t_0}^{t_1} \left(\sum_{i=1}^{n} P_i \dot{Q}_i - K(Q, P, t) \right) \mathrm{d}t = 0. \tag{3.4}$$

From Eqs. 3.3 to 3.4, we conclude that the integrands of the two integrals differ at most by a total time derivative of an arbitrary function F:

$$\sum_{i=1}^{n} p_i \mathrm{d}q_i - H\mathrm{d}t = \sum_{j=1}^{n} P_j \mathrm{d}Q_j - K\mathrm{d}t + \mathrm{d}F. \tag{3.5}$$

Such a function is called a generating function for the canonical transformation f. It is, a priori, a function of both the old and the new variables and time. The two sets of coordinates being connected by the $2n$ equations, namely, $f(q, p, t) = (Q, P, t)$, F can be reduced to a function of $2n+1$ variables among the $4n+1$. Hence, we can define 4^n generating functions that have n variables in $\mathcal{P}_1$ and n in $\mathcal{P}_2$. Among these are the four kinds defined by Goldstein [9]:

$$F_1(q_1, \ldots, q_n, Q_1, \ldots, Q_n, t), \quad F_2(q_1, \ldots, q_n, P_1, \ldots, P_n, t),$$
$$F_3(p_1, \ldots, p_n, Q_1, \ldots, Q_n, t), \quad F_4(p_1, \ldots, p_n, P_1, \ldots, P_n, t).$$

Let us first consider the generating function $F_1(q, Q, t)$. The total time derivative of F_1 reads:

$$\mathrm{d}F_1(q, Q, t) = \sum_{i=1}^{n} \frac{\partial F_1}{\partial q_i} \mathrm{d}q_i + \sum_{j=1}^{n} \frac{\partial F_1}{\partial Q_i} \mathrm{d}Q_i + \frac{\partial F_1}{\partial t} \mathrm{d}t. \tag{3.6}$$

Hence Eq. (3.5) yields:

$$\sum_{i=1}^{n} \left(p_i - \frac{\partial F_1}{\partial q_i} \right) \mathrm{d}q_i - H\mathrm{d}t = \sum_{j=1}^{n} \left(P_j + \frac{\partial F_1}{\partial Q_j} \right) \mathrm{d}Q_j - K\mathrm{d}t + \frac{\partial F_1}{\partial t} \mathrm{d}t. \tag{3.7}$$

Assume that (q, Q, t) is a set of independent variables. Then Eq. (3.7) is equivalent to:

$$p_i = \frac{\partial F_1}{\partial q_i}(q, Q, t), \tag{3.8}$$

$$P_i = -\frac{\partial F_1}{\partial Q_i}(q, Q, t), \tag{3.9}$$

$$K\left(Q, -\frac{\partial F_1}{\partial Q}, t\right) = H\left(q, \frac{\partial F_1}{\partial q}, t\right) + \frac{\partial F_1}{\partial t}. \tag{3.10}$$

If (q, Q) is not a set of independent variables, we say that F_1 is singular.

Let us consider more general generating functions. Let $(i_1, \ldots, i_s)(i_{s+1}, \ldots, i_n)$ and $(k_1, \ldots, k_r)$ $(k_{r+1}, \ldots, k_n)$ be two partitions of the set $(1, \ldots, n)$ into two non-intersecting parts such that $i_1 < \cdots < i_s$, $i_{s+1} < \cdots < i_n$, $k_1 < \cdots < k_r$, and $k_{r+1} < \cdots < k_n$ and define $I_s = (i_1, \ldots, i_s)$, $\bar{I}_s = (i_{s+1}, \ldots, i_n)$, $K_r = (k_1, \ldots, k_r)$, and $\bar{K}_r = (k_{r+1}, \ldots, k_n)$. If

$$(q_{I_s}, p_{\bar{I}_s}, Q_{K_r}, P_{\bar{K}_r}) = (q_{i_1}, \ldots, q_{i_s}, p_{i_{s+1}}, \ldots, p_{i_n}, Q_{k_1}, \ldots, Q_{k_r}, P_{k_{r+1}}, \ldots, P_{k_n})$$

are independent variables, then we can define the generating function F_{I_s, K_r}:

$$F_{I_s,K_r}(q_{I_s}, p_{\bar{I}_s}, Q_{K_r}, P_{\bar{K}_r}, t) = F(q_{i_1}, \ldots, q_{i_s}, p_{i_{s+1}}, \ldots, p_{i_n}, Q_{k_1}, \ldots, Q_{k_r}, P_{k_{r+1}}, \ldots, P_{k_n}, t). \tag{3.11}$$

Expanding $\mathrm{d}F_{I_s,K_r}$ yields:

$$\mathrm{d}F_{I_s,K_r} = \sum_{a=1}^{p} \frac{\partial F_{I_s,K_r}}{\partial q_{i_a}} \mathrm{d}q_{i_a} + \sum_{a=p+1}^{n} \frac{\partial F_{I_s,K_r}}{\partial p_{i_a}} \mathrm{d}p_{i_a} + \sum_{a=1}^{r} \frac{\partial F_{I_s,K_r}}{\partial Q_{k_a}} \mathrm{d}Q_{k_a} + \sum_{a=r+1}^{n} \frac{\partial F_{I_s,K_r}}{\partial P_{k_a}} \mathrm{d}P_{k_a} + \frac{\partial F_{I_s,K_r}}{\partial t} \mathrm{d}t \tag{3.12}$$

and rewriting Eq. (3.5) as a function of the linearly independent variables leads to:

$$\sum_{a=1}^{p} p_{i_a} \mathrm{d}q_{i_a} - \sum_{a=p+1}^{n} q_{i_a} \mathrm{d}p_{i_a} - H\mathrm{d}t = \sum_{a=1}^{r} P_{k_a} \mathrm{d}Q_{k_a} - \sum_{a=r+1}^{n} Q_{k_a} \mathrm{d}P_{k_a} - K\mathrm{d}t + \mathrm{d}F_{I_s,K_r}, \tag{3.13}$$

where

$$F_{I_s,K_r} = F_1 + \sum_{a=r+1}^{n} Q_{k_a} P_{k_a} - \sum_{a=p+1}^{n} q_{i_a} p_{i_a}. \tag{3.14}$$

Eq. (3.14) is often referred to as the *Legendre transformation*, it allows one to transform one generating function into another.

We then substitute Eq. (3.12) into Eq. (3.13):

$$\sum_{a=1}^{r} \left(P_{k_a} + \frac{\partial F_{I_s,K_r}}{\partial Q_{k_a}} \right) \mathrm{d}Q_{k_a} + \sum_{a=r+1}^{n} \left(\frac{\partial F_{I_s,K_r}}{\partial P_{k_a}} - Q_{k_a} \right) \mathrm{d}P_{k_a} - K\mathrm{d}t + \frac{\partial F_{I_s,K_r}}{\partial t} \mathrm{d}t = \sum_{a=1}^{p} \left(p_{i_a} - \frac{\partial F_{I_s,K_r}}{\partial q_{i_a}} \right) \mathrm{d}q_{i_a} - \sum_{a=p+1}^{n} \left(q_{i_a} + \frac{\partial F_{I_s,K_r}}{\partial p_{i_a}} \right) \mathrm{d}p_{i_a} - H\mathrm{d}t, \tag{3.15}$$

and obtain the set of equations that characterizes F_{I_s,K_r}:

$$p_{I_s} = \frac{\partial F_{I_s,K_r}}{\partial q_{I_s}} (q_{I_p}, p_{\bar{I}_p}, Q_{Kr}, P_{\bar{K}_r}, t), \tag{3.16}$$

$$q_{\bar{I}_s} = -\frac{\partial F_{I_s,K_r}}{\partial q_{\bar{I}_s}} (q_{I_p}, p_{\bar{I}_p}, Q_{Kr}, P_{\bar{K}_r}, t), \tag{3.17}$$

$$P_{K_r} = -\frac{\partial F_{I_s,K_r}}{\partial Q_{K_r}} (q_{I_p}, p_{\bar{I}_p}, Q_{Kr}, P_{\bar{K}_r}, t), \tag{3.18}$$

$$Q_{\bar{K}_r} = \frac{\partial F_{I_s,K_r}}{\partial P_{\bar{K}_r}} (q_{I_p}, p_{\bar{I}_p}, Q_{Kr}, P_{\bar{K}_r}, t), \tag{3.19}$$

$$K\left(Q_{K_r}, \frac{\partial F_{I_s,K_r}}{\partial P_{\bar{K}_r}}, -\frac{\partial F_{I_s,K_r}}{Q_{K_r}}, P_{\bar{K}_r}, t \right) = H\left(q_{I_s}, -\frac{\partial F_{I_s,K_r}}{\partial p_{\bar{I}_s}}, \frac{\partial F_{I_s,K_r}}{\partial q_{I_s}}, p_{\bar{I}_s}, t \right) + \frac{\partial F_{I_s,K_r}}{\partial t}. \tag{3.20}$$

For a generating function to be well-defined, we need to make the assumption that its variables are linearly independent. Later we see that this hypothesis is often not satisfied. The following property grants us that at least one of the generating function is well-defined at every instant (the proof of this result is given by Arnold [3]).

Proposition 3.2.3. *Let $f : \mathcal{P}_1 \times \mathbb{R} \to \mathcal{P}_2 \times \mathbb{R}$ be a canonical transformation. Using the above notations, there exist at least two partitions I_s and K_r such that $(q_{I_p}, p_{\bar{I}_p}, Q_{Kr}, P_{\bar{K}_r}, t)$ are linearly independent.*

3.2.2 The phase flow and its generating functions

The Hamilton–Jacobi theory has found many applications over the years but was first used to integrate the equations of motion of integrable Hamiltonian systems [10, 11]. This approach consists of finding a canonical map that transforms the system into an easily integrable one. Once the system is reduced to a trivial one, integration of the equations of motion are easily carried out. However, the search for such a map remains difficult and this aspect limits the use of the Hamilton–Jacobi theory in practice. Instead, in the present research we focus on a single transformation, the one induced by the phase flow that maps the system to its initial state. Under this transformation, the system is in equilibrium, every point in phase space is an equilibrium point. In general, we cannot compute this transformation (if we were able to find this transformation, it would mean that we could integrate the equations of motion) and so we focus on the generating functions that generate this transformation. In particular, we prove that they solve two-point boundary value problems.

Consider a Hamiltonian system and let Φ_t be its flow:

$$\begin{aligned} &\Phi_t : P \to P \\ &(q_0, p_0) \mapsto (\Phi_t^1(q_0, p_0) = q(q_0, p_0, t), \Phi_t^2(q_0, p_0) = p(q_0, p_0, t)) . \end{aligned} \tag{3.21}$$

Φ_t induces a transformation ϕ on $\mathcal{P} \times \mathbb{R}$ as follows:

$$\phi : (q_0, p_0, t) \mapsto (\Phi_t(q_0, p_0), t) . \tag{3.22}$$

ϕ^{-1} transforms the state of the system at time t to its state at the initial time while preserving the time. Let us now prove that ϕ, and a fortiori ϕ^{-1}, are canonical transformations.

Proposition 3.2.4. *The transformation ϕ induced by the phase flow is canonical.*

Proof. From the theory of differential equations[1], ϕ is an isomorphism. Moreover, the solution ϕ is by definition a diffeomorphism mapping from a symplectic space to a symplectic space, preserving the time, and trivially preserving the Hamiltonian. Thus, by Def. 3.2.1, ϕ is canonical. □

[1] Uniqueness of solutions of ordinary differential equations.

The inverse solution ϕ^{-1} maps the Hamiltonian system to equilibrium, i.e., its conditions at an initial epoch which are constant for all time along that trajectory. Therefore, the associated generating functions, F_{I_s,K_r}, verify the Hamilton–Jacobi equation (Eq. (3.20)). In addition, they must also verify Eqs. (3.16)–(3.19), where (Q, P) now denotes the initial state (q_0, p_0) and K is a constant that can be chosen to be 0:

$$p_{I_s} = \frac{\partial F_{I_s,K_r}}{\partial q_{I_s}}(q_{I_p}, p_{\bar{I}_p}, q_{0_{K_r}}, p_{0_{\bar{K}_r}}, t), \tag{3.23}$$

$$q_{\bar{I}_s} = -\frac{\partial F_{I_s,K_r}}{\partial p_{\bar{I}_s}}(q_{I_p}, p_{\bar{I}_p}, q_{0_{K_r}}, p_{0_{\bar{K}_r}}, t), \tag{3.24}$$

$$p_{0_{K_r}} = -\frac{\partial F_{I_s,K_r}}{\partial q_{0_{K_r}}}(q_{I_p}, p_{\bar{I}_p}, q_{0_{K_r}}, p_{0_{\bar{K}_r}}, t), \tag{3.25}$$

$$q_{0_{\bar{K}_r}} = \frac{\partial F_{I_s,K_r}}{\partial p_{0_{\bar{K}_r}}}(q_{I_p}, p_{\bar{I}_p}, q_{0_{K_r}}, p_{0_{\bar{K}_r}}, t), \tag{3.26}$$

$$0 = H\left(q_{I_s}, -\frac{\partial F_{I_s,K_r}}{\partial p_{\bar{I}_s}}, \frac{\partial F_{I_s,K_r}}{\partial q_{I_s}}, p_{\bar{I}_s}, t\right) + \frac{\partial F_{I_s,K_r}}{\partial t}. \tag{3.27}$$

The last equation is often referred to as the Hamilton–Jacobi equation. For the case where the partitions are $(1, \ldots, n)()$ and $(1, \ldots, n)()$ (i.e., $s = n$ and $r = n$), we recover the generating function F_1, which now verifies the following equations (note that the subscripts are suppressed in the following):

$$p = \frac{\partial F_1}{\partial q}(q, q_0, t), \tag{3.28}$$

$$p_0 = -\frac{\partial F_1}{\partial p_0}(q, q_0, t), \tag{3.29}$$

$$0 = H\left(q, \frac{\partial F_1}{\partial q}, t\right) + \frac{\partial F_1}{\partial t}. \tag{3.30}$$

The case $s = n$ and $r = 0$ corresponds to the generating function of the second kind:

$$p = \frac{\partial F_2}{\partial q}(q, p_0, t), \tag{3.31}$$

$$q_0 = \frac{\partial F_2}{\partial p_0}(q, p_0, t), \tag{3.32}$$

$$0 = H\left(q, \frac{\partial F_2}{\partial q}, t\right) + \frac{\partial F_2}{\partial t}. \tag{3.33}$$

If $s=0$ and $r=n$, we recover the generating function of the third kind, F_3:

$$q = -\frac{\partial F_3}{\partial p}(p, q_0, t), \tag{3.34}$$

$$p_0 = -\frac{\partial F_3}{\partial q_0}(p, q_0, t), \tag{3.35}$$

$$0 = H\left(-\frac{\partial F_3}{\partial p}, p, t\right) + \frac{\partial F_3}{\partial t}. \tag{3.36}$$

Finally, if $s=0$ and $r=0$, we obtain F_4:

$$q = \frac{\partial F_4}{\partial p}(p, p_0, t), \tag{3.37}$$

$$q_0 = -\frac{\partial F_4}{\partial p_0}(p, p_0, t), \tag{3.38}$$

$$0 = H\left(\frac{\partial F_4}{\partial p}, p, t\right) + \frac{\partial F_4}{\partial t}. \tag{3.39}$$

To compute the generating functions, one needs boundary conditions to solve the Hamilton–Jacobi equation. At the initial time, the flow induces the identity transformation, and thus the generating functions should also do so. In other words, at the initial time,

$$q(t_0) = q_0, \quad p(t_0) = p_0, \tag{3.40}$$

that is, $(q(t_0), p_0)$ and $(p(t_0), q_0)$ are the only sets of independent variables that contain n initial conditions and n components of the state vector at the initial time. As a consequence, all the generating functions save F_2 and F_3 are singular at the initial time, i.e., they are not defined as functions (they do not map a point into a single point). We will give deeper insight into this notion later, and we will especially show that singularities corresponds to multiple solutions from Eqs. (3.23) to (3.27).

Example 3.2.5. Let us look, for example, at the generating function of the first kind, $F_1(q, q_0, t)$. At the initial time, q is equal to q_0 whatever values the associated momenta p and p_0 take. We conclude that F_1 is singular.

We now focus on the boundary conditions for the F_2 and F_3 generating functions. At the initial time we must have:

$$\begin{cases} p_0 = \frac{\partial F_2}{\partial q}(q = q_0, p_0, t_0), \\ q_0 = \frac{\partial F_2}{\partial p_0}(q = q_0, p_0, t_0), \end{cases} \qquad \begin{cases} q_0 = -\frac{\partial F_3}{\partial p}(p = p_0, q_0, t_0), \\ p_0 = -\frac{\partial F_3}{\partial q_0}(p = p_0, q_0, t_0). \end{cases}$$

Due to the non-commutativity of the derivative operator and the operator that assigns the value t_0 at t, solutions to these equations are not unique. As a result, the boundary conditions verified by F_2 and F_3 may not be uniquely defined as well. For instance, they may be chosen to be:

$$F_2(q, p_0, t) = \langle q, p_0 \rangle, \quad F_3(p, q_0, t) = -\langle p, q_0 \rangle, \tag{3.41}$$

or

$$F_2(q, p_0, t) = \frac{1}{t-t_0} e^{(t-t_0)\langle q, p_0 \rangle}, \quad F_3(p, q_0, t) = -\frac{1}{t-t_0} e^{(t-t_0)\langle p, q_0 \rangle}, \tag{3.42}$$

where $\langle , \rangle$ is the inner product. One can readily verify that Eqs. (3.41) and (3.42) generate the identity transformation (3.40) at the initial time $t = t_0$.

The singularity at the initial time of all but two generating functions is a major issue: it prevents us from initializing the integration, i.e., from solving the Hamilton–Jacobi equation for most generating functions. The algorithm we use circumvents this problem by specifying boundary value conditions for every generating functions at a later time (see Section 3.4).

Lemma 3.2.6. *Generating functions solve two-point boundary value problems.*

Consider two points in phase space, $X_0 = (q_0, p_0)$ *and* $X_1 = (q, p)$, *and two partitions of* $(1, \ldots, n)$ *into two non-intersecting parts,* $(i_1, \ldots, i_s)(i_{s+1}, \ldots, i_n)$ *and* $(k_1, \ldots, k_r)(k_{r+1}, \ldots, k_n)$. *A two-point boundary value problem is formulated as follows:*

Given $2n$ *coordinates* $(q_{i_1}, \ldots, q_{i_s}, p_{i_{s+1}}, \ldots, p_{i_n})$ *and* $(q_{0_{k_1}}, \ldots, q_{0_{k_r}}, p_{0_{k_{r+1}}}, \ldots, p_{0_{k_n}})$, *find the remaining* $2n$ *variables such that a particle starting at* X_0 *will reach* X_1 *in* T *units of time.*

From the relationship defined by Eqs. (3.23–3.26), we see that the generating function F_{I_s, K_r} *solves this problem. This remark is of prime importance since it provides us with a very general technique to solve any Hamiltonian boundary value problems.*

Example 3.2.7. Lambert's problem is a particular case of boundary value problem where the partitions of $(1, \ldots, n)$ are $(1, \ldots, n)()$ and $(1, \ldots, n)()$. Though, given two positions q_f and q_0 and a transfer time T, the corresponding momentum vectors are found from Eqs (3.23) to (3.26)

$$\begin{aligned} p_i &= \frac{\partial F_1}{\partial q_i}(q, q_0, T), \\ p_{0_i} &= -\frac{\partial F_1}{\partial q_{0_i}}(q, q_0, T). \end{aligned} \tag{3.43}$$

3.2.3 Linear systems theory

In this section, we particularize the theory developed above to linear systems. Specifically, we reduce the Hamilton–Jacobi equation to a set of four matrix ordinary differential equations. Then, we relate the state transition matrix and generating functions. We show that properties of one may be deduced from properties of the other. The theory we present has implications in the study of relative motion and in optimal control theory for instance [12, 26–28].

3.2.3.1 Hamilton–Jacobi equation

To study the relative motion of two particles, one often linearizes the dynamics about the trajectory (called the reference trajectory) of one of the particles. Then one uses this linear approximation to study the motion of the other particle relative to the reference trajectory (perturbed trajectory). Thus, the dynamics of relative motion reduces at first order to a time-dependent linear Hamiltonian system, i.e., a system with a quadratic Hamiltonian function without any linear terms:

$$\mathrm{H}^h = \frac{1}{2}\mathbf{X}^{hT}\begin{pmatrix} H_{qq}(t) & H_{qp}(t) \\ H_{pq}(t) & H_{pp}(t) \end{pmatrix}\mathbf{X}^h, \tag{3.44}$$

where $\mathbf{X}^h = \binom{\Delta q}{\Delta p}$ is the relative state vector.

Lemma 3.2.8. *The generating functions associated with the phase flow transformation of the system defined by Eq. (3.44) are quadratic without linear terms.*

The proof of this lemma is trivial once we understand the link between the generating functions and the state transition matrix (see later in the section). From a heuristic perspective, we note that a linear term in the generating function would correspond to a non-homogenous term in the solution to the linear equation, which must equal zero for the dynamical system considered above.

From the above lemma, a general form for F_2 is:

$$F_2 = \frac{1}{2}\mathbf{Y}^T\begin{pmatrix} F_{11}^2(t) & F_{12}^2(t) \\ F_{21}^2(t) & F_{22}^2(t) \end{pmatrix}\mathbf{Y}, \tag{3.45}$$

where $\mathbf{Y} = \begin{pmatrix} \Delta q \\ \Delta p_0 \end{pmatrix}$ and $\begin{pmatrix} \Delta q_0 \\ \Delta p_0 \end{pmatrix}$ is the relative state vector at the initial time. We point out that both matrices defining H^h and F_2 are symmetric by definition. Then Eq. (3.31) reads:

$$\begin{aligned}\Delta p &= \frac{\partial F_2}{\partial \Delta q} \\ &= \left(F_{11}^2(t)\ F_{12}^2(t)\right)\mathbf{Y},\end{aligned}$$

Substituting into Eq. (3.33) yields:

$$\mathbf{Y}^T\left\{\begin{pmatrix} \dot{F}_{11}^2(t) & \dot{F}_{12}^2(t) \\ \dot{F}_{12}^2(t)^T & \dot{F}_{22}^2(t) \end{pmatrix} + \begin{pmatrix} I & F_{11}^2(t)^T \\ 0 & F_{12}^2(t)^T \end{pmatrix}\begin{pmatrix} H_{qq}(t) & H_{qp}(t) \\ H_{pq}(t) & H_{pp}(t) \end{pmatrix}\begin{pmatrix} I & 0 \\ F_{11}^2(t) & F_{12}^2(t) \end{pmatrix}\right\}\mathbf{Y} = 0. \tag{3.46}$$

Though the above equation has been derived using F_2, it is also valid for F_1 (replacing $\mathbf{Y} = \begin{pmatrix} \Delta q \\ \Delta p_0 \end{pmatrix}$ by $\mathbf{Y} = \begin{pmatrix} \Delta q \\ \Delta q_0 \end{pmatrix}$) since F_1 and F_2 solve the same Hamilton–Jacobi equation (Eqs. (3.30) and (3.33)). Eq. (3.46) is equivalent to the following four matrix equations:

$$\begin{aligned}
&\dot{F}_{11}^{1,2}(t) + H_{qq}(t) + H_{qp}(t)F_{11}^{1,2}(t) + F_{11}^{1,2}(t)H_{pq}(t) + F_{11}^{1,2}(t)H_{pp}(t)F_{11}^{1,2}(t) = 0,\\
&\dot{F}_{12}^{1,2}(t) + H_{qp}(t)F_{12}^{1,2}(t) + F_{11}^{1,2}(t)H_{pp}(t)F_{12}^{1,2}(t) = 0,\\
&\dot{F}_{21}^{1,2}(t) + F_{21}^{1,2}(t)H_{pq}(t) + F_{21}^{1,2}(t)H_{pp}(t)F_{11}^{1,2}(t) = 0,\\
&\dot{F}_{22}^{1,2}(t) + F_{21}^{1,2}(t)H_{pp}(t)F_{12}^{1,2}(t) = 0,
\end{aligned} \tag{3.47}$$

where we replaced F_{ij}^2 by $F_{ij}^{1,2}$ to signify that these equations are valid for both F_1 and F_2. We also recall that $F_{21}^{1,2} = F_{12}^{1,2^T}$. A similar set of equations can be derived for any generating function F_{I_s,K_r}. However, in this section we only give the equations verified by F_3 and F_4:

$$\begin{aligned}
&\dot{F}_{11}^{3,4}(t) + H_{pp}(t) - H_{pq}(t)F_{11}^{3,4}(t) - F_{11}^{3,4}(t)H_{qp}(t) + F_{11}^{3,4}(t)H_{qq}(t)F_{11}^{3,4}(t) = 0,\\
&\dot{F}_{12}^{3,4}(t) - H_{pq}(t)F_{12}^{3,4}(t) + F_{11}^{3,4}(t)H_{qq}(t)F_{12}^{3,4}(t) = 0,\\
&\dot{F}_{21}^{3,4}(t) - F_{21}^{3,4}(t)H_{qp}(t) + F_{21}^{3,4}(t)H_{qq}(t)F_{11}^{3,4}(t) = 0,\\
&\dot{F}_{22}^{3,4}(t) + F_{21}^{3,4}(t)H_{qq}(t)F_{12}^{3,4}(t) = 0.
\end{aligned} \tag{3.48}$$

The first equations of Eqs. (3.47) and (3.48) are Riccati equations. The second and third are non-homogeneous, time varying, linear equations once the Riccati equations are solved and are equivalent to each other (i.e., transform into each other under transpose). The last are just a quadrature once the previous equations are solved.

3.2.3.2 Initial conditions

Although F_1 and F_2 (or more generally F_{I_s,K_r} and $F_{I_s,K_{r'}}$ for all r and r') verify the same Hamilton–Jacobi partial differential equation, these generating functions are different. We noticed earlier that this difference is characterized by the boundary conditions. At the initial time, the flow induces the identity transformation, thus the generating functions should also do so. In other words, at the initial time,

$$\Delta q(t_0) = \Delta q_0, \quad \Delta p(t_0) = \Delta p_0 .$$

In terms of generating functions this translates for F_2 to:

$$\frac{\partial F_2}{\partial \Delta q}(\Delta q_0, \Delta p_0, t_0) = \Delta p_0, \quad \frac{\partial F_2}{\partial \Delta p_0}(\Delta q_0, \Delta p_0, t_0) = \Delta q_0,$$

i.e.,

$$F_{11}^2 \Delta q + F_{12}^2 \Delta p_0 = \Delta p_0, \quad F_{21}^2 \Delta q + F_{22}^2 \Delta p_0 = \Delta q_0,$$

or equivalently:

$$F^2_{11} = F^2_{22} = 0, \quad F^2_{12} = F^2_{21} = Identity\,.$$

On the other hand, F_1 is ill-defined at the initial time. Indeed, at the initial time Eqs. (3.28) and (3.29) read:

$$F^1_{11}\Delta q + F^1_{12}\Delta q_0 = \Delta p_0, \quad F^1_{21}\Delta q + F^1_{22}\Delta q_0 = \Delta p_0\,.$$

These equations do not have any solutions. This was expected since we saw earlier that $(\Delta q, \Delta q_0)$ are not independent variables at the initial time $(\Delta q = \Delta q_0)$.

3.2.3.3 Perturbation matrices

Another approach to the study of relative motion at linear order relies on the state transition matrix. This method was developed by Battin in Ref. [4] for the case of a spacecraft moving in a point mass gravity field. Let Φ be the state transition matrix which describes the relative motion:

$$\begin{pmatrix} \Delta q \\ \Delta p \end{pmatrix} = \Phi \begin{pmatrix} \Delta q_0 \\ \Delta p_0 \end{pmatrix},$$

where $\Phi = \begin{pmatrix} \Phi_{qq} & \Phi_{qp} \\ \Phi_{pq} & \Phi_{pp} \end{pmatrix}$. Battin [4] defines the fundamental perturbation matrices C and $\tilde{C}$ as:

$$\tilde{C} = \Phi_{pq}\Phi_{qq}^{-1},$$
$$C = \Phi_{pp}\Phi_{qp}^{-1}\,.$$

That is, given $\Delta p_0 = 0$, $\tilde{C}\Delta q = \Delta p$ and given $\Delta q_0 = 0$, $C\Delta q = \Delta p$. He shows that for the relative motion of a spacecraft about a circular trajectory in a point mass gravity field the perturbation matrices verify a Riccati equation and are therefore symmetric. Using the generating functions for the canonical transformation induced by the phase flow, we immediately recover these properties. We also generalize these results to any linear Hamiltonian system.

Using the notations of Eq. (3.45), Eqs. (3.31) and (3.32) read:

$$\begin{aligned} \Delta p &= \frac{\partial F_2}{\partial \Delta q} \\ &= F^2_{11}\Delta q + F^2_{12}\Delta p_0, \\ \Delta q_0 &= \frac{\partial F_2}{\partial \Delta p_0} \\ &= F^2_{21}\Delta q + F^2_{22}\Delta p_0\,. \end{aligned}$$

We solve for $(\Delta q, \Delta p)$:

$$\Delta q = {F^2_{21}}^{-1}\Delta q_0 - {F^2_{21}}^{-1}F^2_{22}\Delta p_0,$$

$$\Delta p = F^2_{11}{F^2_{21}}^{-1}\Delta q_0 + (F^2_{12} - F^2_{11}{F^2_{21}}^{-1}F^2_{22})\Delta p_0,$$

and identify the right-hand side with the state transition matrix:

$$\begin{cases} \Phi_{qp} = -F^{2}_{21}{}^{-1} F^{2}_{22}, \\ \Phi_{qq} = F^{2}_{21}{}^{-1}, \\ \Phi_{pp} = F^{2}_{12} - F^{2}_{11} F^{2}_{21}{}^{-1} F^{2}_{22}, \\ \Phi_{pq} = F^{2}_{11} F^{2}_{21}{}^{-1}. \end{cases}$$

We conclude that

$$\tilde{C} = \Phi_{pq} \Phi_{qq}^{-1} = F^{2}_{11}. \tag{3.49}$$

In the same manner, but using F_1, we can show that:

$$C = \Phi_{pp} \Phi_{qp}^{-1} = F^{1}_{11}. \tag{3.50}$$

Thus, C and $\tilde{C}$ are symmetric by nature (as $F^{1,2}_{11}$ is symmetric by definition) and they verify the Riccati equation given in Eq. (3.47).

3.2.3.4 Singularities of generating functions and their relation to the state transition matrix

In the first part of this paper, we studied the local existence of generating functions. We proved that at least one of the generating functions is well-defined at every instant (Prop. 3.2.3). In general we can notice that each of them can become singular at some point, even for simple systems. As an example let us look at the harmonic oscillator.

Example 3.2.9. The Hamiltonian for the harmonic oscillator is given by:

$$H(q, p) = \frac{1}{2m} p^2 + \frac{k}{2} q^2,$$

The F_1 generating function for the phase flow canonical transformation can be found to be:

$$F_1(q, q_0, t) = \frac{1}{2}\sqrt{km} \csc(\omega t) \left[-2qq_0 + (q^2 + q_0^2)\cos(\omega t)\right],$$

where $\omega = \sqrt{\frac{k}{m}}$. One can readily verify that F_1 is a solution of the Hamilton–Jacobi equation (Eq. (3.30)). Although it is well-defined most of the time, at $T = m\pi/\omega$, $m \in \mathbb{Z}$, F_1 becomes singular in that the values of the coefficients of the q's and q_0's increase without bound. To understand these singularities, recall the general solution to the equations of motion:

$$q(t) = q_0 \cos(\omega t) + p_0/\omega \sin(\omega t),$$

$$p(t) = -q_0 \omega \sin(\omega t) + p_0 \cos(\omega t).$$

At $t = T$, $q(T) = q_0$, that is q and q_0 are not independent variables. Therefore the generating function F_1 is undefined at this instant. We say that it is singular at $t = T$.

However, F_1 may be defined in the limit: at $t = T$, $q = q_0$, and thus F_1 behaves as $m\frac{(q-q_0)^2}{2(t-T)}$ as $t \mapsto T$. Finally, at $t = T$, $q = q_0$ any values of p and p_0 are possible, i.e., singularities correspond to multiple solutions to the boundary value problem that consists of going from q_0 to $q = q_0$ in $T = 0$ unit of time.

The harmonic oscillator is a useful example. Since the flow is known analytically, we are able to explicitly illustrate the relationship between the generating functions and the flow ϕ. We can go a step further by noticing that both the state transition matrix and the generating functions generate the flow. Therefore, singularities of the generating functions should be related to properties of the state transition matrix:

$$\begin{aligned}\Delta p &= \frac{\partial F_2}{\partial \Delta p}\\ &= F_{11}^2 \Delta q + F_{12}^2 \Delta p_0,\end{aligned}$$

but we also have

$$\Delta p = \Phi_{pq}\Phi_{qq}^{-1}\Delta q + (\Phi_{pp} - \Phi_{pq}\Phi_{qq}^{-1}\Phi_{qp})\Delta p_0 . \tag{3.51}$$

Similarly,

$$\begin{aligned}\Delta q_0 &= \frac{\partial F_2}{\partial \Delta p_0}\\ &= F_{21}^2 \Delta q + F_{22}^2 \Delta p_0,\end{aligned}$$

but we also have

$$\Delta q_0 = \Phi_{qq}^{-1}\Delta q - \Phi_{qq}^{-1}\Phi_{qp}\Delta p_0 . \tag{3.52}$$

A direct identification yields:

$$F_{11}^2 = \Phi_{pq}\Phi_{qq}^{-1}, \tag{3.53}$$

$$F_{12}^2 = \Phi_{pp} - \Phi_{pq}\Phi_{qq}^{-1}\Phi_{qp}, \tag{3.54}$$

$$F_{21}^2 = \Phi_{qq}^{-1}, \tag{3.55}$$

$$F_{22}^2 = \Phi_{qq}^{-1}\Phi_{qp} . \tag{3.56}$$

Thus, F_2 is singular when and only when Φ_{qq} is not invertible. This relation between singularities of F_2 and invertibility of a sub-matrix of the state transition matrix readily generalizes to other kind of generating functions. For such linear systems in particular, we can show that

- F_1 is singular when Φ_{qp} is singular,
- F_2 is singular when Φ_{qq} is singular,
- F_3 is singular when Φ_{pp} is singular,
- F_4 is singular when Φ_{pq} is singular.

To extend these results to other generating functions, we must consider other block decompositions of the state transition matrix. Every $n \times n$ block of the state transition matrix is associated with a different generating function. Since the determinant of the state transition matrix is 1, there exists at least one $n \times n$ sub-matrix that must have a non-zero determinant. The generating function associated with this block is non-singular, and we recover Prop. 3.2.3 for linear systems.

3.2.4 Non-linear systems theory

We have shown the local existence of generating functions and mentioned that they may not be globally defined. Using linear systems theory we are also able to predict where the singularities are and to interpret their meaning as multiple solutions to the two-point boundary value problem. In this section we generalize these results to singularities of non-linear systems.

The following proposition relates singularities of the generating functions to the invertibility of sub-matrices of the Jacobi matrix of the canonical transformation.

Proposition 3.2.10. *The generating function F_{I_s,K_r} for the canonical transformation ϕ is singular at time t if and only if*

$$\det\left(\frac{\partial \phi_i}{\partial z_j}\right)_{i\in I, j\in J} = 0, \tag{3.57}$$

where $I = \{i \in I_s\} \bigcup \{n+i, i \in \bar{I}_s\}$, $J = \{j \in \bar{K}_r\} \bigcup \{n+j, j \in K_r\}$, and $z = (q_0, p_0)$ is the state vector at the initial time.

Proof. For the sake of clarity, let us prove this property for F_1. In that case, $I = [1, n]$ and $J = [n+1, 2n]$. First we remark that

$$\left(\frac{\partial \phi_i}{\partial z_j}\right)_{i\in I, j\in J} = \left(\frac{\partial q_i}{\partial p_{0_j}}\right).$$

Thus, from the inversion theorem, if $\det\left(\frac{\partial \phi_i}{\partial z_j}\right)_{i\in I, j\in J} = 0$, there is no open set in which we can solve p_0 as a function of q and q_0.

On the other hand, suppose that F_1 is non-singular. Then, from Eq. (3.29), we have:

$$p_0 = -\frac{\partial F_1}{\partial q_0}(q, q_0, t), \tag{3.58}$$

that is, we can express p_0 as a function of (q, q_0). This is in contradiction with the result obtained from the local inversion theorem. Therefore, F_1 is singular. □

Example 3.2.11. From the above proposition, we conclude that the F_1 generating function associated with the phase flow of the harmonic oscillator is singular if and only if:

$$\det\left(\frac{\partial \phi_i}{\partial z_j}\right)_{i\in I, j\in J} = 0. \tag{3.59}$$

In this example, $I = 1$, $J = 2$, and $\phi = (q_0 \cos(\omega t) + p_0/\omega \sin(\omega t), -q_0\omega \sin(\omega t) + p_0 \cos(\omega t))$. Therefore F_1 is singular if and only if $\sin(\omega t) = 0$, i.e., $t = 2\pi/\omega + 2k\pi$. We recover previous results obtained by direct computation of F_1.

Prop. 3.2.10 generalizes to non-linear systems the relation between singularities and non-uniqueness of the solutions to boundary value problems. Indeed, F_{I_s,K_r} is singular if and only if $z_J \mapsto \phi_I(t, z)$ is not an isomorphism. In other words, singularities arise when there exist multiple solutions to the boundary value problem.

To study the singularities of non-linear systems, we need to introduce the concept of Lagrangian submanifolds. The theory of Lagrangian submanifolds goes far beyond the results we present in this section: "Some believe that the Lagrangian submanifold approach will give deeper insight into quantum theories than does the Poisson algebra approach. In any case, it gives deeper insight into classical mechanics and classical field theories" (Abraham and Marsden [1]). We refer to Abraham and Marsden [1], Marsden [23] and Weinstein [32] and references given therein for further information on these subjects.

3.2.4.1 Lagrangian submanifolds and the study of caustics

Consider an arbitrary generating function F_{I_s,K_r}. Then the graph of dF_{I_s,K_r} defines a $2n$-dimensional submanifold called a canonical relation [32] of the $4n$-dimensional symplectic space $(\mathcal{P}_1 \times \mathcal{P}_2, \Omega = \pi_1^*\omega_1 - \pi_2^*\omega_2)$. On the other hand, since the variables (q_0, p_0) do not appear in the Hamilton–Jacobi equation (Eq. (3.27)), we may consider them as parameters. In that case the graph of $(q_{I_s}, p_{\bar{I}_s}) \mapsto dF_{I_s,K_r}$ defines an n-dimensional submanifold of the symplectic space $(\mathcal{P}_1, \omega_1)$ called a Lagrangian submanifold [32]. The study of singularities can be achieved using either canonical relations [1] or Lagrangian submanifolds [3, 23]. In the following we assumed t fixed.

Theorem 3.2.12. *The generating function F_{I_s,K_r} is singular if and only if the local projection of the canonical relation $\mathcal{L}$ defined by the graph of dF_{I_s,K_r} onto $(q_{I_s}, p_{\bar{I}_s}, q_{0_{K_r}}, p_{0_{\bar{K}_r}})$ is not a local diffeomorphism.*

Definition 3.2.2. The projection of a singular point F_{I_s,K_r} onto $(q_{I_s}, p_{\bar{I}_s}, q_{0_{K_r}}, p_{0_{\bar{K}_r}})$ is called a caustic.

If one works with Lagrangian submanifolds then the previous theorem becomes:

Theorem 3.2.13. *The generating function*[2] *F_{I_s,K_r} is singular if the local projection of the Lagrangian submanifold defined by the graph of $(q_{I_s}, p_{\bar{I}_s}) \mapsto dF_{I_s,K_r}$ onto $(q_{I_s}, p_{\bar{I}_s})$ is not a local diffeomorphism.*

These theorems are the geometric formulation of Prop. 3.2.10. If the projection of the canonical relation defined by the graph of dF_{I_s,K_r} onto $(q_{I_s}, p_{\bar{I}_s}, q_{0_{K_r}}, p_{0_{\bar{K}_r}})$ is not

[2] We consider here that the generating function is a function of n variables only, and has n parameters.

a local diffeomorphism, then there exists multiple solutions to the problem of finding (q_0, p_0, q, p) knowing $(q_{I_s}, p_{\bar{I}_s}, q_{0_{K_r}}, p_{0_{\bar{K}_r}})$. From the local inversion theorem, this is equivalent to Prop. 3.2.10.

In the light of these theorems, we can give a geometrical interpretation to Thm. 3.2.3 on the existence of generating functions. Given a canonical relation $\mathcal{L}$ (or a Lagrangian submanifold) defined by a canonical transformation, there exists a $2n$-dimensional (or n-dimensional) submanifold $\mathcal{M}$ of $\mathcal{P}_1 \times \mathcal{P}_2$ (or $\mathcal{P}_1$) such that the local projection of $\mathcal{L}$ onto $\mathcal{M}$ is a local diffeomorphism.

To study caustics two approaches, at least, are possible depending on the problem. A good understanding of the physics may provide information very easily. For instance, consider the two-body problem in two dimensions, and the problem of going from a point A to a point B, symmetrically placed on a single line on either side of the body, in a certain lapse of time, T. For certain values of T, the trajectory that links A to B is an ellipse whose perigee and apogee are A and B. Therefore, there are two solutions to this problem depending upon which way the particle is going. In terms of generating functions, we deduce that F_3 is non-singular (there is a unique solution once the final momentum is given) but F_1 is singular (existence of two solutions).

Another method for studying caustics consists of using a known non-singular generating function to define the Lagrangian submanifold $\mathcal{L}$ and then study its projection. A very illustrative example is given by Ehlers and Newman [8]. Using the Hamilton–Jacobi equation they treat the evolution of an ensemble of free particles whose initial momentum distribution is $p = (1/(1+q^2))$. They identify a time t_1 at which F_1 is singular. Then, using a closed-form expression of F_3, they find the equations defining the Lagrangian submanifold at t_1. Its projection can be studied and they eventually find that the caustic is two folds. Nevertheless, such an analysis is not always possible as solutions to the Hamilton–Jacobi equation are usually found numerically, not analytically. To illustrate this method, let us consider the following example.

Example 3.2.14. (Motion about the Libration point L_2 in the Hill three-body problem.) Consider a spacecraft moving about and staying close to the Libration point L_2 in the normalized Hill three-body problem (see Appendix B for a description of the Hill three-body problem). The algorithm we develop in Section 3.4 computes the generating function for relative motion with respect to L_2 as a Taylor series expansion, of order N, of the exact generating function about L_2. For instance, F_2 reads:

$$\begin{aligned}
F_2(q_x, q_y, p_{0_x}, p_{0_y}, t) = {} & f_{11}^2(t)q_x^2 + f_{12}^2(t)q_x q_y + f_{13}^2(t)q_x p_{0_x} + f_{14}^2(t)q_x p_{0_y} + f_{22}^2(t)q_y^2 \\
& + f_{23}^2(t)q_y p_{0_x}(t) + f_{24}^2(t)q_y p_{0_y} + f_{33}^2(t)p_{0_x}^2 + f_{34}^2(t)p_{0_x}p_{0_y} \\
& + f_{44}^2(t)p_{0_y}^2 + r(q_x, q_y, p_{0_x}, p_{0_y}, t),
\end{aligned}$$

where (q, p, q_0, p_0) are relative position and momenta of the spacecraft with respect to L_2 at t and t_0, the initial time, and r is a polynomial of degree N in its spatial variables with time-dependent coefficients and without any quadratic terms. At $T = 1.6822$, F_1 is singular but F_2 is not. Eqs. (3.31) and (3.32) reads:

$$p_x = 2f_{11}^2(T)q_x + f_{12}^2(T)q_y + f_{13}^2(T)p_{0_x} + f_{14}^2(T)p_{0_y} + D_1 r(q_x, q_y, p_{0_x}, p_{0_y}, T), \quad (3.60)$$

$$p_y = f_{12}^2(T)q_x + 2f_{22}^2(T)q_y + f_{23}^2(T)p_{0_x} + f_{24}^2(T)p_{0_y} + D_2 r(q_x, q_y, p_{0_x}, p_{0_y}, T), \quad (3.61)$$

$$q_{0_x} = f_{13}^2(T)q_x + f_{23}^2(T)q_y + 2f_{33}^2(T)p_{0_x} + f_{34}^2(T)p_{0_y} + D_3 r(q_x, q_y, p_{0_x}, p_{0_y}, T), \quad (3.62)$$

$$q_{0_y} = f_{14}^2(T)q_x + f_{24}^2(T)q_y + f_{34}^2(T)p_{0_x} + 2f_{44}^2(T)p_{0_y} + D_4 r(q_x, q_y, p_{0_x}, p_{0_y}, T), \quad (3.63)$$

where $D_i r$ represents the derivative of r with respect to its ith variable. Eqs. (3.60)–(3.63) define a canonical relation $\mathcal{L}$. By assumption F_1 is singular, therefore the projection of $\mathcal{L}$ onto (q, q_0) is not a local diffeomorphism and there exists a caustic.

Let us now study this caustic. Eqs. (3.60)–(3.63) provide p and q_0 as a function of (q, p_0), but to characterize the caustic we need to study the projection of the Lagrangian manifold on[3] (q, q_0). Hence, we must express p and p_0 as a function of (q, q_0). F_1 being singular, there are multiple solutions to the problem of finding p and p_0 as a function of (q, q_0), and one valuable piece of information is the number k of such solutions. To find p and p_0 as a function of (q, q_0) we first invert Eqs. (3.62) and (3.63) to express p_0 as a function of (q, q_0). Then we substitute this relation into Eqs. (3.60) and (3.61). The first step requires a series inversion that can be carried out using the technique developed by Moulton [25]. Let us rewrite Eqs. (3.62) and (3.63):

$$2f_{33}^2(T)p_{0_x} + f_{34}^2(T)p_{0_y} = q_{0_x} - f_{13}^2(T)q_x - f_{23}^2(T)q_y - D_3 r(q_x, q_y, p_{0_x}, p_{0_y}, T), \quad (3.64)$$

$$f_{34}^2(T)p_{0_x} + 2f_{44}^2(T)p_{0_y} = q_{0_y} - f_{14}^2(T)q_x - f_{24}^2(T)q_y - D_4 r(q_x, q_y, p_{0_x}, p_{0_y}, T). \quad (3.65)$$

The determinant of the coefficients of the linear terms on the left-hand side is zero (otherwise there is a unique solution to the series inversion) but each of the coefficients is non-zero, i.e., we can solve for p_{0_x} as a function of $(p_{0_y}, q_{0_x}, q_{0_y})$ using Eq. (3.64). Then we substitute this solution into Eq. (3.65) and we obtain an equation of the form

$$R(p_{0_y}, q_{0_x}, q_{0_y}) = 0, \quad (3.66)$$

that contains no terms in p_{0_y} alone of the first degree. In addition, R contains a non-zero term of the form $\alpha p_{0_y}^2$, where α is a real number. In this case, Weierstrass proved that there exist two solutions, $p_{0_y}^1$ and $p_{0_y}^2$, to Eq. (3.66).

[3] Since F_1 is a function of (q, q_0).

In the same way, we can study the singularity of F_1 at the initial time. At $t = 0$, F_2 generates the identity transformation, hence $f_{33}^2(0) = f_{34}^2(0) = f_{43}^2(0) = f_{44}^2(0) = 0$. This time there is no non-zero first minor, and we find that there exists infinitely many solutions to the series inversion. Another way to see this is to use the Legendre transformation:

$$F_1(q, q_0, t) = F_2(q, p_0, t) - q_0 p_0,$$

As t tends toward 0, (q, p) goes to (q_0, p_0) and F_2 converges toward the identity transformation $\lim_{t\to 0} F_2(q, p_0, t) = qp_0 \underset{t\to 0}{\longrightarrow} q_0 p_0$. Therefore, as t goes to 0, F_1 also goes to 0, i.e., the projection of $\mathcal{L}$ onto (q, q_0) reduces to a point.

The use of series inversion to quantify the number of solutions to the boundary value problem is a very efficient technique for systems with polynomial generating functions. From the series inversion theory we know that the uniqueness of the inversion is determined by the linear terms whereas the number of solutions (if many) depends on properties of non-linear terms (we illustrated this property in the above example). In addition, this technique allows us to study the projection of the canonical relation at the cost of a single matrix inversion only.

In the case where generating functions are (or can be approximated by a) polynomial, we can recover the phase flow (or its approximation) as a polynomial too. For instance, from

$$p_0 = \frac{\partial F_1}{\partial q_0}(q, q_0, t),$$

we can find $q(q_0, p_0)$ at the cost of a series inversion. Then, $q(q_0, p_0)$ together with $p = \frac{\partial F_1}{\partial q}(q, q_0, t)$ define the flow (or its polynomial approximation). On the other hand, generating functions are well-defined if and only if the transformation from the flow to the generating function has a unique solution (Prop. 3.2.10). From series inversion theory, we conclude that generating functions are well-defined if and only if the inversion of the linear approximation of the flow has a unique solution. Therefore, we have the following property:

Proposition 3.2.15. *Singularities of* ***polynomial*** *generating functions correspond to degeneracy of sub-matrices of the state transition matrix as in the linear case. In other words, using our previous notation,*

- F_1 *is singular when* $\det(\Phi_{qp}) = 0$,
- F_2 *is singular when* $\det(\Phi_{qq}) = 0$,
- F_3 *is singular when* $\det(\Phi_{pp}) = 0$,
- F_4 *is singular when* $\det(\Phi_{pq}) = 0$.

Using other block decompositions of the state transition matrix, these results can be extended to the generating function F_{I_s,K_r}.

Example 3.2.16. (Singularities of the generating functions in the Hill three-body problem.) To illustrate Prop. 3.2.11, let us determine the singularities of F_1 and F_2 in the normalized Hill three-body problem linearized about L_2.

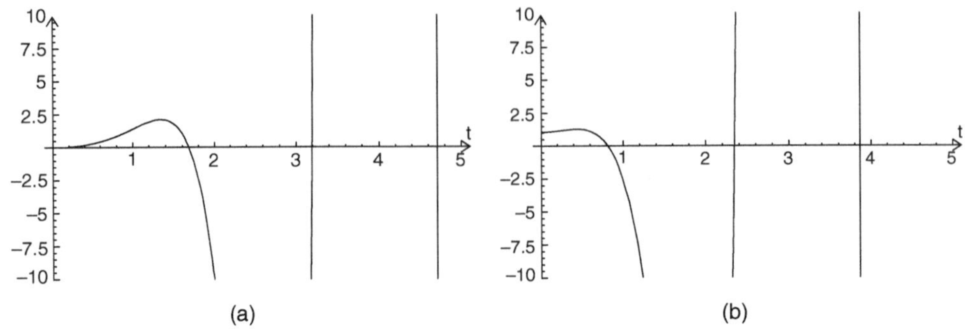

Fig. 3.1. Determinants of Φ_{qq} and Φ_{qp}.

The state transition matrix for this problem satisfies (Appendix B, Eq. (B.8)):

$$\dot{\phi}(t) = \begin{pmatrix} -8 & 0 & 0 & -1 \\ 0 & 4 & 1 & 0 \\ 0 & 1 & 1 & 0 \\ -1 & 0 & 0 & 1 \end{pmatrix} \phi(t), \quad \phi(0) = Identity.$$

We use the *Mathematica*© built in function *DSolve* to compute a symbolic expression of the state transition matrix. We plot in Figure 3.1 the determinant of Φ_{qq} and Φ_{qp} as a function of time. As noticed before F_1 is singular at the initial time and at $t = \{1.6821, 3.1938, 4.710\}$ and F_2 is singular at $t = \{0.809, 2.3443, 3.86\}$. The singularity at $t = 1.6821$ was studied above.

3.3 Hamilton's principal function

Though generating functions are used in the present research to solve boundary value problems, they were introduced by Jacobi, and mostly used thereafter, as fundamental functions which can solve the equations of motion by simple differentiations and eliminations, without integration. Nevertheless, it was Hamilton who first hit upon the idea of finding such a fundamental function. He first proved its existence in geometrical optics (i.e., for time-independent Hamiltonian systems) in 1834 and called it the characteristic function [19]. One year later he published a second essay [20] on systems of attracting and repelling points in which he showed that the evolution of dynamical systems is characterized by a single function called Hamilton's principal function:

> The former Essay contained a general method for reducing all the most important problems of dynamics to the study of one characteristic function, one central or radical relation. It was remarked at the close of that Essay, that many eliminations required by this method in its first conception, might be avoided by a general transformation, introducing the time explicitly into a part S of the whole characteristic function V; and it is now proposed to fix the attention chiefly on this part S, and to call it the Principal

Function. (William R. Hamilton, in the introductory remarks of "Second essay on a General Method in Dynamics" [20]).

Although Hamilton's principal function has been introduced to derive solutions to the equations of motion, it may also be used to solve boundary value problems, similar to the generating functions. Therefore, in the next section we introduce Hamilton's principal function and prove that it solves two-point boundary value problems. Then we discuss how it compares to the generating functions.

3.3.1 *Existence of the Hamilton principal function*

Similarly to the generating functions, Hamilton's principal function may be derived using the calculus of variations. Consider the extended action integral:

$$A = \int_{\tau_0}^{\tau_1} (pq' + p_t t')\mathrm{d}\tau, \tag{3.67}$$

under the auxiliary condition $K(q, t, p, p_t) = 0$, where $q' = \mathrm{d}q/\mathrm{d}\tau$, p_t is the momentum associated with the generalized coordinates t and $K = p_t + H$.

Define a line element[4] $\mathrm{d}\sigma$ for the extended configuration space (q, t) by

$$\mathrm{d}\sigma = L\mathrm{d}t = Lt'\mathrm{d}\tau.$$

where L is the Lagrangian. Then, we can connect two points (q_0, t_0) and (q_1, t_1) of the extended configuration space by a shortest line γ and measure its length from:

$$A = \int_\gamma \mathrm{d}\sigma = \int_\gamma Lt'\mathrm{d}\tau.$$

The distance we obtain is a function of the coordinates of the end-points and, by definition, is given by the Hamilton principal function: $W(q_0, t_0, q_1, t_1)$.

From the calculus of variations (see e.g., Lanczos [22]) we know that the variation of the action A can be expressed as a function of the boundary terms if we vary the limits of the integral:

$$\delta A = p_1\delta q_1 + p_{t_1}\delta t_1 - p_0\delta q_0 - p_{t_0}\delta t_0 .$$

On the other hand, we have:

$$\delta A = \delta W(q_0, t_0, q_1, t_1) = \frac{\partial W}{\partial q_0}\delta q_0 + \frac{\partial W}{\partial t_0}\delta t_0 + \frac{\partial W}{\partial q_1}\delta q_1 + \frac{\partial W}{\partial t_1}\delta t_1,$$

i.e.,

$$p_0 = -\frac{\partial W}{\partial q_0}(q_0, t_0, q_1, t_1), \tag{3.68}$$

$$p_1 = \frac{\partial W}{\partial q_1}(q_0, t_0, q_1, t_1), \tag{3.69}$$

[4] The geometry established by this line element is not Riemannian [22].

and

$$-\frac{\partial W}{\partial t_0}(q_0, t_0, q_1, t_1) + H\left(q_0, -\frac{\partial W}{\partial q_0}, t_0\right) = 0, \tag{3.70}$$

$$\frac{\partial W}{\partial t_1}(q_0, t_0, q_1, t_1) + H\left(q_1, \frac{\partial W}{\partial q_1}, t_1\right) = 0, \tag{3.71}$$

where K has been replaced by $p_t + H$. As with generating functions of the first kind, Hamilton's principal function solves boundary value problems of Lambert's type through Eqs. (3.68–3.69). To find W, however, we need to solve a system of two partial differential equations (Eqs. (3.70) and (3.71)).

3.3.2 *Hamilton's principal function and generating functions*

In this section, we highlight the main differences between generating functions associated with the phase flow and Hamilton's principal function. For sake of simplicity we compare $F_1(q, q_0, t)$ and $W(q, t, q_0, t_0)$.

3.3.2.1 *Calculus of variations*

Even if both functions are derived from the calculus of variations, there are fundamental differences between them. To derive generating functions the time t is considered as an independent variable in the variational principle. In contrast, we increase the dimensionality of the system by adding the time t to the generalized coordinates to derive Hamilton's principal function. As a consequence, generating functions generate a transformation between two points in the phase space, i.e., they act without passage of time. On the other hand, Hamilton's principal function generates a transformation between two points in the extended phase space, i.e., between two points in the phase space with different times. This difference may be viewed as follows: Generating functions allow us to characterize the phase flow given an initial time, t_0 (i.e., to characterize all trajectories whose initial conditions are specified at t_0), whereas Hamilton's principal function does not impose any constraint on the initial time. The counterpart being that Hamilton's principal function must satisfy two partial differential equations (Eq. (3.70) defines W as a function of t_0 and Eq. (3.71) defines W as a function of t_1) whereas generating functions satisfy only one.

Moreover, to derive the generating functions fixed endpoints are imposed, i.e., we impose the trajectory in both sets of variables to verify the principle of least action. On the other hand, the variation used to derive Hamilton's principal function involves moving endpoints and an energy constraint. This difference may be interpreted as follows: Hamilton's principal function generates a transformation which maps a point of a given energy surface to another point on the same energy surface and is not defined for points that do not lie on this surface. As a consequence of the energy constraint, we have [22]:

$$\left| \frac{\partial^2 W}{\partial q_0 \partial q_1} \right| = 0. \tag{3.72}$$

As noticed by Lanczos [22], "this is a characteristic property of the W-function which has no equivalent in Jacobi's theory". On the other hand, generating functions map any point of the phase space into another one, the only constraint is imposed through the variational principle (or equivalently by the definition of canonical transformation): we impose the system in both sets of coordinates to be Hamiltonian with Hamiltonian functions H and K, respectively.

3.3.2.2 Fixed initial time

In the derivation of Hamilton's principal function dt_0 may be chosen to be zero, i.e., the initial time is imposed. Then Hamilton's principal function loses its dependence with respect to t_0. Eq. (3.70) is trivially verified and Eq. (3.72) does not hold anymore, meaning that W and F_1 become equivalent.

Finally, in Ref. [20] Hamilton also derives another principal function $Q(p_0, t_0, p_1, t_1)$ which compares to W as F_4 compares to F_1. The derivation being the same we will not go through it.

To conclude, Hamilton's principal function appears to be more general than the generating functions for the canonical transformation induced by the phase flow. On the other hand, the initial and final times are usually specified when solving two-point boundary problems and therefore, any of these functions will identically solve the problem. However, to find Hamilton's principal function we need to solve two partial differential equations whereas only one needs to be solved to find the generating functions. For these reasons, generating functions are more appropriate to address the problem of solving two-point boundary value problems.

3.4 Local solutions of the Hamilton–Jacobi equation

In this section, we provide a detailed discussion of a method we use to solve for the generating functions of the solution flow. This method only applies for systems with polynomial generating functions. This case obviously includes systems with polynomial Hamiltonian such as the double well potential. It also includes systems describing the relative motion of two particles moving in a Hamiltonian vector field and more generally, the motion of a particle in the vicinity of an equilibrium point or of a known trajectory. In the following we focus our discussion on the problem of relative motion between particles.

3.4.1 Direct solution for the generating function

Suppose we are interested in the relative motion of a particle whose coordinates are (q, p) with respect to another one on a known reference trajectory whose coordinates are (q^0, p^0), both moving in an Hamiltonian field. If both particles stay "close" to each other, we can expand (q, p) as a Taylor series about the reference trajectory. The dynamics of the relative motion is described by the Hamiltonian function H^h [16]:

$$\mathrm{H}^h(X^h, t) = \sum_{p=2}^{\infty} \sum_{\substack{i_1,\ldots,i_{2n}=0,\\ i_1+\cdots+i_{2n}=p}}^{p} \frac{1}{i_1!\cdots i_{2n}!} \frac{\partial^p H}{\partial q_1^{i_1}\cdots\partial q_n^{i_n}\partial p_1^{i_{n+1}}\cdots\partial p_n^{i_{2n}}}(q^0, p^0, t) X_1^{h\,i_1}\ldots X_{2n}^{h\,i_{2n}}, \tag{3.73}$$

where $\mathbf{X}^h = \begin{pmatrix} q \\ p \end{pmatrix}$ is the relative state vector. Since H^h has infinitely many terms, we are usually not able to solve the Hamilton–Jacobi equation but we can approximate the dynamics by truncating the series H^h in order to only keep finitely many terms. Suppose N terms are kept, then we say that we describe the relative motion using an approximation of order N. Clearly, the greater N is, the better our approximation is to the non-linear motion of a particle about the reference trajectory. When an approximation of order N is used, we look for a generating function F_{I_s,K_r} as a polynomial of order N in its spatial variables with time-dependent coefficients. It is important to note that even for a Hamiltonian with a finite expansion, the generating functions for that Hamiltonian will in general be analytic functions with infinite series expansions. Thus, truncation of a generating function at order N is not equivalent to a solution of the generating function of the order N Hamiltonian system, but is always only an approximation to it. Substituting the expansions into the Hamilton–Jacobi equation, it is reduced to a set of ordinary differential equations that can be integrated numerically. Once F_{I_s,K_r} is known, we find the other generating functions from the Legendre transformation, at the cost of a series inversion. If a generating function is singular, the inversion does not have a unique solution and the number of solutions characterizes the caustic.

Recall the Hamilton–Jacobi equation (Eq. (3.27)):

$$H\left(q_{I_s}, -\frac{\partial F_{I_s,K_r}}{\partial p_{\bar{I}_s}}, \frac{\partial F_{I_s,K_r}}{\partial q_{I_s}}, p_{\bar{I}_s}, t\right) + \frac{\partial F_{I_s,K_r}}{\partial t} = 0. \tag{3.74}$$

Since H is a Taylor series in its spatial variables, we look for a solution of the same form, that is, we assume that generating functions are Taylor series as well:

$$F_{I_s,K_r}(y, t) = \sum_{q=0}^{\infty} \sum_{\substack{i_1,\ldots,i_{2n}=0\\ i_1+\cdots+i_{2n}=q}}^{q} \frac{1}{i_1!\cdots i_{2n}} f^{p,r}_{q,i_1,\ldots,i_{2n}}(t) y_1^{i_1}\cdots y_{2n}^{i_{2n}}, \tag{3.75}$$

where $y = (q_{I_s}, p_{\bar{I}_s}, q_{0_{K_r}}, p_{0_{\bar{K}_r}})$. We substitute this expression into Eq. (3.74). The resulting equation is an ordinary differential equation that has the following structure:

$$P\left(y, f^{p,r}_{q,i_1,\ldots,i_{2n}}(t), \dot{f}^{p,r}_{q,i_1,\ldots,i_{2n}}(t)\right) = 0, \tag{3.76}$$

where P is a series in y with time-dependent coefficients. An explicit expression of P up to order 3 is given in Appendix A. Equation (3.76) holds for all y if and only if all the coefficients of P are zero. In this manner, we transform the ordinary differential equation (Eq. (3.76)) into a set of ordinary differential equations whose solutions are the coefficients of the generating function F_{I_s,K_r}.

Now it remains to specify initial conditions for the integration. We have seen before that only F_2 and F_3 can generate the identity transformation, the other generating functions being singular. Let us look more closely at F_2 and F_3, and especially at the coefficients[5] $f^2_{q,i_1,\ldots,i_{2n}}(t_0)$ and $f^3_{q,i_1,\ldots,i_{2n}}(t_0)$. At the initial time we have:

$$\begin{aligned} p_0 &= p \\ &= \frac{\partial F_2}{\partial q}, \end{aligned}$$

and

$$\begin{aligned} q &= q_0 \\ &= \frac{\partial F_2}{\partial p_0}. \end{aligned}$$

Within the radius of convergence, the Taylor series defining the generating functions (Eq. (3.75)) converge normally. Therefore, we can invert the summation and the derivative operator. We obtain a unique set of initial conditions:

$$f^2_{q,i_1,\ldots,i_{2n}}(t_0) = \begin{cases} 1 & \text{if } q=2,\ i_k = i_{k+n} = 1,\ i_{l\neq\{k,k+n\}} = 0,\ \forall (k,l) \in [1,n]\times[1,2n], \\ 0 & \text{otherwise.} \end{cases}$$

Similarly, we obtain for F_3:

$$f^3_{q,i_1,\ldots,i_{2n}}(t_0) = \begin{cases} -1 & \text{if } q=2,\ i_k = i_{k+n} = 1,\ i_{l\neq\{k,k+n\}} = 0,\ \forall (k,l) \in [1,n]\times[1,2n], \\ 0 & \text{otherwise.} \end{cases}$$

These initial conditions allow one to integrate two generating functions among the 4^n, but what about the other ones? This issue on singular initial conditions is similar to the one on singularity avoidance during the integration. In the next section we propose a technique to handle these problems based on the Legendre transformation. But before going further, one remark needs to be made. After we proceed with the integration, one must always verify that the series converge and that they describe the true dynamics[6] in some open set. If these two conditions are verified we can identify the generating functions with their Taylor series.

---

[5] We change our notation for convenience: f^2 stands for $f^{n,0}$, i.e., represents the coefficients of the Taylor series of F_2. We do the same for all four kinds of generating functions F_1, F_2, F_3 and F_4.

[6] Remember that even if a function is C^∞ and has a converging Taylor series, it may not equal its Taylor series. As an example take $f(x) = \exp(1/x^2)$ if $x \neq 0$, $f(0) = 0$, it is C^∞ and its Taylor series at $x = 0$ is 0, and therefore converges. However, f is not identically zero.

3.4.1.1 Singularity avoidance

We have seen that most of the generating functions are singular at the initial time. Moreover solutions to the Hamilton–Jacobi equations often develop caustics. These two issues prevent numerical integration. The goal of this section is to introduce a technique to overcome this difficulty.

We first need to recall the Legendre transformation, which allows one to derive one generating function from another (Eq. (3.14)). Suppose F_2 is known, then we can find F_1 from:

$$F_1(q, q_0, t) = F_2(q, p_0, t) - \langle q_0, p_0 \rangle, \tag{3.77}$$

where p_0 is viewed as a function of (q, q_0). Obviously, the difficulty in proceeding with a Legendre transformation lies in finding p_0 as a function of (q, q_0). To find such an expression we use Eq. (3.32):

$$q_0 = \frac{\partial F_2}{\partial p_0}(q, p_0, t), \tag{3.78}$$

and then solve for $p_0(q, q_0)$.

For the class of problems we consider, F_2 is a Taylor series. Therefore we need to perform a series inversion to eventually find p_0 as a Taylor series of (q, q_0). Series inversion is a classical problem, we adopt the procedure developed by Moulton [25]. We first suppose that there exists a series expansion of p_0 as a function of q and q_0. Then we insert this expression into Eq. (3.78) and balance terms of the same order. We obtain a set of linear equations, whose solution is found at the cost of a $n \times n$ matrix inversion (we recall that n is the dimension of the configuration space, it is small in general). If its rank is $n - p$, Weierstrass proved that the series inversion has $p + 1$ solutions (for instance, if $p = 1$, there are two solutions to the problem). This is the linear version of Prop. 3.2.10 for the F_1 generating function.

Let us return to the problem of singularity avoidance. So far, we were able to integrate generating functions of the second and third kinds since they have well-defined initial conditions. To integrate other generating functions, say F_{I_s,K_r}, we need to specify boundary conditions. Using the Legendre transformation, we can find the value of F_{I_s,K_r} at $t_1 > 0$ from the value of F_2 or F_3. This value can in turn be used to initialize the integration of the Hamilton–Jacobi equation for F_{I_s,K_r} which can be continued forward or backward in time until it encounters a singularity.

Now suppose F_2 is singular at t_2. Let us see how we can take advantage of the Legendre transformation to integrate F_2 for $t > t_2$. Proposition 3.2.3 tells us that at least one of the generating functions is non-singular at t_2. Without loss of generality, we assume that F_1 is non-singular at t_2. At $t_1 < t_2$ we carry out a Legendre transformation to find F_1 from F_2, then we integrate F_1 over $[t_1, t_3 > t_2]$ and carry out another Legendre transformation to recover F_2 at t_3. Once the value of F_2 is found at t_3, the integration of the Hamilton–Jacobi equation can be continued. Finally, we recall from Proposition 3.2.15 that we can predict the locus of the singularities using the state transition matrix.

We have described an algorithm to solve the Hamilton–Jacobi equation and developed techniques to continue the integration despite singularities. In the next section, we introduce an indirect approach to compute the generating functions based on the initial value problem. This approach naturally avoids singularities but requires more computations (see Section 3.4.3).

3.4.2 An indirect approach

By definition, generating functions implicitly define the canonical transformation they are associated with. Hence, we may compute the generating functions from the canonical transformation, i.e., compute the generating functions associated with the flow from knowledge of the flow. In this section, we develop an algorithm based on these remarks.

Recall Hamilton's equations of motion:

$$\begin{pmatrix} q \\ p \end{pmatrix} = J\nabla H(q, p, t)\,. \tag{3.79}$$

Suppose that $q(q_0, p_0, t)$ and $p(q_0, p_0, t)$ can be expressed as series in the initial conditions (q_0, p_0) with time-dependent coefficients, truncate the series to order N and substitute these into Eq. (3.79). Hamilton's equations reduce to an ordinary differential equation of a form that is polynomial in (q_0, p_0). As before, we balance terms of the same order and transform Hamilton's equations into a set of ordinary differential equations whose variables are the time-dependent coefficients defining q and p as series of q_0 and p_0. Using $q(q_0, p_0, t_0) = q_0$ and $p(q_0, p_0, t_0) = p_0$ as initial conditions for the integration, we are able to compute an approximation of order N of the phase flow. Once the flow is known, we recover the generating functions by performing a series inversion.

Example 3.4.1. Suppose we want to compute F_1 at $t = T$. From $q = q(q_0, p_0, T)$ we carry out a series inversion to eventually find $p_0 = p_0(q, q_0, T)$. Then $p_0 = p_0(q, q_0, T)$ together with $p = p(q_0, p_0, T)$ defines the gradient of F_1:

$$\begin{aligned} \frac{\partial F_1}{\partial q}(q, q_0, T) &= p \\ &= p(q_0, p_0(q, q_0, T)), \end{aligned} \tag{3.80}$$

$$\begin{aligned} \frac{\partial F_1}{\partial q_0}(q, q_0, T) &= -p_0 \\ &= -p_0(q, q_0, T). \end{aligned} \tag{3.81}$$

We recover F_1 from its gradient by performing two quadratures over the polynomial terms. If one uses traditional numerical integrators to integrate the phase flow, Eqs. (3.80) and (3.81) are not integrable due to numerical round off

$$\left(\frac{\partial p(q_0, p_0(q, q_0, T))}{\partial q_0} \neq -\frac{\partial p_0(q, q_0, T)}{\partial q} \right).$$

Using symplectic algorithms to compute the approximate phase flow, we preserve the Hamiltonian structure of the flow and thus are assured that Eqs. (3.80) and (3.81) are integrable [13].

3.4.3 A comparison of the direct and indirect approach

We have introduced two algorithms that compute the generating functions associated with the phase flow. In this section, we highlight the advantages and drawbacks of each method. In addition, we show that by combining them we obtain a robust and powerful algorithm.

3.4.3.1 Method specifications

The direct approach

The direct approach provides us with a closed form approximation of the generating functions over a given time interval. However, there are inherent difficulties as generating functions may develop singularities which prevent the integration from going further in time. The technique we developed to bypass this problem results in additional computations. It requires us to first identify the times at which generating functions become singular, and then to find a non-singular generating function at each of these times. Over a long time simulation, this method reaches its limits as many singularities may need to be avoided.

The indirect approach

The main advantage of the indirect method is that it never encounters singularities, as the flow is always non-singular. On the other hand, this method requires us to solve many more equations than the direct approach (see below). Furthermore, a major drawback of the indirect approach is that it computes an expression for the generating functions at a given time only, the time at which the series inversion is performed. To generate solutions to a two-point boundary value problem over a range of times then requires that a series inversion be performed at each point in time.

The curse of dimensionality

In this paragraph, we point out a difficulty inherent to both methods, namely the "curse of dimensionality". As we solve the generating functions to higher and higher orders, the number of variables grows dramatically. This problem is the limiting factor for computation: typically on a $2GHz$ Linux computer with $1G$ RAM, we have trouble solving the generating functions to order 7 and up for a 6-dimensional Hamiltonian system.

Computation of the generating functions using the direct approach requires us to find all the coefficients of a $2n$-dimensional series with no linear terms. At order N, a $2n$-dimensional Taylor series has x terms, where

$$x = \binom{2n-1+N}{N} = \frac{(2n-1+N)!}{N!(2n-1)!}.$$

In the indirect approach we express the $2n$-dimensional state vector as Taylor series with respect to the $2n$ initial conditions. Therefore, we need to compute the coefficients of $2n$ $2n$-dimensional Taylor series.

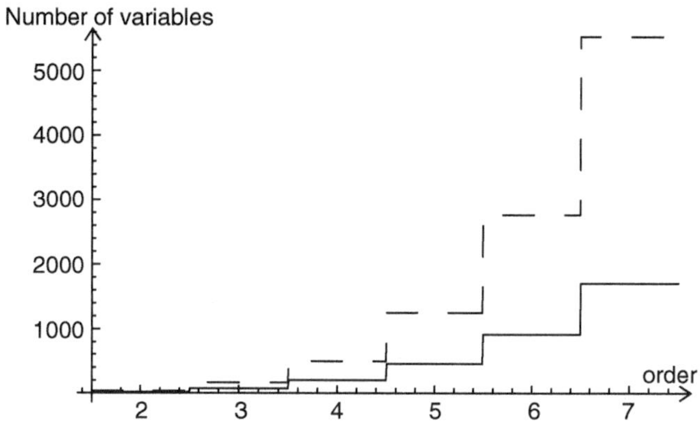

Fig. 3.2. Number of variables in the indirect (dashed) and direct (solid) methods.

To summarize, an approximation of order N of the generating functions is found by solving:

- $\sum_{k=2}^{N} \frac{(2n-1+k)!}{k!(2n-1)!}$ ordinary differential equations using the direct approach,
- $2n \sum_{k=1}^{N-1} \frac{(2n-1+k)!}{k!(2n-1)!}$ ordinary differential equations using the indirect approach[7].

In Figure 3.2, the solid line and dotted line indicate the numbers of equations that needs to be solved with the direct and indirect methods for a 6-dimensional Hamiltonian system.

3.4.3.2 A combined algorithm

In practice, to solve boundary value problems over a long time span it is most convenient to combine both methods. Typically, we first solve the initial value problem (indirect method) up to a time of interest, say T. Then we solve the Hamilton–Jacobi equation (direct approach) about T, with initial conditions equal to the values of the generating functions at T found using the indirect approach. This approach has been applied in [17, 18] to solve two-point boundary value problems over week-long time spans in low Earth orbit.

3.4.4 Convergence and existence of solutions

We now study the convergence properties of our algorithm. In particular, we provide a criterion to evaluate the domain in which the approximation of order N of the generating functions is valid. An example to illustrate this criterion is given.

[7] The summation goes from 1 to $N-1$ because the indirect approach computes the gradient of the generating functions.

3.4.4.1 Theoretical considerations

Recall the general form of a generating function (Eq. (3.75)):

$$F_{I_s,K_r}(q_{I_p}, p_{\bar{I}_p}, q_{O_{K_r}}, p_{O_{\bar{K}_r}}, t) = \sum_{q=0}^{\infty} \sum_{\substack{i_1,\ldots,i_{2n}=0 \\ i_1+\cdots+i_{2n}=q}}^{q} \frac{1}{i_1!\cdots i_{2n}!} f^{p,r}_{q,i_1,\ldots,i_{2n}!}(t) y_1^{i_1} \cdots y_{2n}^{i_{2n}}.$$

Definition 3.4.1. (Radius of convergence.) The radius of convergence of the multi-variable series defining F_{I_s,K_r} at t is the real number R_t such that the sum:

$$\sum_{q=0}^{\infty} \left(\sum_{\substack{i_1,\ldots,i_{2n}=0 \\ i_1+\cdots+i_{2n}=q}}^{q} \frac{1}{i_1!\cdots i_{2n}!} f^{p,r}_{q,i_1,\ldots,i_{2n}}(t) \right) \eta^q$$

converges absolutely $\forall \eta,\ 0 < \eta < R_t$ and diverges $\forall \eta > R_t$.

The following proposition, whose proof can be found in many textbooks, concerns the normal convergence of the series. Earlier, we used this result for finding the initial conditions to integrate the Hamilton–Jacobi equation.

Proposition 3.4.1. *Let R_t be the radius of convergence of the multi-variable series defining F_{I_s,K_r} at the time t. Then for all $\eta < R_t$ the series converges normally in $\{y \in \mathbb{R}^{2n} : \|y\| \leq \eta\}$ at t.*

The radius of convergence is not appropriate for studying series of functions as it is a function of time. To remove the time dependency, we define the domain of convergence, a domain $\mathcal{D}$ in $\mathbb{R} \times \mathbb{R}^{2n}$ in which the series converge uniformly.

Definition 3.4.2. (Domain of convergence.) The domain of convergence $\mathcal{D}$ is a region in $\mathbb{R} \times \mathbb{R}^{2n}$ in which the series

$$\sum_{q=0}^{\infty} \sum_{\substack{i_1,\ldots,i_{2n}=0 \\ i_1+\cdots+i_{2n}=q}}^{q} \frac{1}{i_1!\cdots i_{2n}!} f^{p,r}_{q,i_1,\ldots,i_{2n}}(t) y_1^{i_1} \cdots y_{2n}^{i_{2n}}$$

converges uniformly.

In contrast with the radius of convergence, the domain of convergence is not uniquely defined. The spatial domain depends on the time interval and vice versa. For instance, $\sum_n t^n y^n$ converges if and only if $ty < 1$. $\mathcal{D} = \{(t, y) \in [0, 2] \times [0, 0.5]\}$ and $\mathcal{D} = \{(t, y) \in [0, 0.5] \times [0, 2]\}$ are two well-defined domains of convergence.

In Def. 3.4.2, the uniform convergence of the series is of prime importance. It allows one to bound the error between the true series and its truncation. Indeed, by definition we have:

$$\forall \epsilon > 0, \; \exists N > 0, \; \forall (t, y) \in \mathcal{D},$$

$$F_{I_s,K_r}(q_{I_p}, p_{\bar{I}_p}, q_{0_{K_r}}, p_{0_{\bar{K}_r}}, t) - \sum_{q=0}^{N} \sum_{\substack{i_1,\ldots,i_{2n}=0 \\ i_1+\cdots+i_{2n}=q}}^{q} \frac{1}{i_1! \cdots i_{2n}} f^{p,r}_{q,i_1,\ldots,i_{2n}!}(t) y_1^{i_1} \cdots y_{2n}^{i_{2n}} < \epsilon . \tag{3.82}$$

In other words, given a domain of convergence and a precision goal ϵ, there exists a positive integer N such that the truncated Taylor series of order N approximates the true function within ϵ in the domain.

3.4.4.2 *Practical considerations*

In practice, for most of the problems we are interested in, we are only able to compute finitely many terms in the series. As a result, it is impossible to estimate a domain of convergence. Worse, we cannot theoretically guarantee that the generating functions can be expressed as Taylor series. In fact, we have seen earlier that even if the Taylor series of F_{I_s,K_r} converges on some open set and F_{I_s,K_r} is smooth, then F_{I_s,K_r} may not be equal to its Taylor series. One can readily verify that the function $f(x) = \exp(1/x^2)$ if $x \neq 0$, $f(0) = 0$ is smooth and has a converging Taylor series at 0. However, f is not equal to its Taylor series. In the following we make two realistic assumptions in order to develop a practical tool for estimating a domain of convergence.

We first assume that the flow may be expressed as a Taylor series in some open set. This is a very common assumption when studying dynamical systems. For example, we make this hypothesis when we approximate the flow by the state transition matrix at linear order. We noticed in the indirect approach that the generating functions may be computed from the flow at the cost of a series inversion. From the series inversion theory (see e.g., Moulton [25]), we conclude that the generating functions can also be expressed as Taylor series (when they are not singular). Thus, for almost every t, there exists a non-zero radius of convergence. In addition, the concept of domain of convergence is well-defined.

The second assumption we make is also reasonable. We assume that there exists a domain in which the first-order terms of the series defining F_{I_s,K_r} are dominant. In other words, we assume that there exists a domain in which the linear order is the largest, followed by the second order, third order, etc. This is again a very common assumption for dynamical systems. When approximating the flow with the state transition matrix, we implicitly assume that the linear term is dominant. However, in the present case, there is a subtlety due to the presence of singularities. We observe that this assumption no longer holds as we get closer to a singularity. Let us look at an example to illustrate this phenomenon.

Example 3.4.2. The Taylor series in x of $f(x,t) = (1-t)^x$ for $t \in (0,1)$ is

$$\sum_{r=0}^{\infty} a_n x^n, \text{ where } a_n = \frac{\log(1-t)^n}{n!}.$$

Its radius of convergence is $R_t = \infty$ for all $t \in (0,1)$ and it is singular at $t = 1$. In Figure 3.3, we plot the four first terms of the series as a function of x for different times. Clearly, as t gets closer to 1, the first-order terms are less and less dominant. Equivalently, the x-interval in which the first-order terms are dominant shrinks as t goes to 1. In Figure 3.4, we plot $(1-t)^x - \sum_{r=0}^{3} \frac{\log(1-t)^n}{n!} x^n$. One can readily verify that given a prescribed error margin, the domain in which the order 4 approximates f within this margin shrinks as t gets closer to 1. This is a very common behavior that motivates the need for a new criterion.

Suppose that the fourth-order approximation of f is to be used for solving a given problem where the time evolves from 0 to 0.6. We know that such an approximation is relevant if the first-order terms are dominant, i.e., $a_0 > a_1 > a_2 > a_3$. From Figure 3.3, we infer that this condition is satisfied if and only if $\|x\| \leq 1$. We call the domain $\mathcal{D}_u = \{[0,1], [0, 0.6]\}$ the domain of use.

Let us formalize the concept of *domain of use*.

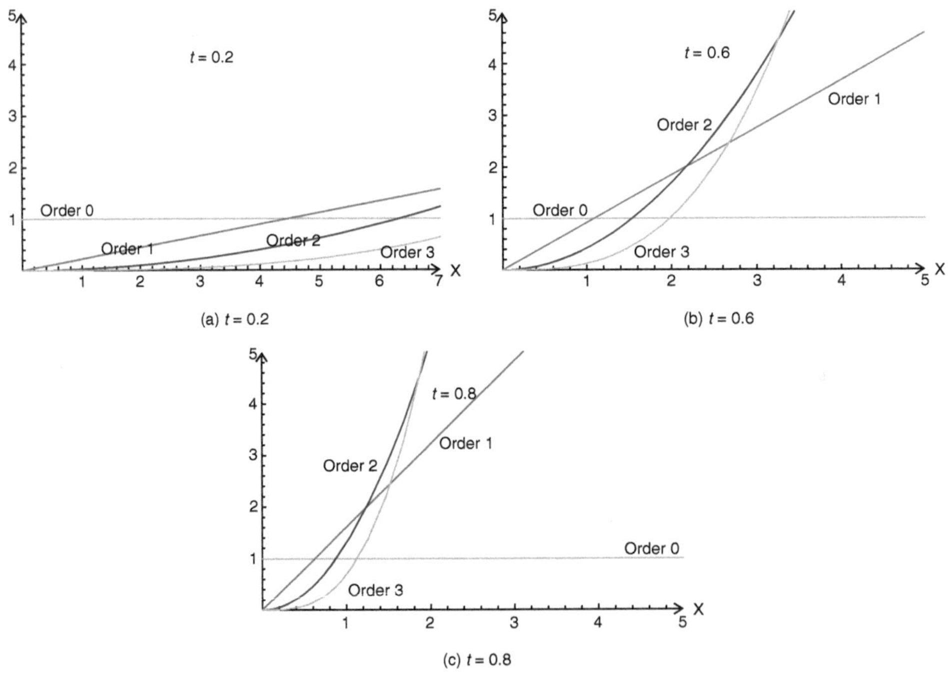

Fig. 3.3. Contribution of the first four terms in the Taylor series of $(1-t)^x$.

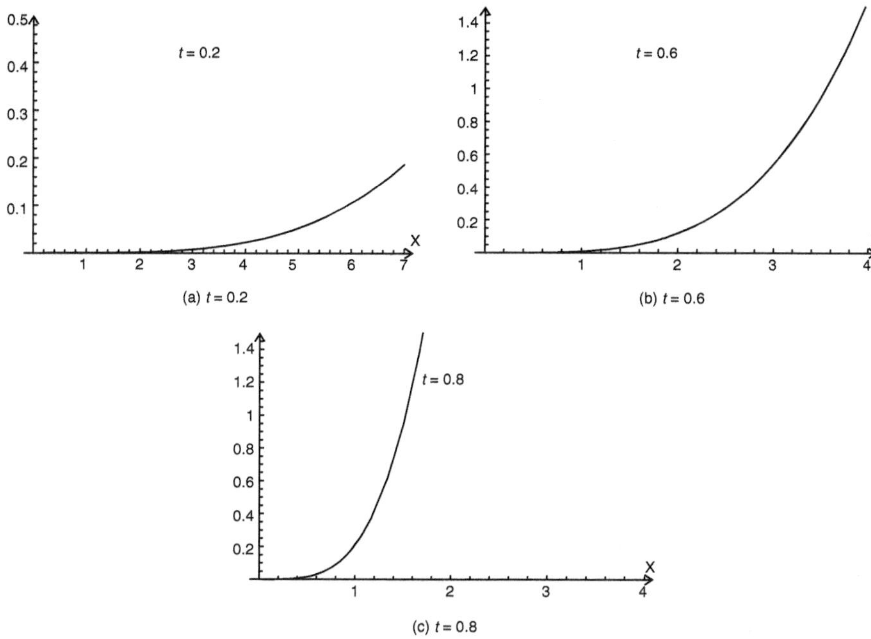

Fig. 3.4. $(1-t)^x - \sum_{r=0}^{3} \frac{\log(1-t)^n}{n!} x^n$.

Definition 3.4.3. (Domain of use.) The domain of use $\mathcal{D}_u$ is a domain in $\mathbb{R} \times \mathbb{R}^{2n}$ in which

$$\left(\sum_{\substack{i_1,\ldots,i_{2n}=0 \\ i_1+\cdots+i_{2n}=q}}^{q} \frac{1}{i_1!\cdots i_{2n}} f^{p,r}_{q,i_1,\ldots,i_{2n}}(t) y_1^{i_1} \cdots y_{2n}^{i_{2n}} \right)_q$$

is a decreasing sequence.

This definition is very conservative but very easy to work with. For a given problem, we identify a time interval (or a spatial domain) in which we want to use the generating functions. Then we compute the spatial domain (or the time interval) in which our solution is valid. Once we have identified the domain of use, one can safely work with the solution within this domain. Let us illustrate the use of the above tool with an example.

3.4.5 Examples

We consider the following fictional space mission: A formation of spacecraft is flying about the Libration point L_2 in the Hill three-body problem and we wish to use F_1 to solve the position to position boundary value problem in order to design a reconfiguration.

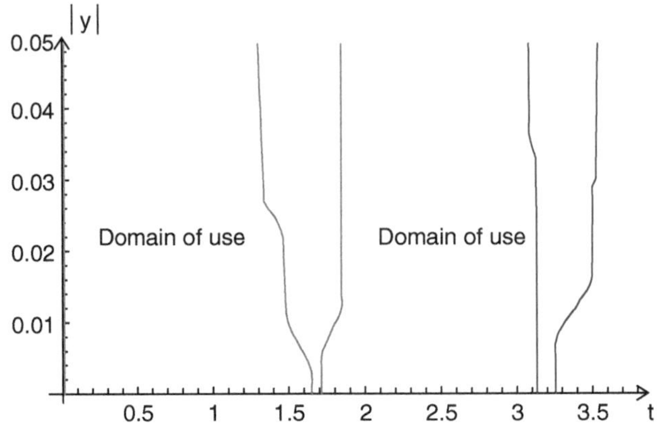

Fig. 3.5. Domain of use.

The mission specifications impose the spacecraft to stay within 0.05 units of length from the equilibrium point L_2 (i.e., 107 500 km in the Earth–Sun system). The normalized Hamiltonian describing the Hill's dynamics is (Appendix B):

$$H(q, p) = \frac{1}{2}(p_x^2 + p_y^2) + (q_y p_x - q_x p_y) - \frac{1}{\sqrt{q_x^2 + q_y^2}} + \frac{1}{2}(q_y^2 - 2q_x^2), \tag{3.83}$$

where $q_x = x$, $q_y = y$, $p_x = \dot{x} - y$, and $p_y = \dot{y} + x$. Using this coordinate system, L_2 is an equilibrium point with coordinates $(q_x, q_y) = (3^{-1/3}, 0)$. To use the above algorithm, H must be expressed as a Taylor series in the spatial variables. Hence, since we want to study the dynamics about L_2, we linearize H about L_2. Then, we use the algorithm to solve F_1 up to order 5 in the time interval $(0, 3.5)$. Using the direct approach, this is equivalent to solving 121 ordinary differential equations. We encounter a number of singularities for F_1 at $t = 0$, $t = 1.68$, and $t = 3.19$. In Figure 3.5, we plot the maximum value of $\|y\|$ so that the first five terms are in decreasing order[8]. We notice that as we get closer to the singularity, the maximum value of $\|y\|$ goes to 0. To find the domain of use, we only need to intersect this plot with $\|y\| = 0.05$.

3.4.5.1 Error in the approximation

We can verify a posteriori that the Taylor series expansion found for the generating function F_1 approximates the true dynamics within this domain. To do so, we again use the example from before and set $q(T) = q_1$ and q_0, and find $p(T) = p_1$ and p_0 from Eqs. (3.28) to (3.29). Then we integrate the trajectory whose initial condition is (q_0, p_0) to find $(q(T), p(T)) = (q_2, p_2)$. The error in the approximation is defined as the norm of

[8] Some terms may change sign and therefore may be very small. In that case we ignore these terms so that the decreasing condition can be satisfied (For instance if the order 2 term goes to 0, it will be smaller than any other terms and therefore must be ignored).

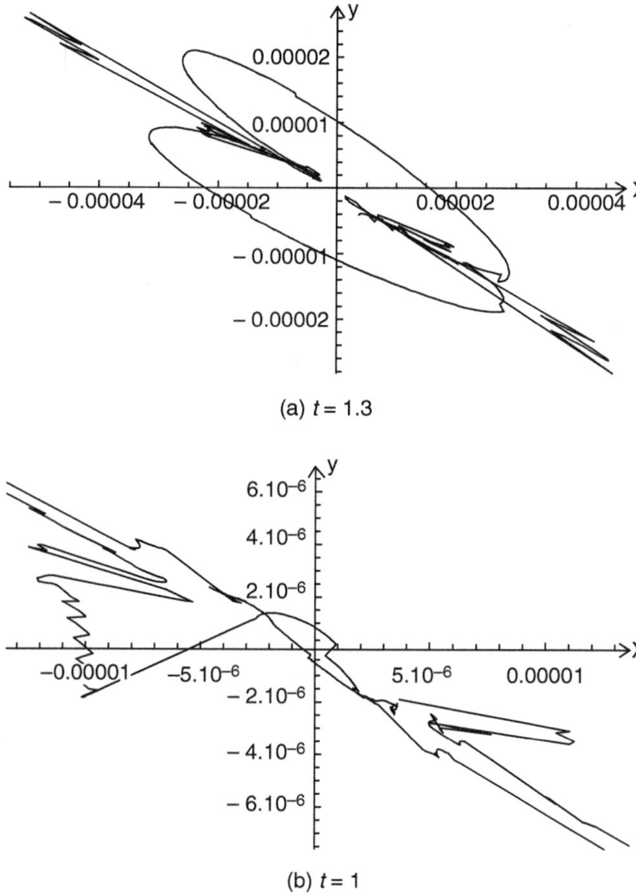

Fig. 3.6. Difference between the true and the approximate dynamics.

$(q_2 - q_1, p_2 - p_1)$. In Figure 3.6 we plot this error for $q_0 = 0$ and q_1 that takes values on the circle centered at L_2 of radius 0.05 for different values of t. The solution is checked at points along the circle, generating the error curves in the figures. We observe that the truncated series provide a good approximation of the true dynamics.

We also point out that since the series is converging and the magnitude of each order decreases in the domain of use, the accuracy must always increase if an additional order is taken into account. In Figure 3.7, we observe that the order two solution provides a poor approximation to the initial momentum because the error ranges up to 4.5×10^{-3} units of length (i.e., 9615 km in the Earth–Sun system). Order three and four give order of magnitude improvements, the error is less than 2.2×10^{-4} units of length (480 km) for order three and less than 3.5×10^{-5} units of length (77 km) for order four, over two orders of magnitude better than the order two solution.

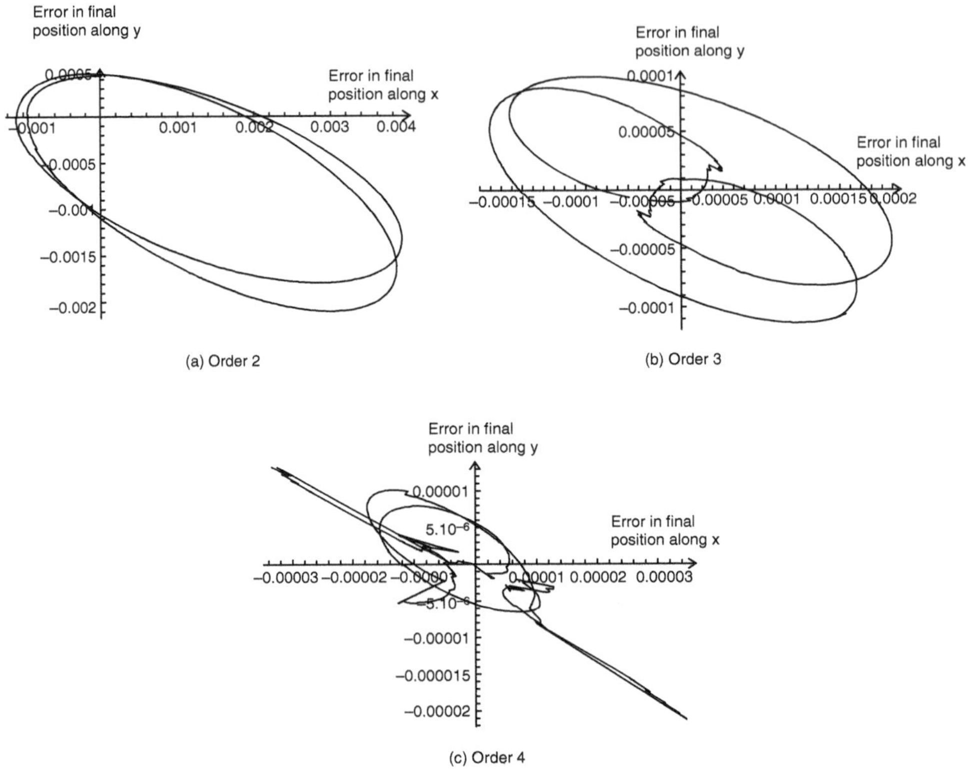

Fig. 3.7. Error in the normalized final position for $t = 0.9$.

3.5 Applications

We now illustrate the use of our novel approach to solving two-point boundary value problems. We consider the normalized Hill three-body problem (see Appendix B). Using our algorithm outlined previously we compute the Taylor series expansion of the generating function F_1 about the Libration point L_2 up to order N (N is determined by the accuracy we wish to achieve). In other words, F_1 can be expressed as a polynomial of order N with time-dependent coefficients whose values are known. The gradient of F_1 is then a polynomial of order $N-1$. Hence, solutions to any position to position boundary value problem (i.e., Lambert's type of boundary value problem) is solved by evaluating a polynomial of order $N-1$. Once F_1 is known, we can solve many problems. In this section, we choose to focus on the study of periodic orbits and on optimal control problems. The study of periodic orbits has been presented in Ref. [15] and the optimal control problems can be found in Ref. [12]. The method has also been applied to the formation flight of spacecraft, we refer the interested reader to Refs. [16–18].

3.5.1 The search for periodic orbits

3.5.1.1 Methodology

To find periodic orbits using the theory we developed above, we need to characterize them as solutions to two-point boundary value problems.

Periodic orbits in a $2n$-dimensional Hamiltonian dynamical system are characterized by the following equations:

$$q(T) = q_0, \tag{3.84}$$

$$p(T) = p_0, \tag{3.85}$$

where T is the period of the orbit, (q_0, p_0) are the initial conditions at time $t_0 = 0$ and $(q(t), p(t))$ verifies Hamilton's equations:

$$\dot{q} = \frac{\partial H}{\partial p}(q, p, t), \quad \dot{p} - \frac{\partial H}{\partial q}(q, p, t). \tag{3.86}$$

In the most general case, the search for periodic orbits consists of solving the $2n$ equations (3.84) and (3.85) for the $2n+1$ unknowns (q_0, p_0, T). Simple methods that solve this problem take a set of initial conditions (q_0, p_0), and integrate Hamilton's equations. If there exists a time $t = T$ such that Eqs. (3.84) and (3.85) are verified, then a periodic orbit is found. Else, other initial conditions need to be guessed. In the approach we propose in this chapter, instead of looking at the initial conditions and the period as the only variables of the problem, we suppose that the period, n initial conditions as well as n components of the state vector at time T are unknowns. Then the search for periodic orbits reduces to solving the $2n$ equations (3.84) and (3.85) for these $2n+1$ unknowns.

For instance, if $(q(T), q_0, T)$ are taken to be the $2n+1$ unknowns, then the search for periodic orbits consists of solving the $2n$ equations (3.84) and (3.85) for $(q(T), q_0, T)$. Let us now find all periodic orbits of a given period. In other words, T is given and we need to find $(q(T), q_0)$ such that $q(T) = q_0$ and $p(T) = p_0$. This is a boundary value problem with constraints that can be solved with the generating function F_1. Combining Eqs. (3.28) and (3.29) and Eqs. (3.84) and (3.85) we obtain:

$$\begin{aligned} p(T) &= \frac{\partial F_1}{\partial q}(q, q_0, T), \qquad q(T) = q_0, \\ p_0 &= -\frac{\partial F_1}{\partial q_0}(q, q_0, T), \qquad p(T) = p_0, \end{aligned} \tag{3.87}$$

i.e.,

$$\frac{\partial F_1}{\partial q}(q = q_0, q_0, T) + \frac{\partial F_1}{\partial q_0}(q = q_0, q_0, T) = 0, \tag{3.88}$$

$$p = p_0 = \frac{\partial F_1}{\partial q}(q = q_0, q_0, T). \tag{3.89}$$

Eqs. (3.88) and (3.89) define necessary and sufficient conditions for the existence of periodic orbits. Therefore, the search for all periodic orbits of a given period is reduced to solving n equations (3.88) for n variables, the q_0's, and then evaluate n equations (3.89) to compute the corresponding momenta. Most importantly, once F_1 is known, solutions of these conditions are computed using algebraic manipulations, no integration is required.

Similarly, if we want to find all periodic orbits going through a given point in the position space, we set q_0 in Eq. (3.88) and solve for T. However, instead of solving the n equations defined by Eq. (3.88) for one variable, T, we may combine them in the following way:

$$\left\| \frac{\partial F_1}{\partial q}(q = q_0, q_0, T) + \frac{\partial F_1}{\partial q_0}(q = q_0, q_0, T) \right\| = 0, \tag{3.90}$$

where $\|\cdot\|$ is a norm. This equation can be easily solved numerically or even graphically.

3.5.1.2 Examples

In the following we use a truncated Taylor series of F_1 of order $N = 5$ to study periodic orbits about the Libration point L_2.

First, let us find all periodic orbits going through $q_0 = (0.01, 0)$. To solve this problem, we use the necessary and sufficient condition defined by Eq. (3.90). In Figure 3.8 we plot the left-hand side of Eq. (3.90) as a function of the normalized time. We observe that the norm vanishes only at $t = T = 3.03353$. Therefore, there exists only one periodic orbit going through q_0, and its period is T (there may be additional periodic orbits of period $T > 3.2$, but we cannot see them in this figure). Again, these results are in agreement with known results on periodic orbits about L_2. One can show that any point in the vicinity of L_2 belongs to a periodic orbit. The periods of these orbits increase as their distances from L_2 increase. In the limit, as the distance between periodic orbits and L_2 goes to zero, the period goes to $T = T_{linear} = 3.033019$.

Another problem is to find all periodic orbits of a given period $T = 3.0345$. To solve this problem, we use Eq. (3.88) which defines two equations with two unknowns that can be solved graphically. In Figure 3.9, we plot the solutions to each of these two equations and then superimpose them to find their intersection. The intersection corresponds to

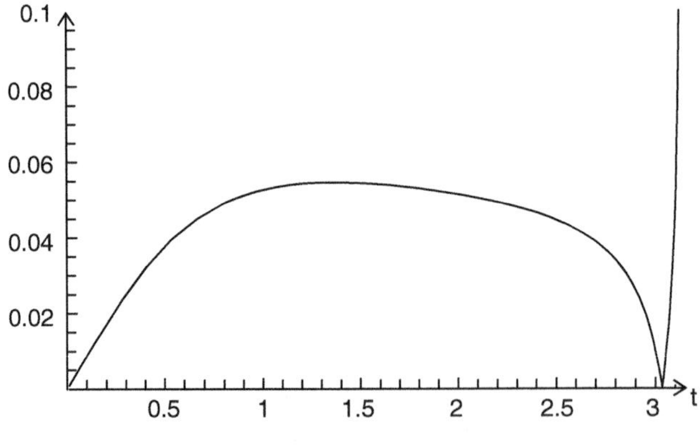

Fig. 3.8. Plot of $\|\frac{\partial F_1}{\partial q}(q = q_0, q_0, T) + \frac{\partial F_1}{\partial q_0}(q = q_0, q_0, T)\|$ where $q_0 = (0.01, 0)$.

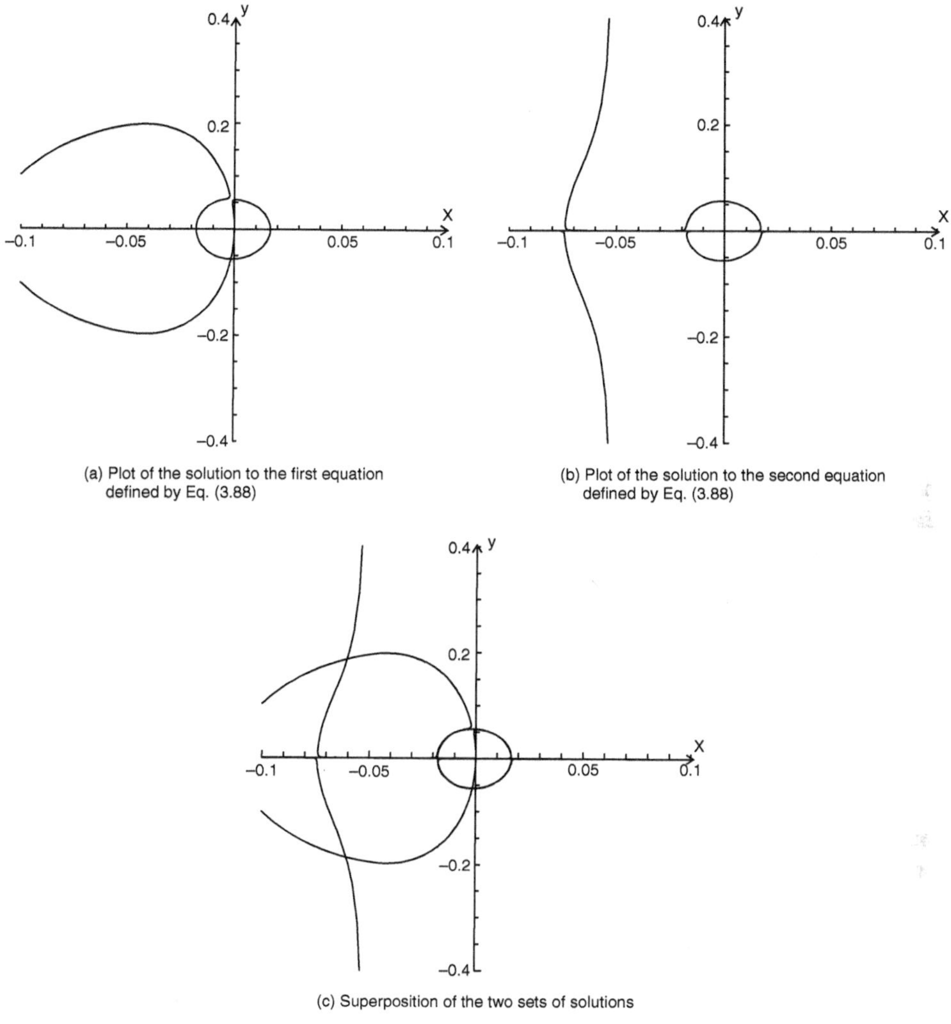

Fig. 3.9. Periodic orbits for the non-linear motion about a Libration point.

solutions to Eq. (3.90), i.e., to the set of points that belongs to periodic orbits of period T. We observe that the intersection is composed of a circle and two points whose coordinates are $(q_x, q_y) = (-0.0603795, \pm 0.187281)$. The circle is obviously a periodic orbit but the two points are not equilibrium points, and rather correspond to out-of-plane periodic orbits[9].

[9] In the Hill three-body problem these out-of-plane orbits do not exist. At that distance of L_2, our approximation of the dynamics of order 5 is no more valid, therefore these two points do not have any physical meaning. In practice, we can evaluate a domain in which an approximation of order N of the dynamics is valid. We refer to Section 3.4.4 and Ref. [12] for more details.

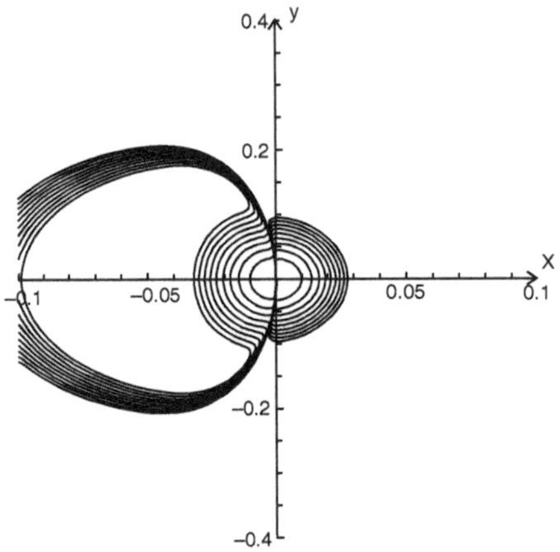

(a) Plot of the solution to the first equation defined by Eq. (3.88) for $t = 3.033 + 0.0005n$ $n \in \{1 \ldots 10\}$

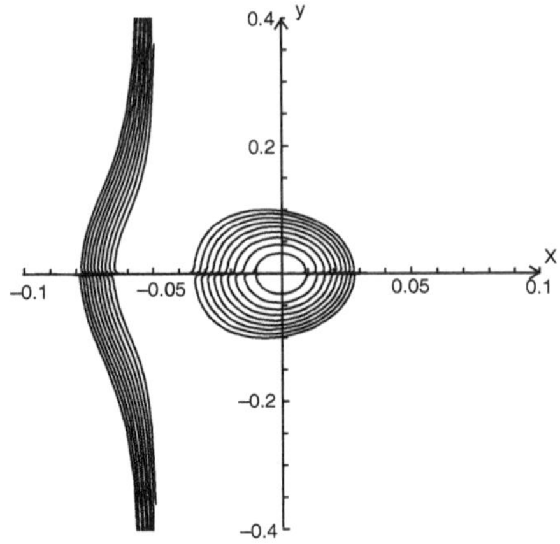

(b) Plot of the solution to the second equation defined by Eq. (3.88) for $t = 3.033 + 0.0005n$ $n \in \{1 \ldots 10\}$

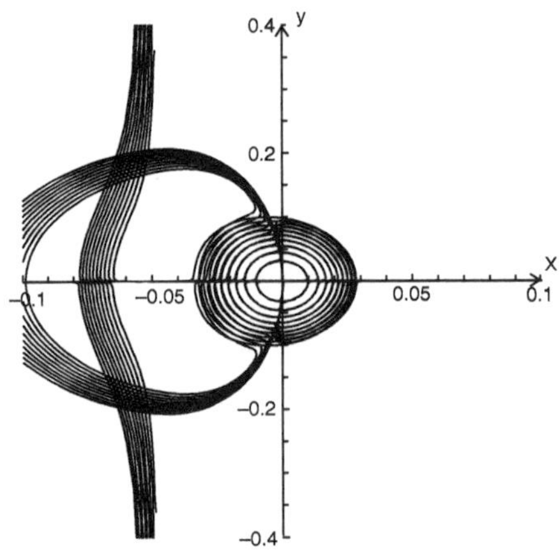

(c) Superposition of the two sets of solutions for $t = 3.033 + 0.0005n$ $n \in \{1 \ldots 10\}$

Fig. 3.10. Periodic orbits for the non-linear motion about a Libration point.

By plotting the intersection for different periods T, we generate a family of periodic orbits around the Libration point. In Figure 3.10 we represent the solutions to Eq. (3.88) for $t = 3.033 + 0.0005n,\ \ n \in \{1 \cdots 10\}$. For $t = 3.033$ (which is less that the periodic orbit period of the linearized system), the intersection only contains the origin, which is why there are only 9 periodic orbits shown around the origin. We note that at larger values of $x^2 + y^2$ the curves do not overlay precisely, indicating that higher order terms are needed.

The method we propose allows us to search for periodic orbits anywhere in the phase space or in the time domain without requiring any initial guess or knowledge of a periodic orbit that belongs to the family. This is a major advantage compared to traditional methods.

Most important, we reduce the search for periodic orbits to solving a non-linear system of equations. Once the generating functions are known, no integration is required to find periodic orbits of different periods and/or going through different points in the phase space. This is a fundamental property of the generating functions; once the generating functions are known, any two-point boundary value problem can be solved at the cost of a single function evaluation.

3.5.2 Optimal control and mission planning

To illustrate the use of the generating functions to solve non-linear optimal control problems we now consider a targeting problem in the two-dimensional Hill three-body problem (Appendix B). We consider a spacecraft away from the Libration point L_2 and want to find the control sequence that moves the spacecraft at the equilibrium point L_2 while minimizing the fuel consumption. Specifically, this optimal control problem formulates as follows:

We want to minimize the cost function $J = \frac{1}{2}\int_{t=0}^{t=t_f} \left(u_x^2 + u_y^2\right) \mathrm{d}t$ subject to the constraints

$$\begin{cases} \dot{x_1} = x_3, \\ \dot{x_2} = x_4, \\ \dot{x_3} = 2x_4 - \frac{x_1}{(x_1^2+x_2^2)^{3/2}} + 3x_1 + u_x, \\ \dot{x_4} = -2x_3 - \frac{x_2}{(x_1^2+x_2^2)^{3/2}} + u_y, \end{cases} \tag{3.91}$$

and the boundary conditions:

$$X(t=0) = X_0, \quad X(t=t_f) = X_{L_2} = (3^{-1/3}, 0, 0, 0), \tag{3.92}$$

where $r^2 = x^2 + y^2$ and $X = (x_1, x_2, x_3, x_4) = (x, y, \dot{x}, \dot{y})$. Define the Hamiltonian:

$$\begin{aligned} H(X, P, U) =& p_1 x_3 + p_2 x_4 + p_3 \left(2x_4 - \frac{x_1}{(x_1^2+x_2^2)^{3/2}} + 3x_1 + u_x\right) \\ &+ p_4 \left(-2x_3 - \frac{x_2}{(x_1^2+x_2^2)^{3/2}} + u_y\right) + \frac{1}{2}u_x^2 + \frac{1}{2}u_y^2, \end{aligned}$$

where $P = (p_1, p_2, p_3, p_4)$ and $U = (u_x, u_y)$. Then, from $\frac{\partial H}{\partial U} = 0$, we find the optimal control feedback law:

$$u_x = -p_3, \quad u_y = -p_4.$$

Substituting $U = (u_x, u_y)$ into H yields:

$$\begin{aligned} \bar{H}(X, P) =& p_1 x_1 + p_2 x_2 + p_3 \left(2x_4 - \frac{x_1}{(x_1^2+x^2)^{3/2}} + 3x_1 - p_3\right) \\ &+ p_4 \left(-2x_3 - \frac{x_2}{(x_1^2+x^2)^{3/2}} - p_4\right) + \frac{1}{2}p_3^2 + \frac{1}{2}p_4^2. \end{aligned} \tag{3.93}$$

We deduce the necessary conditions for optimality:

$$\dot{X} = \frac{\partial \bar{H}}{\partial P}, \tag{3.94}$$

$$\dot{P} = -\frac{\partial \bar{H}}{\partial X}, \tag{3.95}$$

$$X(t=0) = X_0, \quad X(t=t_f) = (0,0,0,0).$$

This is a position to position boundary value problem that can be solved using F_1. In this example, we compute F_1 at order 4 and use this approximation together with Eqs. (3.28) and (3.29) to find the value of the co-state P at the initial and final times. Then, the optimal trajectory is found by integrating Hamilton's equations (Eqs. (3.94) and (3.95)).

In Figure 3.11, we plot the trajectories for different final times. As t_f increases, the trajectory tends to wrap around the Libration point so that the spacecraft takes advantage of the geometry of the Libration point (Appendix B). On the other hand, if the transfer time is small, the trajectory is almost a straight line, it completely ignores the dynamics. In Figure 3.12 the associated control laws are represented. As expected, the longer the transfer time is, the smaller the magnitude of the control. We emphasize that we only need to evaluate the gradient of F_1 (which is a polynomial of order 3) seven times and integrate Eqs. (3.94) and (3.95) seven times to obtain the seven curves in Figure 3.11. Similarly, in Figure 3.13, at the cost of sixteen evaluations of the gradient of F_1, we are able to represent the optimal trajectories of spacecraft starting at $X_0 = (r\cos(\theta), r\sin(\theta))$ where $r = 10\,700$ km and $\theta = k\pi/8$, and ending at L_2 in 145 days. In Figure 3.14 the corresponding optimal control law is represented.

Further, if different types of boundary conditions are imposed (for instance, the terminal state is not fully specified) then we need to perform a Legendre transform to find the generating function that solves this new boundary value problem. There is no need

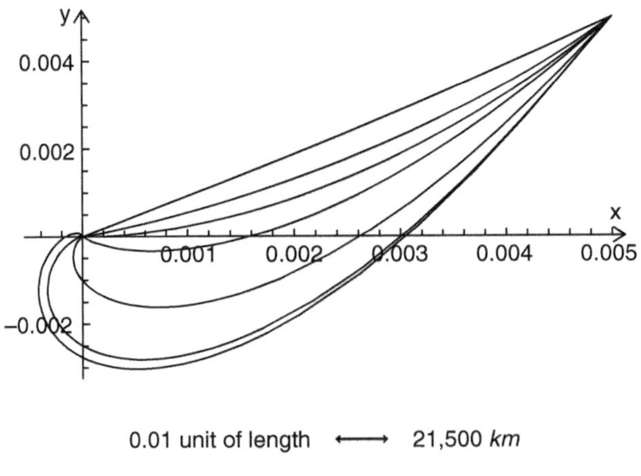

Fig. 3.11. Optimal trajectories of the spacecraft for different transfer times.

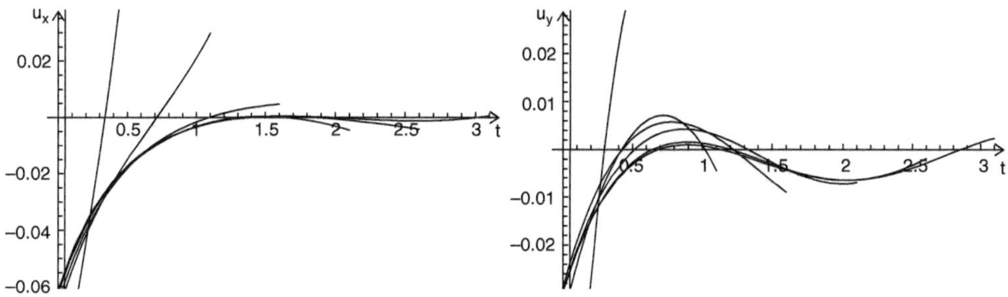

(a) Time history of u_x

(b) Time history of u_y

1 unit of time ⟷ 58 *days*,

10^{-3} unit of control ⟷ $1.36 \cdot 10^{-2}$ *m·s*$^{-2}$

Fig. 3.12. Time history of the control laws.

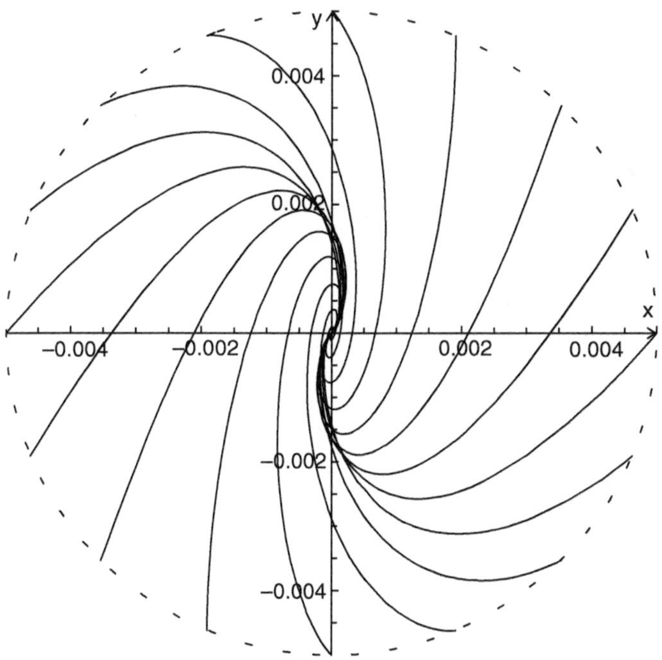

0.01 unit of length ↔ 21,500 *km*

Fig. 3.13. Optimal trajectories of the spacecraft as a function of the initial position.

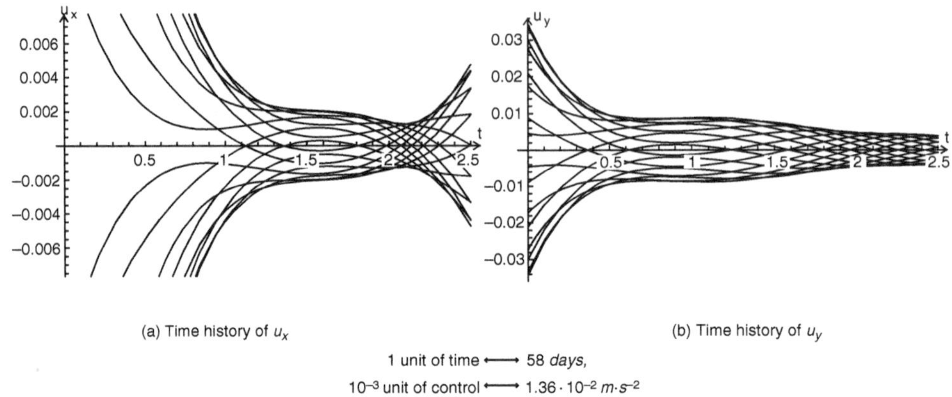

Fig. 3.14. Time history of the control laws.

to resolve the optimal control problem. This is an intrinsic property of the generating functions that opens the doors to truly reconfigurable optimal control. In [27, 28] the application of generating functions to solve optimal control problems is studied in detail, and a number of strong results are discovered for the necessary and sufficient conditions for optimal control.

Finally, we point out that this method is very efficient for mission planning. By varying the transfer time for instance, we can find times for which the optimal transfer requires less fuel expenditure. For more details on this topic we refer to Ref. [16]. Another major application of the theory presented in this chapter is spacecraft formation design. We noticed in the introduction that the reconfiguration of a formation of N spacecraft requires us to solve $N!$ boundary value problems. Using the generating functions this task can be achieved at the cost of $N!$ function evaluations. This problem and others relating to formation flight are solved in Refs. [17, 18].

3.6 Conclusions

The method we develop in this chapter is based on the Hamilton–Jacobi theory. We have observed that the generating functions associated with the phase flow readily solve any Hamiltonian two-point boundary value problems. This observation has many consequences that we now re-state. Above all, it provides a very general methodology for solving boundary value problems for Hamiltonian systems. Whereas traditional methods solve boundary value problems about an initial guess only, our approach gives a "full picture". In particular, traditional methods completely ignore the number of solutions to the boundary value problem. Our approach, however, indicates the presence of multiple solutions as singularities of generating functions. In turn, we proved and illustrated that these singularities can be studied and the number of solutions may be determined.

In linear systems theory, it is well-known that perturbation matrices solve boundary value problems. These matrices have distinctive properties that are studied in the literature.

Using generating functions we have recovered and extended some of these properties. Most importantly, we have proved that they correspond to coefficients of the generating functions. As a result, our approach naturally contains the theory of perturbation matrices. The relation between perturbation matrices and generating functions may also be investigated using the state transition matrix. In this respect, we have shown that state transition matrix and generating functions are closely related. One of the main consequences, is that we can predict singularities of the generating functions using the state transition matrix. This result broadens to some extent to non-linear systems.

In non-linear systems theory, there is no equivalent of the perturbation matrices. Thus, the approach we have proposed is the first to define functions, namely the generating functions, that directly relate boundary conditions. Obviously, no results as general as the ones derived for linear systems may be gleaned in this case. However, for polynomial generating functions we have established that singularities of the generating functions may still be predicted from the state transition matrix. As a result, the existence of multiple solutions to two-point boundary value problems is fully predicted by the linear dynamics. The number of solutions, however, depends on the non-linear dynamics.

The solution of the Hamilton–Jacobi equation for a generating function is extremely difficult in general. Thus, to make our observations and theory realizable it is essential that we be able to construct solutions for some general class of problems. In this chapter we detail an explicit set of algorithms that allows us to do so, found by performing a Taylor series expansion of the generating function and the Hamiltonian function about a nominal solution to the dynamical system. This provides a solution for the generating function which is analytic in its spatial variables, and can incorporate high-order non-linearities, where the coefficients of the expansion are found from numerically integrating a series of ordinary differential equations in time. Through some specific examples we demonstrate some of the convergence properties of this algorithm and establish that it provides an accurate representation of the generating function in the neighborhood of a solution.

Finally, the approach we have presented to solve two-point boundary value problems applies to any Hamiltonian systems. It is therefore not surprising that it has implications in several fields. In particular, it allows us to develop new methods to study the phase space structure, solve optimal control problems and design spacecraft formations.

Appendix A. The Hamilton–Jacobi equation at higher orders

In this appendix, we give an explicit expression of P as defined by Eq. (3.76). We assume a $2n$-dimensional Hamiltonian system with polynomial Hamiltonian function and polynomial generating functions. We have seen that the Hamilton–Jacobi partial differential equation reduces to an ordinary differential equation of the form

$$P(y, f^{p,r}_{i_1,\dots,i_{2n}}(t), \dot{f}^{p,r}_{i_1,\dots,i_{2n}}(t)) = 0. \tag{A.1}$$

In the following we use tensor notation in order to derive an explicit expression of P. In tensor notation, a Taylor series expansion writes as:

$$f(x,t) = f^0(t) + f^1(t)\cdot\vec{x} + (f^2(t)\cdot\vec{x})\cdot\vec{x} + ((f^3(t)\cdot\vec{x})\cdot\vec{x})\cdot\vec{x} + \cdots. \tag{A.2}$$

Applying this formula to $H(\vec{x}, t)$ and $F_2 = F(\vec{y}, t)$ yields:

$$H(\vec{x}) = h_{i,j}(t)x_i x_j + h_{i,j,k}(t)x_i x_j x_k + \cdots, \tag{A.3}$$

$$F(\vec{y}) = f_{i,j}(t)y_i y_j + f_{i,j,k}(t)y_i y_j y_k + \cdots, \tag{A.4}$$

where we assume the summation convention. Let us now express $\vec{x} = (\Delta q, \Delta p)$ as a function of $\vec{y} = (\Delta q, \Delta p_0)$ (we drop the time dependence in the notation, i.e., we shall write $h_{i,j}$ instead of $h_{i,j}(t)$). For all $a \leq n$ and $j = n + a$

$$x_a = y_a, \tag{A.5}$$

$$x_j = \frac{\partial F}{\partial y_a} \tag{A.6}$$

$$= f_{a,k}y_k + f_{k,a}y_k + f_{a,k,l}y_k y_l + f_{k,a,l}y_k y_l + f_{k,l,a}y_k y_l + \cdots, \tag{A.7}$$

where n is the dimension of the configuration space. The Hamilton–Jacobi equation becomes:

$$\dot{f}_{i,j}y_i y_j + \dot{f}_{i,j,k}y_i y_j y_k + \cdots + h_{i,j}x_i x_j + h_{i,j,k}x_i x_j x_k + \cdots = 0. \tag{A.8}$$

Replacing $\vec{x}$ by $\vec{y}$ in Eq. (A.8) using Eq. (A.7), and keeping only terms of order less than 3 yields:

$$\begin{aligned}
0 = {} & \dot{f}_{i,j}y_i y_j + \dot{f}_{i,j,k}y_i y_j y_k + h_{a,b}y_a y_b + h_{a,b,c}y_a y_b y_c \\
& + (h_{a,n+b} + h_{n+b,a})y_a(f_{b,k}y_k + f_{k,b}y_k + f_{b,k,l}y_k y_l + f_{l,b,k}y_k y_l + f_{k,l,b}y_k y_l) \\
& + h_{n+a,n+b}(f_{a,k}y_k + f_{k,a}y_k + f_{a,k,l}y_k y_l + f_{l,a,k}y_k y_l + f_{k,l,a}y_k y_l) \\
& \times (f_{b,m}y_m + f_{m,b}y_m + f_{b,m,p}y_m y_p + f_{p,b,m}y_m y_p + f_{m,p,b}y_m y_p) \\
& + (h_{n+a,b,c} + h_{c,n+a,b} + h_{b,c,n+a})y_b y_c(f_{a,k}y_k + f_{k,a}y_k) \\
& + (h_{n+a,n+b,c} + h_{n+b,c,n+a} + h_{c,n+a,n+b})y_c(f_{a,k}y_k + f_{k,a}y_k)(f_{b,l}y_l + f_{l,b}y_l) \\
& + h_{n+a,n+b,n+c}(f_{a,k}y_k + f_{k,a}y_k)(f_{b,l}y_l + f_{l,b}y_l)(f_{c,m}y_m + f_{m,c}y_m).
\end{aligned} \tag{A.9}$$

Eq. (A.9) is the expression of P up to order 3 as defined by Eq. (3.76). It is a polynomial equation in the y_i variables with time-dependent coefficients and holds if every coefficient is zero. The equations of order 3 reads:

$$\begin{aligned}
& \dot{f}_{i,j,k}y_i y_j y_k + (A_{i,j,k} + B_{i,j,k} + C_{i,j,k})y_i y_j y_k + (D_{a,i,j} + E_{a,i,j})y_a y_i y_j \\
& + G_{a,b,i}y_a y_b y_i + h_{a,b,c}y_a y_b y_c = 0,
\end{aligned} \tag{A.10}$$

where

$$A_{i,j,k} = h_{n+a,n+b,n+c}(f_{a,i} + f_{i,a})(f_{b,j} + f_{j,b})(f_{c,k} + f_{k,c}),$$

$$B_{i,j,k} = h_{n+a,n+b}(f_{a,i} + f_{i,a})(f_{b,j,k} + f_{j,k,b} + f_{k,b,j}),$$

$$C_{i,j,k} = h_{n+a,n+b}(f_{b,i} + f_{i,b})(f_{a,j,k} + f_{j,k,a} + f_{k,a,j}),$$

$$
\begin{aligned}
D_{a,i,j} &= (h_{a,n+b,n+c} + h_{n+c,a,n+b} + h_{n+b,n+c,a})(f_{b,i} + f_{i,b})(f_{c,j} + f_{j,c}),\\
E_{a,i,j} &= (h_{a,n+b} + h_{n+b,a})(f_{b,i,j} + f_{j,b,i} + f_{i,j,b}),\\
G_{a,b,i} &= (h_{a,b,n+c} + h_{b,n+c,a} + h_{n+c,a,b})(f_{c,i} + f_{i,c})\,.
\end{aligned} \tag{A.11}
$$

We deduce the coefficients of $y_i y_j y_k$:

- Coefficients of $y^3_{i\le n}$

$$A_{i,i,i} + B_{i,i,i} + C_{i,i,i} + D_{i,i,i} + E_{i,i,i} + \dot{f}_{i,i,i} + G_{i,i,i} + h_{i,i,i} = 0\,. \tag{A.12}$$

- Coefficients of $y^3_{i>n}$

$$A_{i,i,i} + B_{i,i,i} + C_{i,i,i} + \dot{f}_{i,i,i} = 0\,. \tag{A.13}$$

- Coefficients of $y^2_{i\le n} y_{j\le n}$

$$(A + B + C + D + E + \dot{f} + G + h)_{\tau(i,i,j)} = 0\,, \tag{A.14}$$

where $\tau(i, j, k)$ represents all the distinct permutations of (i, j, k), that is

$$A_{\tau(i,j,k),l} = A_{i,j,k,l} + A_{i,k,j,l} + A_{k,i,j,l} + A_{k,j,i,l} + A_{j,k,i,l} + A_{j,i,k,l}$$

but

$$A_{\tau(i,i,j),l} = A_{i,i,j,l} + A_{i,j,i,l} + A_{j,i,i,l}\,.$$

- Coefficients of $y^2_{i\le n} y_{j>n}$

$$(A + B + C + \dot{f})_{\tau(i,i,j)} + (D + E)_{i,\tau(i,j)} + G_{i,i,j} = 0\,. \tag{A.15}$$

- Coefficients of $y_{i\le n} y_{j\le n} y_{k\le n}$:

$$(A + B + C + D + E + \dot{f} + G + h)_{\tau(i,j,k)} = 0\,. \tag{A.16}$$

- Coefficients of $y_{i\le n} y_{j\le n} y_{k>n}$

$$(A + B + C + \dot{f})_{\tau(i,j,k)} + (D + E)_{i,\tau(j,k)} + (D + E)_{j,\tau(i,k)} + G_{\tau(i,j),k} = 0\,. \tag{A.17}$$

- Coefficients of $y^2_{i>n} y_{j\le n}$

$$(A + B + C + \dot{f})_{\tau(i,i,j)} + (E + D)_{j,i,i} = 0\,. \tag{A.18}$$

- Coefficients of $y^2_{i>n} y_{j>n}$

$$(A + B + C + \dot{f})_{\tau(i,i,j)} = 0\,. \tag{A.19}$$

- Coefficients of $y_{i\le n} y_{j>n} y_{k>n}$

$$(A + B + C + \dot{f})_{\tau(i,j,k)} + (D + E)_{i,\tau(j,k)} = 0\,. \tag{A.20}$$

- Coefficients of $y_{i>n} y_{j>n} y_{k>n}$

$$(A + B + C + \dot{f})_{\tau(i,j,k)} = 0\,. \tag{A.21}$$

Eqs. (A.12)–(A.21) allow us to solve for F_2 (and F_1 since they both verify the same Hamilton-Jacobi equation, only the initial conditions being different). The process of deriving equations for the generating functions can be continued to arbitrarily high order using a symbolic manipulation program (we have implemented and solved the expansion to order 8 using *Mathematica*©).

Appendix B: The Hill three-body problem

The three-body problem describes the motion of three-point mass particles under their mutual gravitational interactions. This is a classical problem that covers a large range of situations in astrodynamics. An instance of such situations is the motion of the Moon about the Earth under the influence of the Sun. However, this problem does not have a general solution and thus we usually consider simplified formulations justified by physical reasoning. In this chapter, we consider three simplifications. We assume that:

1. One of the three bodies has negligible mass compared to the other two bodies (for instance a spacecraft under the influence of the Sun and the Earth).
2. One of the two massive bodies is in circular orbit about the other one.
3. One of the two massive bodies has larger mass than the other one (the Sun compared to the Earth for instance).

Under these three assumptions, the Hamiltonian for this system reads:

$$H(q, p) = \frac{1}{2}(p_x^2 + p_y^2) + (q_y p_x - q_x p_y) - \frac{1}{\sqrt{q_x^2 + q_y^2}} + \frac{1}{2}(q_y^2 - 2q_x^2), \tag{B.1}$$

and the equations of motion become:

$$\begin{cases} \dot{q}_x = p_x + q_y, \\ \dot{q}_y = p_y - q_x, \\ \dot{p}_x = p_y + 2q_x - \frac{q_x}{(q_x^2+q_y^2)^{3/2}}, \\ \dot{p}_y = -p_x - q_y - \frac{q_y}{(q_x^2+q_y^2)^{3/2}}, \end{cases} \tag{B.2}$$

where $q_x = x$, $q_y = y$, $p_x = \dot{x} - y$ and $p_y = \dot{y} + x$.

This problem has two equilibrium points, L_1 and L_2 whose coordinates are

$$L_1\left(-\left(\frac{1}{3}\right)^{1/3}, 0\right) \quad \text{and} \quad L_2\left(\left(\frac{1}{3}\right)^{1/3}, 0\right).$$

Using linear systems theory, one can prove that the libration points have a stable, an unstable and two center manifolds (Figure 3.15).

To study the relative motion of a spacecraft about L_2, we use Eq. (3.73) to compute H^h, the Hamiltonian function describing the relative motion dynamics.

$$H^h = \frac{1}{2} X^{h^T} \begin{pmatrix} H_{qq}(t) & H_{qp}(t) \\ H_{pq}(t) & H_{pp}(t) \end{pmatrix} X^h + \cdots, \tag{B.3}$$

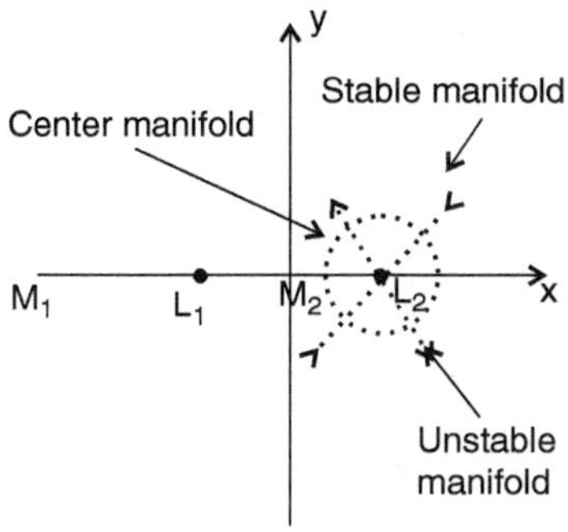

Fig. 3.15. The Libration points in the Hill three-body problem.

where $X^h = \begin{pmatrix} q - q_0 \\ p - p_0 \end{pmatrix} = \begin{pmatrix} \Delta q_x \\ \Delta q_y \\ \Delta p_x \\ \Delta p_y \end{pmatrix}$, $(q_0, p_0) = \left(\left(\frac{1}{3}\right)^{1/3}, 0, 0, \left(\frac{1}{3}\right)^{1/3} \right)$ refers to the state at the equilibrium point L_2 and,

$$H_{qq}(t) = \begin{pmatrix} \frac{1}{(q_{0x}^2+q_{0y}^2)^{3/2}} - \frac{3q_{0x}^2}{(q_{0x}^2+q_{0y}^2)^{5/2}} - 2 & -\frac{3q_{0x}q_{0y}}{(q_{0x}^2+q_{0y}^2)^{5/2}} \\ -\frac{3q_{0x}q_{0y}}{(q_{0x}^2+q_{0y}^2)^{5/2}} & \frac{1}{(q_{0x}^2+q_{0y}^2)^{3/2}} - \frac{3q_{0y}^2}{(q_{0x}^2+q_{0y}^2)^{5/2}} + 1 \end{pmatrix}, \tag{B.4}$$

$$H_{qp}(t) = \begin{pmatrix} 0 & -1 \\ 1 & 0 \end{pmatrix}, \tag{B.5}$$

$$H_{pq}(t) = \begin{pmatrix} 0 & 1 \\ -1 & 0 \end{pmatrix}, \tag{B.6}$$

$$H_{pp}(t) = \begin{pmatrix} 1 & 0 \\ 0 & 1 \end{pmatrix}. \tag{B.7}$$

Substituting (q_0, p_0) by its value yields the expression of H^h at second order:

$$H^h = \frac{1}{2} \left(\Delta q_x \, \Delta q_y \, \Delta p_x \, \Delta p_y \right) \begin{pmatrix} -8 & 0 & 0 & -1 \\ 0 & 4 & 1 & 0 \\ 0 & 1 & 1 & 0 \\ -1 & 0 & 0 & 1 \end{pmatrix} \begin{pmatrix} \Delta q_x \\ \Delta q_y \\ \Delta p_x \\ \Delta p_y \end{pmatrix}. \tag{B.8}$$

At higher order, we find:

$$H^h = \frac{1}{2} \left(\Delta q_x \, \Delta q_y \, \Delta p_x \, \Delta p_y \right) \begin{pmatrix} -8 & 0 & 0 & -1 \\ 0 & 4 & 1 & 0 \\ 0 & 1 & 1 & 0 \\ -1 & 0 & 0 & 1 \end{pmatrix} \begin{pmatrix} \Delta q_x \\ \Delta q_y \\ \Delta p_x \\ \Delta p_y \end{pmatrix}$$
$$+ 3^{4/3} \Delta q_x^3 - \frac{3^{7/3}}{2} \Delta q_x \Delta q_y^2 - 3^{5/3} \Delta q_x^4 + 3^{8/3} \Delta q_x^2 \Delta q_y^2 - \frac{3^{8/3}}{8} \Delta q_y^4 \cdots \tag{B.9}$$

We point out that H^h is time-independent.

Finally, we give in the following table the values of the normalized variables for the Earth–Sun system.

Normalized units		Earth–Sun system
0.01 unit of length	$\longleftrightarrow$	21,500 km
1 unit of time	$\longleftrightarrow$	58 days 2 hours
1 unit of velocity	$\longleftrightarrow$	428 m/s
1 unit of acceleration	$\longleftrightarrow$	$1.38 \cdot 10^{-5}$ m/s^2

References

1. Abraham Ralph, and Marsden, Jerrold E. (1978). Foundations of Mechanics (W.A. Benjamin, ed.) 2nd edition, 1978.
2. Alfriend, K.T., Yan, Hui and Valadi, S.R. (2002). Nonlinear considerations in satellite formation flying. In *Proceedings of the AIAA/AAS Astrodynamics Specialist Conference and Exhibit*. AIAA.
3. Arnold, Vladimir I. (1988). Mathematical Methods of Classical Mechanics. Springer-Verlag, 2nd edition.
4. Battin, Richard H. (1999). An Introduction to the Mathematics and Methods of Astrodynamics. American Institute of Aeronautics and Astronautics, revised edition.
5. Betts, John T. (1998). Survey of numerical methods for trajectory optimization. *Journal of Control, Guidance, and Dynamics*, **21**(2), pp. 193–207.
6. Bloch, A.M., Baillieul, J., Crouch, P.E. and Marsden, J.E. (2003). Nonholonomic Mechanics and Control. Springer.
7. Bryson, Arthur E. and Ho, Yu-Chi (1975). Applied Optimal Control: Optimization, estimation, and control. Halsted Press, revised edition.
8. Ehlers, Juergen and Newman, Ezra T. (2000). The theory of caustics and wavefront singularities with physical applications. *Journal of Mathematical Physics A*, **41**(6), pp. 3344–3378.
9. Goldstein, Herbert (1965). Classical Mechanics. Addison-Wesley.
10. Goldstein, Herbert (1980). Classical Mechanics. Addison-Wesley, 2nd edition.
11. Greenwood, Donald T. (1977). Classical Dynamics. Prentice-Hall.
12. Guibout, Vincent M. (2004). The Hamilton–Jacobi theory for solving two-point boundary value problems: Theory and Numerics with application to spacecraft formation flight, optimal control and the study of phase space structure. PhD thesis, University of Michigan.
13. Guibout, Vincent M. and Bloch, Anthony M. (2004). Discrete variational principles and Hamilton-Jacobi theory for mechanical systems and optimal control problems. *Physica D*, submitted.
14. Guibout, Vincent M. and Scheeres, Daniel J. (2002). Formation flight with generating functions: Solving the relative boundary value problem. In *Proceedings of the AIAA/AAS Astrodynamics Specialist Conference and Exhibit, Monterey, California. Paper AIAA 2002-4639*. AIAA.
15. Guibout, Vincent M. and Scheeres, Daniel J. (2003). Periodic orbits from generating functions. In *Proceedings of the AAS/AIAA Astrodynamics Specialist Conference and Exhibit, Big Sky, Montana. Paper AAS 03-566*. AAS.
16. Guibout, Vincent M. and Scheeres, Daniel J. (2003). Solving relative two-point boundary value problems: Spacecraft formation flight transfers application. *AIAA, Journal of Control, Guidance and Dynamics*, **27**(4), 693–704.
17. Guibout, Vincent M. and Scheeres, Daniel J. (2004). Spacecraft formation dynamics and design. In *Proceedings of the AIAA/AAS Astrodynamics Specialist Conference and Exhibit, Providence, Rhode Island.*
18. Guibout, V.M. and Scheeres, D.J. (2006). Spacecraft formation dynamics and design. *Journal of Guidance, Control, and Dynamics*, **29**(1), 121–133.

19. Hamilton, William Rowan (1834). On a general method in dynamics. *Philosophical Transactions of the Royal Society, Part II*, pp. 247–308.
20. Hamilton, William Rowan (1835). Second essay on a general method in dynamics. *Philosophical Transactions of the Royal Society, Part I*, pp. 95–144.
21. Keller, H.B. (1968). Numerical Methods for Two-point Boundary Value Problems. Blaisdell.
22. Lanczos, Cornelius (1977). The variational principles of mechanics. University of Toronto Press, 4th edition.
23. Marsden, Jerrold E. and Ratiu, Tudor S. (1998). Introduction to Mechanics and Symmetry: A basic exposition of classical mechanical systems. Springer-Verlag, 2nd edition.
24. Melton, Robert G. (2000). Time-explicit representation of relative motion between elliptical orbits. *Journal of Guidance, Control, and Dynamics*, **23**, 604–610.
25. Moulton, Forest R. (1930). Differential equations. The Macmillan company.
26. Park, Chandeok, Guibout, Vincent M. and Scheeres, Daniel J. (2006). Solving optimal continuous thrust rendezvous problems with generating functions. *Journal of Guidance, Control, and Dynamics*, **29**(2), 321–331.
27. Park, Chandeok and Scheeres, Daniel J. (2003). Indirect solutions of the optimal feedback control using Hamiltonian dynamics and generating functions. In *Proceedings of the 2003 IEEE conference on Decision and Control, 2003. Maui, Hawaii*. IEEE.
28. Park, Chandeok and Scheeres, Daniel J. (2004). Solutions of optimal feedback control problems with general boundary conditions using Hamiltonian dynamics and generating functions. In *Proceedings of the American Control Conference, Boston, Massachusetts, June 2004. Paper WeM02.1.*
29. Powers, David L. (1987). Boundary Value Problems. San Diego, Harcourt Brace Jovanovich.
30. Press, William H., Teukolsky, Saul A., Vetterling, William T. and Flannery, Brian P. (1992). Numerical Recipes in C, the Art of Scientific Computing. Cambridge University Press, 2nd edition.
31. Wang, P.K.C. and Hadaegh, F.Y. (1999). Minimum-fuel formation reconfiguration of multiple free-flying spacecraft. *The Journal of the Astronautical Sciences*, **47**(1–2), 77–102.
32. Weinstein, Alain (1977). Lectures on symplectic manifolds. *Regional Conference Series in Mathematics*, 29.

4 Low-Energy Transfers and Applications

EDWARD BELBRUNO
Department of Astrophysical Sciences, Princeton University; N.J. 08544-1000, U.S.A.

Contents

4.1 Introduction

The application of methods of dynamical systems theory to the field of astrodynamics has uncovered new types of low-energy trajectories that have many important applications. In particular, for the purpose of finding transfer trajectories from the Earth to lunar orbit. The mechanism used to obtain a transfer that is 'low energy' is called 'ballistic capture'. This is a process where a spacecraft is captured into lunar orbit without the use of rocket engines to slow down. The resulting transfer to the Moon is called a 'ballistic capture transfer'. Their property of being captured automatically into lunar orbit is completely different than that of the standard Hohmann transfer where substantial fuel must be used. This offers many advantages to the Hohmann transfer. In particular, they are substantially lower cost to use and operationally safer. The dynamical properties of the ballistic capture transfer are much more complicated than that of the Hohmann transfer, utilizing Newtonian four-body dynamics, as opposed to Newtonian two-body dynamics of the Hohmann transfer. The ballistic capture process itself is dynamically sensitive, but can be stabilized with a negligible maneuver.

A ballistic capture transfer was first operationally demonstrated in 1991 by the rescue of the Japanese spacecraft *Hiten* [2]. More recently, another type of ballistic capture transfer was used by ESAs spacecraft *SMART-1* [16, 17]. Their properties from the perspective of dynamical systems theory was investigated in 1994 [4], then by Marsden et al. [13]. Since then, a rigorous proof has been given showing that the ballistic process, in general, is chaotic in nature [4].

The use of low-energy trajectories has an interesting application on the origin of our own Moon. In a theory recently published by Belbruno and Gott [5], a class of low-energy transfers has shed light on the origin of the hypothetical Mars-sized object that slammed into the Earth 4 billion years ago to create the Moon. This is very briefly described in Section 4.5. For details, the reader should consult [5] (See also [10]).

Some popular survey articles on this material are cited in the bibliography. In the subject of astrodynamics, they are Refs. [8, 14, 16], and in the area of dynamical astronomy, Ref. [9]. Reference [4] provides a rigorous theoretical treatment of low-energy transfers. A more popular intuitive approach to the subject of chaos and low-energy transfers is given in [6].

4.2 Capture problem, models, and transfer types

In order to study transfers of spacecraft to the Moon, and other bodies, we define the *capture problem* to facilitate this.

A special four-body problem is generally defined between the spacecraft and three other planetary bodies. We will assume that the only forces acting on the spacecraft are the gravitational forces of the three bodies. We first consider a 'planar elliptic restricted three-body problem' between the particles P_1, P_2, P_3. That is, we assume that the spacecraft, labeled P_3, moves in the same plane as two planetary bodies P_1, P_2. The two planetary bodies move in prescribed mutually uniform elliptical Kelperian orbits about their common center of mass of eccentricity $e_{12} \approx 0$. We assume that P_1, P_2, P_3 move in a coordinate system Q_1, Q_2 which is inertial and centered at P_1 at the origin. The mass of P_1 is $m_1 > 0$; the mass of P_2 is $m_2 > 0$, where $m_2 \ll m_1$; and the mass of $P_3 = 0$. This latter assumption makes sense since P_1, P_2 are planetary sized bodies, and the mass of a spacecraft relative to them will be negligibly small.

A fourth mass point P_4, of mass $m_4 > 0$, is introduced. It is assumed to move about the center of mass point P_{cm} between P_1, P_2 in a uniform Keplerian ellipse of eccentricity $e_{124} \approx 0$. Since $m_2 \ll m_1$, then $P_{cm} \approx (0, 0)$ and P_4 approximately moves about P_1. We assume that the distance of P_4 from the center of mass of P_1, P_2 is much larger than the distance between P_1 and P_2, and that $m_1 \ll m_4$. The zero mass particle P_3 moves in the gravitational field generated by the assumed elliptic motions of P_1, P_2, P_4.

We refer to this model as a *planar elliptic restricted four-body problem*. It is shown in Figure 4.1. It could also be referred to as the *co-elliptic restricted four-body problem*. An example of this type of problem is where P_1 = Earth, P_2 = Moon, P_3 = spacecraft, P_4 = Sun. We will assume this labeling for the remainder of this chapter, although it should be noted that the results are not just limited to this choice of bodies.

If $e_{12} = e_{124} = 0$, then we refer to this as a co-circular restricted four-body problem. When $e_{12} = 0$ and we turn off the gravitational influence of P_4 by setting $m_4 = 0$, then this problem reduces to the *planar circular restricted three-body problem* between P_1, P_2, P_3. When $m_4 > 0$, and we turn off the gravitational influence of P_2 by setting $m_2 = 0$, then the circular restricted problem is obtained between P_1, P_3, P_4.

We define transfers from P_1 to P_2. In Figure 4.2 we just show P_1, P_2, which shows the conditions required for a transfer of P_3 from P_1 to P_2. P_4 is not shown.

Referring to Figure 4.2, the following assumptions are made:

- A1: The spacecraft, P_3, initially moves in a circular orbit about the Earth, P_1, of radius r_{13} as measured from the center of P_1.
- A2: A velocity increment magnitude ΔV_0 at the location $\mathbf{Q}_0$ at time t_0 on the circular orbit is added to the circular velocity $(Gm_1 r_{13}^{-1})^{\frac{1}{2}}$ so P_3 can transfer to the location $\mathbf{Q}_F$ near the Moon, P_2. Note that the vectors $\mathbf{Q}_0, \mathbf{Q}_F$ are in the coordinate system Q_1, Q_2,

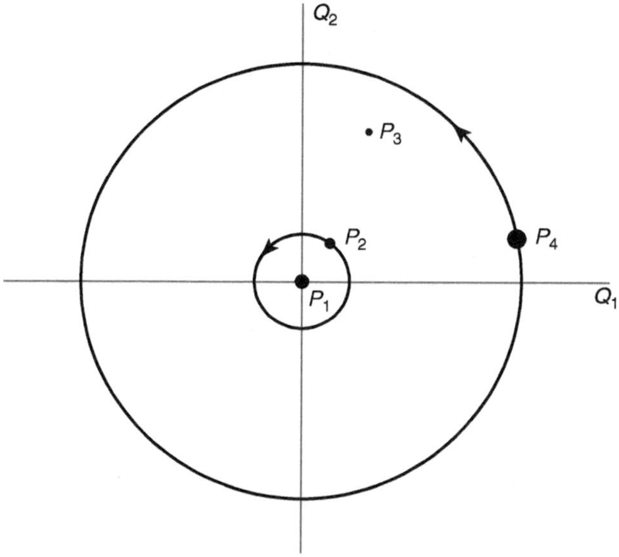

Fig. 4.1. Co-elliptic restricted four-body problem.

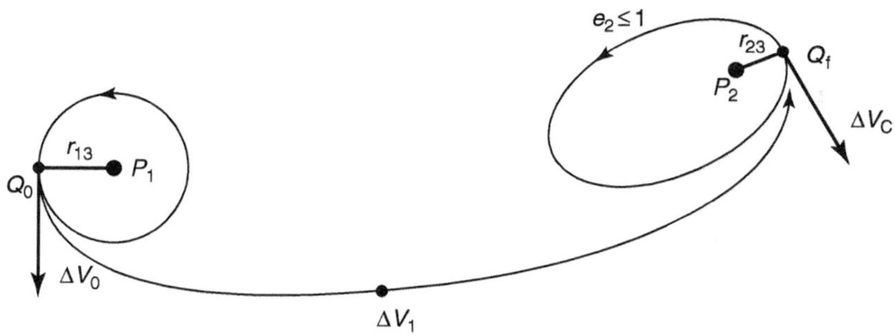

Fig. 4.2. The capture problem in inertial coordinates centered at P_1.

centered at the origin, P_1. G is the Newtonian gravitational constant, and r_{13} is the distance between P_1 and P_3.

- A3: A velocity increment magnitude ΔV_1 is applied at a time t_1, $t_0 < t_1 < t_F$; t_F is the arrival time at $\mathbf{Q}_F$.
- A4: A velocity increment magnitude ΔV_C is applied at $\mathbf{Q}_F$ in order that the two-body Keplerian energy E_2 between P_2, P_3 is negative or zero at $\mathbf{Q}_F$ so that at $t = t_F$ an oscillating ellipse of given eccentricity $0 \leq e_2 \leq 1$ is obtained of periapsis distance r_{23}. This defines an *instantaneous capture* at $\mathbf{Q}_F$ at time $t = t_F > t_0$ into an ellipse or parabola. We regard a parabola as an ellipse of infinite semimajor axis. $\mathbf{Q}_F$ represents the periapsis of the oscillating ellipse with respect to P_2 at distance r_{23}.

4.2.1 Remarks

Remark 1. The velocity increments $\Delta V_0, \Delta V_1, \Delta V_C$ are provided by firing the rocket engines of the spacecraft to impart a thrust, and hence a change in velocity. These increments are called ΔV's or *maneuvers*. In practice they cannot be achieved instantaneously, as we are assuming here, and depend on the magnitude of the ΔV. The engines may need to fire for a duration of a few seconds or several minutes. In general, modeling the ΔV's in an instantaneous or *impulsive* manner yields accurate modeling.

Remark 2. The term instantaneous capture in A4 also implies that for $t > t_F$ E_2 may become positive again. That is, P_3 is ejected right after being captured. It is generally the case that the ellipse shown in Figure 4.2 about the Moon may just exist when $t = t_F$. If ΔV_C is sufficiently large, then the capture ellipses can be stabilized for long times after $t = t_F$. In general, if it is desired to place a spacecraft several hundred kilometers from the surface of the Moon in a circular orbit after applying ΔV_C, then the orbit remains approximately circular for several months. Frequent ΔV's need to be applied by the spacecraft to maintain an orbit about the Moon; in general these are not stable due to nonuniformities of the mass distribution of the Moon and gravitational perturbations due to the Earth and Sun.

Remark 3. The term *osculating ellipse* in A4 means that the elliptical state at $t = t_F$ at $\mathbf{Q}_F$ may be unstable.

The *capture problem* is defined by the problem

$$\min\{\Delta V_0 + \Delta V_1 + \Delta V_C\}, \tag{4.1}$$

where the minimization is taken over all transfers from $\mathbf{Q}_0$ to $\mathbf{Q}_F$ and for assumptions A1–A4.

A solution of the capture problem for simplified assumptions with $\Delta V = 0$ is given by the Hohmann transfer.

4.2.2 Hohmann transfers

It is instructive to consider a typical Hohmann transfer between the Earth (P_1) and Moon(P_2). They have the property that at $\mathbf{Q}_F$, $\Delta V_C \gg 0$, or equivalently $E_2 \gg 0$, i.e., they are substantially hyperbolic with respect to the Moon at $\mathbf{Q}_F$. As we will see later, this is substantially different than ballistic capture transfers where $\Delta V_C = 0$ at $\mathbf{Q}_F$ and $E_2 \leq 0$. We only describe Hohmann transfers briefly here. They are described in detail throughout the astrodynamics literature.

The Hohmann transfer was developed by W. Hohmann in the early 1900s. Although his assumptions are oversimplifying in nature, they nevertheless lead to transfers from $\mathbf{Q}_0$ to $\mathbf{Q}_F$ which are very useful in practice, not just for the case $P_1 =$ Earth, $P_2 =$ Moon, $P_4 =$ Sun, but for transfers from the Earth to the other planets of our solar system.

Here, we discuss the Hohmann transfer that is relevant to Figure 4.2 for the Earth–Moon system. We assume that P_1, P_2 are in mutually circular orbits, i.e. $e_{12} = 0$.

The basic assumptions are the following: First, $m_4 = 0$ so that P_4 is not considered. Second, as P_3 transfers from $\mathbf{Q}_0$ to $\mathbf{Q}_F$, i.e., for $t_0 \leq t \leq t_F$, the gravity of P_2 is ignored, i.e., $m_2 = 0$. This yields a simple two-body problem between P_3, P_1, where then the two-body energy is then minimized. This gives one-half of a Kepler ellipse with periapsis at $\mathbf{Q}_0$ and apoapsis at $\mathbf{Q}_F$. This is the Hohmann transfer from $\mathbf{Q}_0$ to $\mathbf{Q}_F$. This ellipse arc has an eccentricity e_1. Upon arrival at $\mathbf{Q}_F$, the gravity of m_1 is ignored, $m_1 = 0$. m_2 is now assumed to be nonzero, and ΔV_C is computed relative to a two-body problem between P_3, P_2. $\mathbf{Q}_F$ is assumed to be on the far side of P_2 on the P_1–P_2 line. Breaking up a four-body problem into two disjoint two-body problems is an enormous simplification to the capture problem and dynamically is not correct. Nevertheless, these transfers change little when they are applied with full solar system modeling in many useful cases. This is because of the high energy associated with them. Their derivation is elegantly simple, and their usefulness is remarkable. They have paved the way for both human and robotic exploration of our solar system.

Of particular interest for applications considered later in this book is when r_{13}, r_{23} are relatively small numbers. Let km = kilometer, s = second. For example, $r_{13} = r_E + 200$ km, $r_{23} = r_M + 100$ km are typical radial distances used in applications of P_3 from P_1, P_2 at the locations $\mathbf{Q}_0, \mathbf{Q}_\mathbf{F}$, respectively, and at times $t = t_0, t_F$, respectively. The r_E and r_M represent the radii of the Earth and Moon, respectively. We will assume these values of r_{13}, r_{23} for the remainder of this paper for convenience. It is verified that $\Delta V_0 = 3.142$ km/s, $\Delta V_1 = 0$. $\Delta V_C = 0.200\,\text{km/s}, 0.648\,\text{km/s}$ for $e_2 = 0.95, 0$, respectively. Also, $t_F - t_0 = 5$ days. The transfer itself is nearly parabolic where $e_1 = 0.97$. Visually it would appear to be nearly linear. For Hohmann transfers in general, $E_2 \gg 0$ at $\mathbf{Q}_\mathbf{F}$. These values of r_{13}, r_{23} are the values that we desire for a solution of the capture problem.

It is verified that $E_2 > 0$ at $\mathbf{Q}_F$ for P_3, and this causes a large value of ΔV_C to occur. This property of $E_2 > 0$ is satisfied by Hohmann transfers. The reason $E_2 > 0$ follows from the fact that the magnitude V_F of the velocity vector at $\mathbf{Q}_\mathbf{F}$ of P_3 on the transfer at lunar periapsis, where the direction is in the same direction as the Moons orbit about the Earth, has the property that $V_F \ll V_M$, where V_M is the magnitude of the velocity of the Moon about the Earth. It turns out that under the given assumptions, $V_F = 0.176$ km/s and $V_M = 1.019$ km/s. This implies that $E_2 = 0.843\ \text{km}^2/\text{s}^2$. It is the discrepancy between V_F and V_M that yields a large value of ΔV_C of several hundred meters per second, depending on the value of e_2. The calculation of E_2 for a Hohmann transfer is estimated by noting that relative to P_2, the transfer is hyperbolic, with a hyperbolic periapsis at $\mathbf{Q}_\mathbf{F}$. The corresponding velocity at $r_2 = \infty$, called the hyperbolic excess velocity and labeled V_∞, is estimated by $V_\infty = V_M - V_F = 0.843$ km/s yielding $E_2 = (1/2)V_\infty^2$. The calculation of ΔV_C follows from a functional relationship it has with V_∞, or equivalently E_2.

4.2.3 Ballistic capture transfers

A ballistic capture transfer is defined to be a solution of the capture problem where $\Delta V_C = 0$ at $\mathbf{Q}_\mathbf{F}$ for $t = t_F$. It will arrive at periapsis at $\mathbf{Q}_\mathbf{F}$ where E_2 is negative, and therefore it will have no V_∞. This enables capture where $\Delta V_C = 0$. Eliminating the V_∞ is the motivation for the construction of ballistic capture transfers. From our discussion of the Hohmann transfer, this means that a ballistic capture transfer has to arrive at $\mathbf{Q}_F$ where the spacecraft's velocity approximately matches the velocity of the Moon about the Earth. We will see that a ballistic capture transfer going from $\mathbf{Q}_0$ to $\mathbf{Q}_F$ can be constructed with approximately the same value of ΔV_0 for a Hohmann transfer and also with $\Delta V_1 = 0$. We refer to a Hohmann transfer as *high energy* since V_∞ is significantly high, and a ballistic capture transfer is called *low energy* since the V_∞ is eliminated.

The basic idea behind finding ballistic capture transfers to the Moon, or any body, is to find a region about the Moon, in position-velocity space (i.e., phase space), where an object can be ballistically captured. When such a region is found, then one can try to find trajectories from the Earth that go to that region. Ballistic capture enables capture to occur in an natural way, where a spacecraft need not slow down using its engines. A spacecraft moving in this region about the Moon lies in the transition between capture and escape from the Moon. Its motion in this region is very sensitive—being both chaotic and unstable.

In the next section we study in more detail how to determine where ballistic capture can occur about the Moon.

4.3 Ballistic capture regions and transfers

To better understand the process of ballistic capture, and the transfers themselves, it is instructive to consider the planar circular restricted three-body problem between the spacecraft, Earth, and Moon.

This defines the motion of P_3 in the gravitational field generated by the uniform circular motion of P_1, P_2 in an inertial coordinate system. P_3 moves in the same plane of motion as P_1, P_2. The constant frequency ω of motion of P_1, P_2 about their common center of mass at the origin is normalized to 1. It is assumed that $m_3 = 0$, and $m_1 + m_2 = 1$. We set $m_1 = 1 - \mu, m_2 = \mu, \ \mu = m_2/(m_1 + m_2)$. In a rotating coordinate system x_1, x_2 which rotates with the same frequency ω, both P_1, P_2 are fixed. We normalize the distance between P_1, P_2 to be 1. Without loss of generality we place P_1 at $(\mu, 0)$ and P_2 at $(-1+\mu, 0)$. We assume here that $m_2 \ll m_1$, or equivalently $\mu \ll 1$. With P_1 = Earth, P_2 = Moon, $\mu = 0.0123$.

The differential equations of motion for P_3 are given by

$$\begin{aligned} \ddot{x}_1 - 2\dot{x}_2 &= x_1 + \Omega_{x_1} \\ \ddot{x}_2 + 2\dot{x}_1 &= x_2 + \Omega_{x_2}, \end{aligned} \tag{4.2}$$

where $\dot{} \equiv \frac{d}{dt}, \ \Omega_x \equiv \frac{\partial \Omega}{\partial x},$

$$\Omega = \frac{1-\mu}{r_1} + \frac{\mu}{r_2},$$

r_1 = distance of P_3 to $P_1 = [(x_1 - \mu)^2 + x_2^2]^{\frac{1}{2}}$, and r_2 = distance of P_3 to $P_2 = [(x_1 + 1 - \mu)^2 + x_2^2]^{\frac{1}{2}}$, see Figure 4.3. The right-hand side of Eq. (4.3) represents the sum of the radially directed centrifugal force $\mathbf{F_C} = (x, y)$ and the sum $\mathbf{F_G}$ of the gravitational forces due to P_1 and P_2.

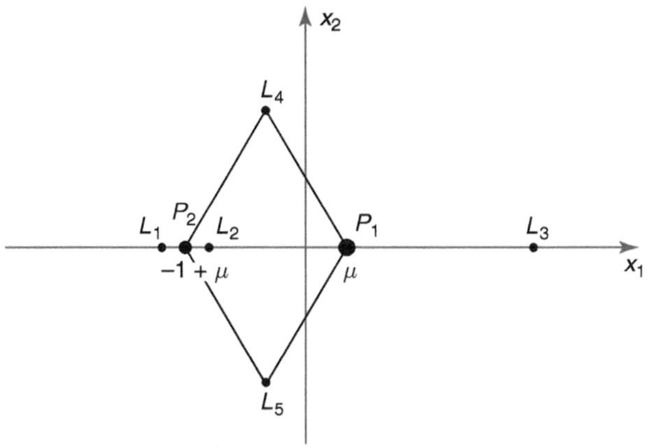

Fig. 4.3. Rotating coordinate system and locations of the Lagrange points.

The x_1 and x_2 are called barycentric rotating coordinates. If the coordinate systems were not rotating, then that defines a barycentric inertial coordinate system Q_1, Q_2. The transformation between x_1, x_2 and Q_1, Q_2 is given by a rotation matrix of rotational frequency 1. For notation we set $x = (x_1, x_2)$ and $Q = (Q_1, Q_2)$. x, Q are understood to be vectors.

An integral of motion for Eq. (4.2) is the Jacobi energy given by

$$J = -|\dot{x}|^2 + |x|^2 + \mu(1-\mu) + 2\Omega. \tag{4.3}$$

Thus

$$J^{-1}(C) = \{(x, \dot{x}) \in \mathbb{R}^4 \mid J = C,\ C \in \mathbb{R}\}$$

is a three-dimensional manifold in phase space for which the solutions of Eq. (4.2) which start on $J^{-1}(C)$ remain on it for all time. C is called the Jacobi constant. The additive term $\mu(1-\mu)$ occurring in Eq. (4.3) is present so that the values of C are normalized.

The manifold $J^{-1}(C)$ projected onto the physical (x_1, x_2)-plane form the *Hill regions*

$$\mathcal{H}(C) = \{x \in \mathbb{R}^2 \mid 2\widetilde{\Omega} - C \geq 0\},$$

where

$$\widetilde{\Omega} = \Omega + \frac{1}{2}|x|^2 + \frac{1}{2}\mu(1-\mu).$$

The particle P_3 is constrained to move in $\mathcal{H}(C)$, see Ref. [4]. The boundary of $\mathcal{H}(C)$ is given by the curves

$$\mathcal{Z}(C) = \{x \in \mathbb{R}^2 \mid 2\widetilde{\Omega} - C = 0\},$$

which are called *zero velocity curves*, because the velocity of P_3 vanishes there. The qualitative appearance of the Hill regions $\mathcal{H}(C)$ for different values of C are shown in

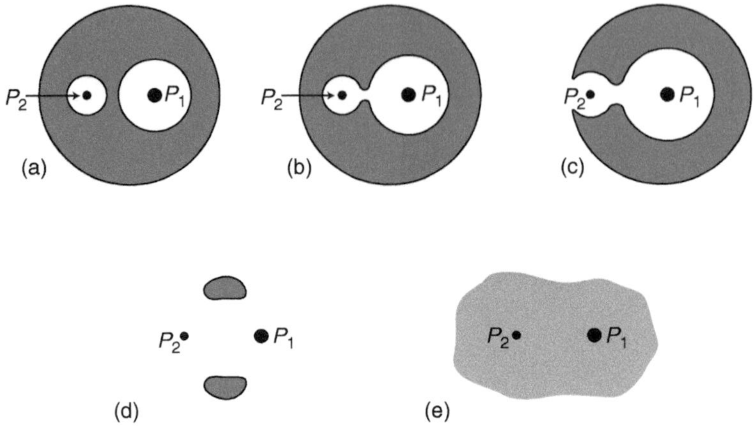

Fig. 4.4. Basic Hill's regions: Starting from the top three figures, left to right, C has the respective values, $C > C_2$, $C_1 < C < C_2$, $C \lesssim C_1$, and then the bottom two figures, left to right, $C_3 < C < C_1, 3 < C < C_3$.

Figure 4.4. The particle P_3 cannot move in the hatched areas. The five values C_i are obtained by evaluating the function J at the five Lagrangian equilibrium points L_i of (4.2). The relative positions of the Lagrange equilibria are shown in Figure 4.3. The ones of interest in this paper are L_1 and L_2, which are unstable saddle center points [4]. The values C_i satisfy

$$C_4 = C_5 = 3 < C_3 < C_1 < C_2.$$

For $C < 3$, the Hill's region becomes the entire x_1, x_2-plane. Thus P_3 can move throughout the entire plane. When $C \lesssim C_2$, P_3 can pass between P_1 and P_2. For $C \geq C_1$ the Hill's region has two components. One is bounded, and the other is unbounded. When $C \lesssim C_1$, P_3 can move between the inner and outer Hill regions. C_2 represents the minimal energy for which P_3 can pass from P_1 to P_2. C_1 represents the minimal energy for which P_3 can pass between the bounded and unbounded components of the Hills regions.

It is noted that an approximation for C_1 and C_2 valid to three digits when $\mu \leq 0.01$, or four digits when $\mu \leq 0.001$, is

$$C_1 \approx 3+9\left(\frac{\mu}{3}\right)^{2/3} - 11\left(\frac{\mu}{3}\right), \quad C_2 \approx 3+9\left(\frac{\mu}{3}\right)^{2/3} - 7\left(\frac{\mu}{3}\right). \tag{4.4}$$

In inertial coordinates Eqs. (4.2) and (4.3), respectively, become,

$$\ddot{Q} = \Omega_Q, \tag{4.5}$$

$$\tilde{J} = -|\dot{Q}|^2 + 2(Q_1\dot{Q}_2 - Q_2\dot{Q}_1) + 2\Omega + \mu(1-\mu), \tag{4.6}$$

where

$$r_1(t) = \sqrt{(Q_1 + \mu c)^2 + (Q_2 + \mu s)^2},$$

$$r_2(t) = \sqrt{(Q_1 - (1-\mu)c)^2 + (Q_2 - (1-\mu)s)^2},$$

where $c \equiv \cos(t)$, $s \equiv \sin(t)$. In these coordinates there is an explicit time dependence which is not the case in rotating coordinates.

Let $\phi(t) = (Q(t), \dot{Q}(t))$ be a solution of Eq. (4.5) for P_3. We assume at time $t = t_0$ it starts at some distance r_1 from P_1 and at time t_1 it is at a distance r_2 from P_2, $t_1 > t_0$. We are viewing $\phi(t)$ in the four-dimensional phase space. In position space it is given by $Q(t)$. As P_3 moves, $\tilde{J}(\phi(t)) = C$. We assume that no collision takes place so that $r_1 > 0, r_2 > 0$ along $\phi(t)$. Let $X = (X_1, X_2)$ be P_2-centered inertial coordinates.

Definition 4.3.1. The two-body Kepler energy of P_3 with respect to P_2 in P_2-centered inertial coordinates is given by

$$E_2(X, \dot{X}) = \frac{1}{2}|\dot{X}|^2 - \frac{\mu}{r_2} \tag{4.7}$$

where $r_2 = |X|$.

Definition 4.3.2. P_3 is *ballistically captured* at P_2 at time $t = t_1$ if

$$E_2(\phi(t_1)) \leq 0. \tag{4.8}$$

$\phi(t)$ is called a *ballistic capture transfer* from $t = t_0$ to $t = t_1$. If $E_2 \gtrsim 0$ at $t = t_1$ then P_3 is *pseudo-ballistically captured* at P_2.

It is noted that the notation $a \gtrsim b$ means that $a > b$ and $a - b = \delta \ll 1$.

When P_3 is ballistically captured with respect to P_2 the capture may lead to ballistic capture for all future time $t \geq t_1$, or this capture may be temporary where at a finite time $t = t_3 > t_2$, $E_2 > 0$. When this occurs then we say that P_3 has *ballistically escaped* P_2 for $t = t_3$. In this case the ballistic capture is *temporary*.

Ballistic capture can be stable or unstable whether it is temporary or not. By stability we mean *orbital stability*. That is, if the orbital elements of the motion change to a significant degree with very small changes in the initial conditions. If infinitesimally small changes in the initial conditions lead to predictably small changes in all the orbital elements for arbitrarily long time, then the motion is called stable, otherwise it is called unstable.

Temporary ballistic capture can be stable, so that although the Kepler energy is changing from negative to positive values, the orbital elements change in a small predictable way for all time. Likewise, ballistic capture for all time need not be stable. Thus whether or not the capture is temporary or not is not a good measure to describe the motion.

The key quantity to measure is the orbital stability. When ballistic capture is unstable we refer to it as *weak ballistic capture* or *weak capture* for brevity. A set where this occurs numerically can be estimated and is described in [4]. It can be analytically approximated by looking at the values of C of the Jacobi integral where the motion of ϕ has sufficiently high energy, where $C < C_1$. Also, we consider those points where $\dot{r}_2 = 0$. This defines a set W on the Jacobi integral surface $J^{-1}(C)$ in the coordinates x, $\dot{x}$,

Definition 4.3.3.

$$W = \{(x, \dot{x}) \in \mathbb{R}^4 | J = C, C < C_1, E_2 \leq 0, \dot{r}_2 = 0\}.$$

W is referred to as the *weak stability boundary.*

As is proven in [4] W on the three-dimensional surface $J^{-1}(C)$ is equivalent to a two-dimensional annular set about P_2. It is described by an explicit functional relationship

$$r_2 = f(\theta_2, e_2)$$

where θ_2 is the polar angle about P_2 in a P_2-centered rotating coordinate system, where f is periodic of period 2π in θ_2, and $0 \leq e_2 \leq 1$. This relationship can be conveniently written explicitly as

$$r_2 \approx \frac{(1 - e_2)\mu^{\frac{1}{3}}}{3^{\frac{5}{3}} - \frac{2}{3}\mu^{\frac{1}{3}}}$$

under the conditions that $C \lesssim C_1$ and $r_2 \gtrsim 0$.

We slightly extend the definition of W for the case of pseudo-ballistic capture and where we need not require $\dot{r}_2 = 0$. This set is labeled W_H, and is given by

$$W_H = \{(x, \dot{x}) \in \mathbb{R}^4 | J = C, C < C_1, E_2 \gtrsim 0 \ (i.e.\ e_2 \gtrsim 1)\}.$$

The set

$$\tilde{W} = W \cup W_H$$

is called the *extended* weak stability boundary. W_H represents points with respect to P_2 which are slightly hyperbolic and have $C < C_1$.

Numerical simulations indicate that the motion of trajectories with initial conditions on $\tilde{W}$ are generally unstable. In Theorem B in the next section shows that this is indeed the case due to the existence of a chaotic motion associated with $\tilde{W}$. W is referred to as the *weak stability boundary*, and $\tilde{W}$ is a hyperbolic extension of that set.

4.3.1 Method of determining ballistic capture transfers, and their properties

A method for finding ballistic capture transfers is to assume that your spacecraft is already at the extended weak stability boundary of the Moon at the desired capture distance r_{23} at the point $\mathbf{Q_F}$. This will give precise values of the velocity the spacecraft will have when it arrives at the Moon. Then one can employ a 'backwards integration method', where the trajectory is integrated backwards in time. Since the capture state at the Moon is unstable, tiny variations in the ballistic capture state can be used to target the trajectory in backwards time to have a periapsis with respect to the Earth at the point $\mathbf{Q_0}$. In another method, one can perform a 'forward algorithm', and start at the Earth at periapsis at $\mathbf{Q_0}$, and by varying specific control variables, target to the desired ballistic capture state at $\mathbf{Q_F}$. The details of this are described in [4].

The first numerical demonstration of the construction of a ballistic capture transfer was in 1986. This was for the Lunar Get Away Special (LGAS) mission study [1]. The numerical simulation was done for a low thrust spacecraft that was designed to be released from low Earth orbit from a Get Away Special cannister in the cargo bay of the space shuttle. After slowly spiraling out of Earth orbit with the solar electric ion engines for 1.5 years (using 3000 spirals), it reached a sufficiently large distance from the Earth where it shut off its engines, and moved on a ballistic capture transfer to the Moon over the north lunar pole, where it arrived in ballistic capture. It then turned its engines back on and took several months to gradually spiral down to the desired altitude at low lunar orbit at 100 km altitude.

The first operational demonstration of a ballistic capture transfer occurred a few years later. In 1991, a special ballistic capture transfer was used by E. Belbruno and J. Miller to resurrect a failed Japanese lunar mission, and get their spacecraft *Hiten* to the Moon since it had almost no fuel [2, 4, 16]. It took three months to reach the Moon instead of the 3 days a Hohmann transfer takes. The spacecraft first goes out about 1.5 million km from the Earth, then falls back to the Moon for ballistic capture. The transfer *Hiten* used is shown in Figure 4.5. The small elliptic orbit shown in the lower third quadrant is just a phasing orbit, and the transfer starts from near the Earth at the end of the phasing orbit.

Another name for a general ballistic capture transfer is a 'weak stability boundary (WSB) transfer'. *Hiten* is using an 'exterior' WSB transfer since it travels outside the orbit of the Moon. If a WSB transfer stays inside the Moon's orbit, it is called an 'interior' WSB transfer. *LGAS* used an interior transfer. ESAs *SMART-1* mission was

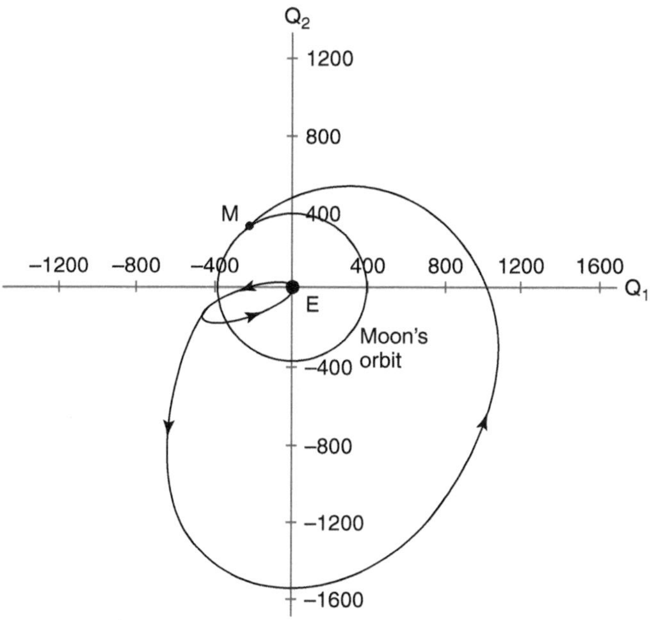

Fig. 4.5. The exterior ballistic capture transfer used by *Hiten*.

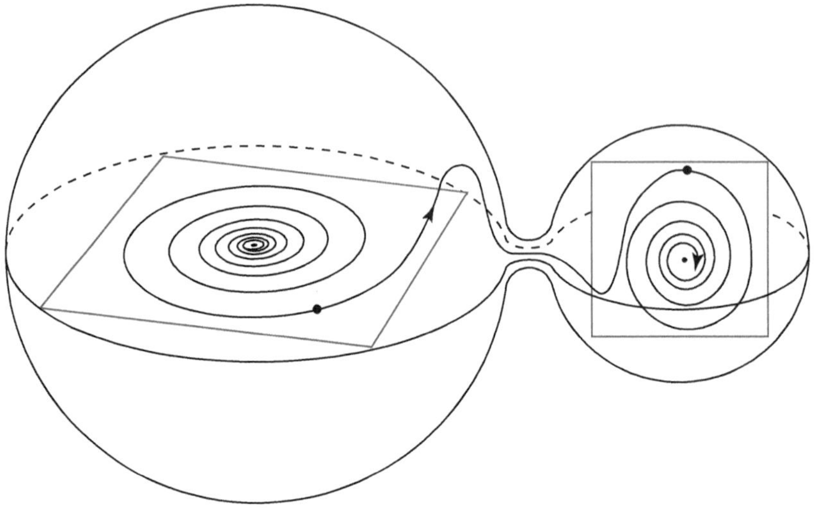

Fig. 4.6. Representation of complete transfer to low circular lunar orbit.

inspired by that design, and their spacecraft arrived at the Moon in November 2004. An illustration, not done accurately to scale, is shown in Figure 4.6. The trajectory lies within a three-dimensional Hills region about the Earth and Moon. The interior transfer itself in Figure 4.6 starts about 100000 from the Earth after the spiraling has stopped and the engines have been shut off. It ends when the transfer arrives at ballistic capture at approximately 30000 km over the north lunar pole. Over a period of a few months, using its engines, it slowly spirals down to low lunar orbit. The interior transfer has the disadvantage that it cannot start closer than approximately 60000 km from the Earth. The exterior transfer solves that problem.

The exterior WSB transfer is particularly important since it can be designed from any altitude from the Earth and go to any altitude at the Moon, and saves substantial fuel as compared to a Hohmann transfer. It promises to have important applications to future lunar missions due to its cost savings in bringing payloads into lunar orbit. The exterior WSB transfer's dynamics is complicated and is described in detail in Ref. [4]. We discuss a few of its properties here because of its interesting dynamics.

The exterior transfer at first appears to resemble a standard Hohmann bi-elliptic transfer. However, the looks are deceiving. Such a transfer, which is analogous to the one *Hiten* used, can leave the Earth at any altitude. If it leaves at low Earth orbit at an altitude of say 200 km, it will need approximately the same ΔV as a Hohmann transfer. But that is where the similarity stops. It takes about 1.5 months to reach the apoapsis at about 1.5 million km. While moving in this region, the gravitational forces of the Earth and Sun approximately balance as the spacecraft moves. It is actually moving in a weak stability boundary region about the Earth with the Sun in this case as the larger perturbing body. As the spacecraft arcs around and falls back towards the Moon, no maneuver is required to fall back towards the Moon. This is completely different than the bi-elliptic transfer which requires a 0.250 km/s maneuver to do this. As the spacecraft falls back towards the Moon, the Sun is positioned in such a way so as to slow down the spacecraft as it approaches the Moon. In this way it can arrive with a velocity that approximately matches the Moons about the Earth. It will approach the Moon from outside the Moon's orbit. If the Jacobi constant C is just slightly less than C_1, the Hill's region opens slightly near the L_1 location, and the trajectory can pass though, passing close to the invariant manifolds associated with the Lyapunov orbit about the L_1 location. This is seen in Figure 4.7. It then passes into the Hill's region about the Moon and to low lunar orbit to weak capture. As is described in [4], and described below, the structure of the phase space where weak capture occurs is very complicated, and consists of an infinite set of intersecting invariant manifolds. It is important to point out that C need not be just be slightly less than C_1. This condition poses a large constraint that the trajectory needs to approach the Moon via the tiny opening near L_1. This condition also generally gives rise to transfers with times of flight on the order of 120 days. If we allow more generally that C could be substantially less than C_1, say $C < 3$, then the spacecraft can move anywhere in the physical space near the Moon, and is not constrained to pass near the location of L_1. This is because the Hill's region becomes the entire physical space and the zero velocity curves bounding the motion, no longer exist. Then the trajectory can be ballistically captured near the Moon by approaching the Moon, in general, from any direction. The time of flight also decreases to approximately 90 days.

Since ballistic capture transfers arrive at P_2 where $E_2 < 0$, they save substantial ΔV required to place P_3 into a capture orbit about P_2 relative to a Hohmann transfer. For

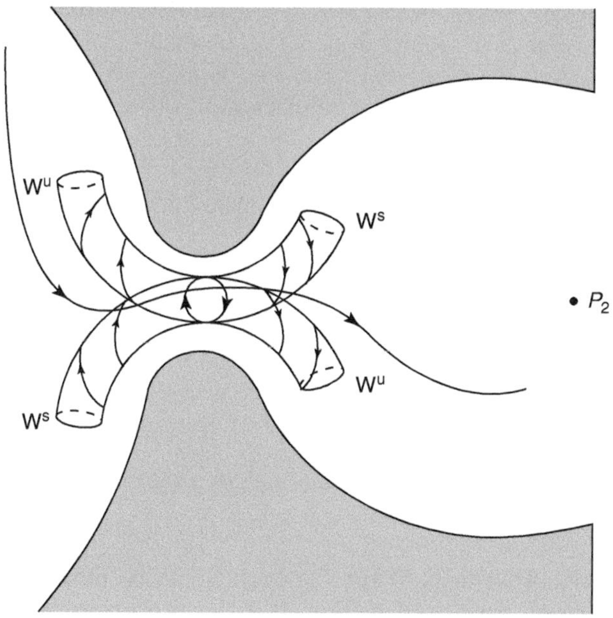

Fig. 4.7. Capture dynamics with $C \lesssim C_1$.

example, in the case of going into circular orbit of 100 km altitude, they save approximately 25% in ΔV, and to achieve an elliptic lunar orbit with a 100 km altitude, they require zero ΔV, where the osculating eccentricity is approximately 0.95 at the time of capture. It is remarked that the savings of 25% in ΔV is very significant, and can double the payload that one can place into low circular lunar orbit. At a cost of approximately 1 million dollars per pound to bring anything into lunar orbit, this savings is significant. Another advantage of WSB transfers is that their capture at the Moon is gradual, and not as risky as the Hohmann transfer which must perform a large capture maneuver in a short time span. A WSB transfer just gradually drifts into capture in a slow fashion and is much less risky.

It is instructive to consider other types of capture and see how ballistic capture may be related to them. A type of capture which is defined in a completely different way than ballistic capture is called permanent capture. This is topologically defined whereas ballistic capture is locally defined analytically.

Definition 4.3.4. P_3 is permanently captured in forward time with respect to both P_1, P_2 if

$$\lim_{t \to -\infty} |Q(t)| = \infty$$

and

$$|Q(t)| < a < \infty$$

as $t \to \infty$ where a is a finite constant. An analogous definition is given for permanent capture in backwards time.

Thus, for permanent capture, the particle P_3 comes in from infinity and remains bound to P_2 *for all time*. Permanent capture does not imply ballistic capture since while P_3 is bound to P_2, E_2 need not be negative. Permanent capture is an unstable process.

From this we can define another type of capture where a particle comes in from infinity, remains bounded for a finite period of time, then goes out to infinity as time goes to infinity. Thus the motion of P_3 is bounded for only a finite period of time. We call this *temporary capture* which is different from 'temporary ballistic capture' we defined previously.

Permanent capture has been studied from a mathematical perspective, and it can be proven to occur only for a set of measure zero in the phase space, and thus it is very unlikely.

It turns out that points exist on $\tilde{W}$ that also lead to permanent capture. Moreover, there exists a region on $\tilde{W}$ whose points lead more generally to chaotic motion of which permanent capture is of one type. This is described in the next section.

4.4 Chaos and weak capture

Consider a solution $Q(t)$ to (4.5) for P_3. $Q(t)$. It can be proven that under special conditions $Q(t)$ will perform chaotic motion. This chaotic motion occurs on a special set in phase space called a hyperbolic invariant set. We describe briefly the types of motions that can occur.

The orbits are near parabolic orbits, and lie between bounded and unbounded motion with respect to P_1 or P_2. Thus, they are in the transition between capture and escape from the P_1, P_2-system. For $\mu = 0$, they are Keplerian parabolic trajectories of P_3 about P_1, with Jacobi energy $|C| = 2\sqrt{2}$. A positive value of C represents direct motion about P_1, and a negative value of C represents retrograde motion.

The orbits start at a reference time $t = 0$ transversal to the Q_1-axis, slightly beyond the Moon, P_2, where $r_2 \gtrsim 0$ and move out to near infinity. For $\mu \ll 1$ the orbits appear nearly parabolic in appearance. They will in general fall back to the Q_1-axis crossing it again for $r_2 \gtrsim 0$ and then move out to infinity again. Then P_3 will fall by P_2 again passing slightly beyond the Moon, etc. This motion can repeat forever. It is also possible that while this oscillatory motion is occurring it is periodic in nature, or it can eventually escape and never return to the Q_1-axis. Or it can start from infinitely far from P_2 and then keep passing slightly beyond P_2 as it crosses the Q_1 axis while repeatably passing out to near infinity at a bounded distance, for all future time. This would correspond to permanent capture. In general many other types of motions can occur that pass slightly beyond the Moon.

The dynamics of this motion can be observed most easily by cutting the near parabolic oscillatory motion by a two-dimensional section Σ_t in phase space, at a given times t, where Σ_t is on the Q_1-axis, where $r_2 \gtrsim 0$, and for $\mu \ll 1$. A given orbit ψ will cut the axis at a sequence of times $t_k, k = 1, \ldots,$ where

$$t_k < t_{k+1},$$

where $Q_2(t_k) = 0$. Set

$$s_k = \left[\frac{t_{k+1} - t_k}{2\pi}\right], \tag{4.9}$$

where $[a]$ is the largest integer $k \le a$, for $a \in \mathbb{R}$. Thus, s_k gives a measure of the number of complete revolutions the primaries P_1, P_2 make (since they have period 2π) in the time it takes P_3 to make two passes through $Q_1 = 0$. The s_k can be used to define bi-infinite sequences

$$s = (\ldots s_{-2}, s_{-1}, s_0; s_1, s_2, \ldots).$$

Let S define the space of all such sequences.

The eventual general pattern of intersection points on Σ_{t_k} from all such orbits ψ takes on the appearance of that shown in Figure 4.8. This is called a *hyperbolic network* associated with a *transverse homoclinic point* r. The intersection of the invariant manifolds W^s, W^u associated to a hyperbolic equilibrium point p of the return map on Σ_t eventually forms a dense set of points, forming a Cantor set. Each of these points has a direction where the motion moves away from the point under iteration and another where it moves towards the point under iteration. This is analogous to a saddle point, except that the points of a hyperbolic network need not be equilibrium points. Thus the motion is very unstable. This Cantor set is called a hyperbolic network we label Λ. A motion defined on a hyperbolic network is called *chaotic*. This network is also called a hyperbolic invariant set [4].

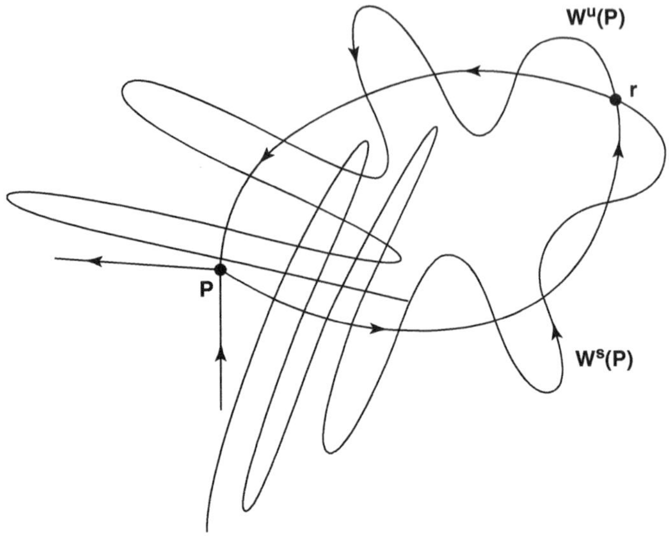

Fig. 4.8. Transverse homoclinic point **r** and hyperbolic network.

The existence of this type of dynamics near parabolic motion in another version of the three-body problem is proven to exist by Moser [15]. The case described above for the restricted problem was proven by Xia [19].

In order to describe the dynamics on Λ, the sequences of S can be used. It is remarkable that it can be proven that prescribing *any* sequence $s \in S$, a motion will exist for the restricted problem passing near to the Moon, where $|C| - 2\sqrt{2} \gtrsim 0$, $\mu \ll 1$. The condition on C means that the motion is near to parabolic for $\mu \ll 1$.

Theorem A. For $|C| - 2\sqrt{2} \gtrsim 0$, $\mu \ll 1$, $r_2 \ll 1$, there exists an integer $m = m(\mu, C)$ such that for any sequence $s \in S$ with $s_k \geq m$ there corresponds a solution of (4.5).

The sequences keep track of the itinerary of the motion of P_3 as it repeatably passes through the points of Λ.

Thus, to every bi-infinite sequence there corresponds an actual solution of the planar circular restricted three-body problem, which is near parabolic and oscillates in a chaotic fashion, repeatably passing very near to P_2 at a distance r_2. This gives an infinite variety of possible motions. Let's see what different sequences say. If the sequence is unbounded, then successive t_k become unbounded. This implies the solution takes so long to come back to $Q_2 = 0$, it in fact is becoming unbounded, but it has infinitely many zeros. This is an unbounded oscillatory solution. A periodic orbit would give rise to a repeating sequence, e.g., $s = \{\ldots, 1, 2, 1, 2, 1, 2, \ldots\}$

A permanent capture orbit which comes in from ∞ corresponds to a sequence terminating on the left with ∞, and then performs infinitely many bounded oscillations for all future time,

$$s = (\infty, \ldots s_k, s_{k+1}, \ldots),$$

where s_k are bounded. Temporary capture is defined by sequences which begin and end with ∞.

Theorem A can be applied to the problem of weak capture. It is very interesting that this is the case, and it is proven [4] that a subset of Λ exists on $\tilde{W}$.

Theorem B. A subset of the hyperbolic network Λ exists on the set W_H of the extended weak stability boundary $\tilde{W}$, which gives rise to the same chaotic motions in the bi-infinite sequence space S as described in Theorem A.

Thus weak capture is chaotic, and can lead to infinitely many possible motions including permanent capture. This fact as interesting applications to the possible construction of permanent capture transfers for spacecraft and also for the permanent capture of small bodies such as comets, asteroids or Kuiper belt objects about the Sun or a planet.

It is remarked that the chaos proven to exist in the weak capture process in Theorem B is associated to near parabolic motion which moves far from P_2. This is done by utilizing the transverse intersection of invariant manifolds associated with parabolic motion. Weak capture can also to studied by studying the possible transverse intersection of the invariant manifolds associated with the Lyapunov orbits about L_1, L_2. This is not yet analytically proven in a general manner; however, there exists an interesting numerically assisted proof of the transverse intersection of these invariant manifolds for some selected parameter values of μ and for $C \lesssim C_1$ [12].

4.5 Origin of the Moon

An outstanding question in astronomy is to understand where the Moon came from. One of the first theories to try to answer this is called the 'sister planet theory'. It proposed that the Moon formed together with the Earth as sister planets, in the solar nebula of gas and dust from which all the planets formed about 4 billion years ago. However, there are some inconsistencies with this. One is the fact that a large iron core is absent in the Moon, and present in the Earth, giving the Earth and Moon different densities, which are 5.5 grams/cm^3 and 3.3 grams/cm^3, respectively. Another theory is that the Moon was formed from beyond the Earth's orbit, and was captured into orbit about the Earth. If this were the case then the Earth and Moon would have different abundances of oxygen isotopes. This is inconsistent with the fact that the Earth and Moon have identical abundances.

A generally accepted theory which explains the differences in iron and the oxygen isotope abundances among other things is called the "impactor theory". It was formulated by W. Hartmann and D. Davis [11], and A. Cameron and W. Ward [7]. It proposes that after the Earth had already formed 4 billion years ago, a giant Mars-sized object smashed into the Earth. When it hit, it formed the Moon from iron poor mantel material debris primarily from the impactor, and also from the Earth, both of which already had iron cores. The Moon coalesced from this material. The iron core of the impactor was deposited into the iron core of the Earth. This explains the iron deficiency of the Moon. This theory also proposes that the impactor formed at the same 1 AU distance that the Earth is from the Sun. This explains the identical oxygen isotope abundances.

Numerical simulations from this theory show conditions of impact that an object the size of Mars would have to have to form the Moon from the resulting debris. They show that this object would have to approach the Earth with a relatively slow velocity—nearly parabolic with respect to the Earth, with velocities only a few hundred meters per second.

A fundamental question to ask is—*Where did this Mars-sized impactor come from?*

In 2001, Richard Gott described his theory to me to explain where the impactor may have come from. He proposed that at the time of the solar nebula from which the Earth was formed, there was so much debris flying around the Sun that it could have settled near the stable equilateral Lagrange points L_4, L_5 with respect to the Earth and Sun. Since these locations are stable, debris arriving there with a small relative velocity could remain trapped there. As more and more debris arrives, it could start to coalesce and a massive body could start to grow. Given several million years a large Mars-sized body could result.

However, here was a problem with his theory. How could it be demonstrated that the impactor could leave the L_4 (or L_5) neighborhood and impact the Earth? Back of the envelope calculations showed that collision would be unlikely since the impactor would likely fly by the Earth at high relative velocities of several kilometers per second, and easily miss the Earth.

The solution was found by calculating a WSB region about L_4 (or L_5) where a small massive object would first move captured in neighborhoods about these points and move in a horseshoe orbit—i.e., moving in an Earth-like orbit and oscillating back and forth, between counterclockwise and clockwise motions (without moving 360 degrees about the Sun), and not moving past the Earth. They would very gradually gain energy, and hence velocity with respect to the Sun, by the resulting interactions with the small planetesimals in the solar nebula at the time. Eventually, the impactor would grow in size, and move beyond the Earth, and the motion would bifurcate from the oscillating horseshoe motion to a non-oscillating cycling motion, repeatably flying closely by the Earth. This cycling motion is called 'breakout', and it is chaotic in nature. This is determined in the restricted problem by fixing a direction of motion at L_4 (or L_5) and gradually increasing the velocity until breakout occurs. This gives a parametric set of critical velocities about L_4 (or L_5) depending on direction, yielding a WSB about L_4 (or L_5). This is seen in Figure 4.9. It is

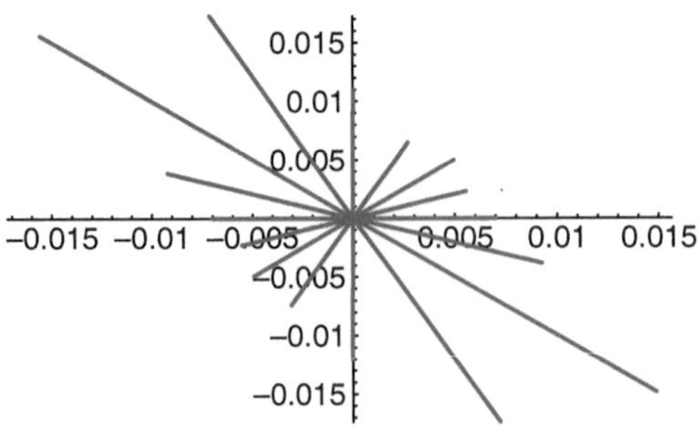

Fig. 4.9. Critical velocity magnitudes as a function of direction at L_4 giving rise to escape, or, equivalently, 'breakout' from L_4. (A velocity value of 1 corresponds to the velocity of the Earth about the Sun.)

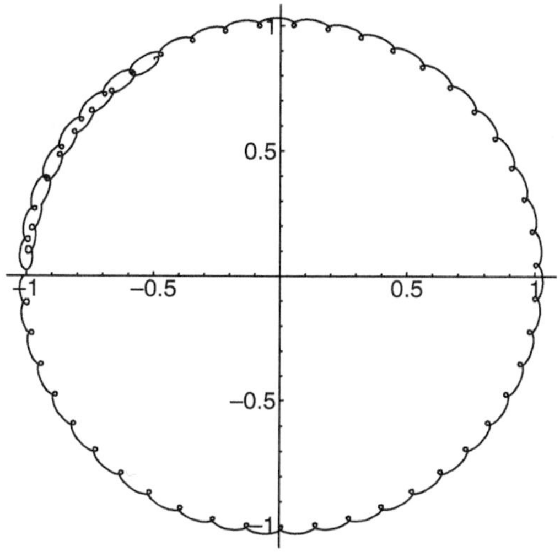

Fig. 4.10. A near parabolic creeping collision orbit emanating from L_4 in the third quadrant, and colliding with the Earth 57.32 years later near -1 on the x-axis. It first moves downward towards the Earth in a counterclockwise direction, then reverses its direction and moves nearly 360 degrees in a clockwise direction about the Sun at the origin, where it collides with the Earth.

shown in Ref. [5] that the likelihood of collision was very high, and that this process is preserved with more accurate modeling of the solar system. An actual collision orbit is shown in Figure 4.10.

The description presented here is very brief. The detailed exposition of this theory is given in Ref. [5]. Also, see Ref. [10]. The theory for the formation of the Mars impactor at L4/L5 as presented in Ref. [5] is currently being applied to the formation of some of the moons of Saturn, and to Saturn's rings in a collaboration with Gott, Vanderbei, and Belbruno [10, 18].

Acknowledgements

I would like to thank Pini Gurfil for helpful suggestions to put this paper into the desired form. This work was supported by grants from NASA.

References

1. Belbruno, E.A. (1987). Lunar Capture Orbits, A Method of Constructing Earth-Moon Trajectories and the Lunar GAS Mission, in: *Proceedings of AIAA/DGLR/JSASS Inter. Propl. Conf.* AIAA Paper No. 87-1054, (May 1987).
2. Belbruno, E.A. and Miller, J. (1993). Sun-perturbed earth-to-moon transfers with ballistic capture. *J. Guid., Control and Dynamics* **16**(4), July–August, 770–775.

3. Belbruno, E.A. (2004). Existence of Chaos Associated with Weak Capture and Applications, in *Astrodynamics, Space Missions, and Chaos* (E. Belbruno, D. Folta, and P. Gurfil eds.), **1017**, *Annals of the New York Academy of Sciences*, pp. 1–10.
4. Belbruno, E.A. (2004). *Capture Dynamics and Chaotic Motions in Celestial Mechanics*, Princeton University Press.
5. Belbruno, E.A. and Gott III, J.R. (2005). Where Did the Moon Come From? *Astronomical Journal*, **129** (4), March, pp. 1724–1745.
6. Belbruno, E.A. (2006). *Fly Me to the Moon*, Princeton University Press (planned for publication on December 15, 2006).
7. Cameron A.G.W. and Ward, W.R. (1976). The Origin of the Moon, in *Proc. Lunar Planet. Sci. Conf. 7th*, pp. 120–122.
8. Case, J. (2004). Celestial Mechanics Theory Meets the Nitty-Gritty of Trajectory Design, (book review of [4]), **37**(6), *SIAM News*, July–August, pp. 1–3.
9. Chown, M. (2004). The Planet that Stalked the Earth, (Featured cover story), *New Scientist*, August 14, pp. 26–30.
10. Gott III, J.R. (2006). Lagrange L4/L5 Points and the Origin of our Moon and Saturn's Moons and Rings, in *Astrodynamics and Its Applications* (E. Belbruno ed.), New York Academy of Sciences, Annals, V 1065.
11. Hartmann, W.K. and Davis, D.R. (1975). Satellite-sized Planetesimals and Lunar Origin, *Icarus*, **24**, pp. 504–515.
12. Koon, W.S., Lo, M.W., Marsden, J.E. and Ross, S.D. (2000). Heteroclinic Connections Between Periodic Orbits and Resonance Transitions in Celestial Mechanics, *Chaos*, **10**, pp. 427–469.
13. Koon, W.S., Lo, M.W., Marsden, J.E. and Ross, S.D. (2000). Shoot the Moon, *AAS/AIAA Astrodynamics Specialist Conference*, Florida, Paper Number AAS 000-166.
14. Klarreich, E. Navigating Celestial Currents (Featured cover story), *Science News*, **167**(16), (April 16, 2005), 250–252.
15. Moser, J. (1973). *Stable and Random Motions in Dynamical Systems*, Princeton University Press, V 77, Annals of Mathematics.
16. Osserman, R. (2005). *Mathematics of the Heavens, Notices of the American Mathematical Society (AMS)*, **52**(4), April, 417–424.
17. Racca, G. (2003). New Challanges to Trajectory Design by the Use of Electric Propulsion and Other Means of Wandering in the Solar System, *Celestial Mechanics and Dynamical Astronomy*, **85**, pp. 1–24.
18. Vanderbei, R. (2006). *Horsing Around on Saturn*, in *Astrodynamics and Its Applications* (E. Belbruno ed.), New York Academy of Sciences, Annals, V 1065.
19. Xia, Z. (1992). Melnikov Method and Transversal homoclinic Points in the Restricted Three-Body problem, *JDE* **96**, pp. 170–184.

5 Set Oriented Numerical Methods in Space Mission Design

MICHAEL DELLNITZ[1] AND OLIVER JUNGE[2]
[1]*Institute for Mathematics, University of Paderborn, Germany*
[2]*Center for Mathematical Sciences, Munich University of Technology, Germany*

Contents

5.1 Introduction

New techniques for the design of energy efficient trajectories for space missions have been proposed which are based on the circular restricted three-body problem as the underlying mathematical model. These techniques exploit the structure and geometry of certain invariant sets and associated invariant manifolds in phase space in order to systematically construct efficient flight paths.

In this chapter we present numerical methods that enable an implementation of this approach. Using a set oriented framework we show how to compute approximations to invariant sets and invariant manifolds and how to detect connecting orbits that might serve as initial guesses for the solution of a more detailed model. We also show how to extend the approach in order to account for a continuously applied control force on the spacecraft as realized by certain low thrust propulsion systems.

All techniques described in this chapter have been implemented within the software package "Global Analysis of Invariant Objects" (`GAIO`) which is available from the authors.

5.2 Dynamical systems and mission design

A new paradigm for the construction of energy efficient trajectories for spacecraft is currently emerging. It heavily bases on concepts and techniques from the theory and numerical treatment of dynamical systems. The basic strategy is the following: Instead of a two-body problem, as in more classical approaches, one considers a restricted three-body problem as the mathematical model for the motion of the spacecraft. This enables one to exploit the intricate structure and geometry of certain invariant sets and their stable and unstable manifolds in phase space—which are not present in two-body problems—as candidate regions for energy efficient trajectories. For example, this approach has recently been used in the design of the trajectory for the *Genesis discovery mission*[1] [29].

Building on this basic concept, techniques have been proposed that synthesize partial orbits from different three-body problems into a single one, yielding energy efficient trajectories with eventually very complicated itineraries [26, 27]. In Ref. [27], a *petit grand tour* among the moons of Jupiter has been constructed by this approach. The idea of the technique is as follows: One computes the intersection of parts of the stable resp. unstable manifold of two specific periodic orbits in the vicinity of two moons, respectively, with a suitably chosen surface. After a transformation of these two curves into a common coordinate system one identifies points on them that lie close to each other—ideally one searches for intersection points. Typically, however, these two curves will not intersect in the chosen surface, so a certain (impulsive) maneuver of the spacecraft will be necessary in order to transit from the part of the trajectory on the unstable manifold to the one on the stable manifold. In a final step this "patched 3-body approximation" to a trajectory is used as an initial guess for standard local solvers using the full n-body dynamics of the solar system as the underlying model (as, e.g., the differential corrector implemented in the JPL-tool LTool [32]).

5.2.1 The circular restricted three-body problem

Let us briefly recall the basics of the *(planar) circular restricted three body problem (PCR3BP)*—for a more detailed exposition and a description of the full spatial model see Refs. [2, 31, 38]. The PCR3BP models the motion of a particle of very small mass within the gravitational field of two heavy bodies (e.g., The Sun and The Earth). Those two *primaries* move in a plane counterclockwise on circles about their common center of mass with the same constant angular velocity. One assumes that the third body does not influence the motion of the primaries while it is only influenced by the gravitational forces of the primaries.

In a normalized rotating coordinate system the origin is the center of mass and the two primaries are fixed on the x-axis at $(-\mu, 0)$ and $(1-\mu, 0)$, respectively, where $\mu = m_1/(m_1+m_2)$ and m_1 and m_2 are the masses of the primaries. The equations of motion for the spacecraft with position (x_1, x_2) in rotating coordinates are given by

$$\ddot{x}_1 - 2\dot{x}_2 = \Omega_{x_1}(x_1, x_2), \qquad \ddot{x}_2 + 2\dot{x}_1 = \Omega_{x_2}(x_1, x_2) \tag{5.1}$$

[1] `http://genesismission.jpl.nasa.gov.`

with

$$\Omega(x_1, x_2) = \frac{x_1^2 + x_2^2}{2} + \frac{1-\mu}{r_1} + \frac{\mu}{r_2} + \frac{\mu(1-\mu)}{2}$$

and

$$r_1 = \sqrt{(x_1+\mu)^2 + x_2^2}, \qquad r_2 = \sqrt{(x_1 - 1 + \mu)^2 + x_2^2}.$$

The system posesses five equilibrium points (the *Lagrange points*): the collinear points L_1, L_2 and L_3 on the x-axis and the equilateral points L_4 and L_5. The equations (5.1) have a first integral, the *Jacobi integral*, given by

$$C(x_1, x_2, \dot{x}_1, \dot{x}_2) = -(\dot{x}_1^2 + \dot{x}_2^2) + 2\Omega(x_1, x_2). \tag{5.2}$$

The three-dimensional manifolds of constant C-values are invariant under the flow of (5.1), their projection onto position-space, the *Hill's region*, determines the allowed region for the motion of the spacecraft (Figure 5.1(a)).

5.2.2 *Patching three-body problems*

The idea of constructing energy efficient trajectories via coupling three-body problems essentially relies on two key observations:

1. For suitable energy values (i.e., values of the Jacobi integral (5.2)) there exist periodic solutions, the *Lyapunov orbits* (Figure 5.1(a)), of (5.1) in the vicinity of the equilibrium points L_1 and L_2 that are unstable in both time directions. Their unstable resp. stable

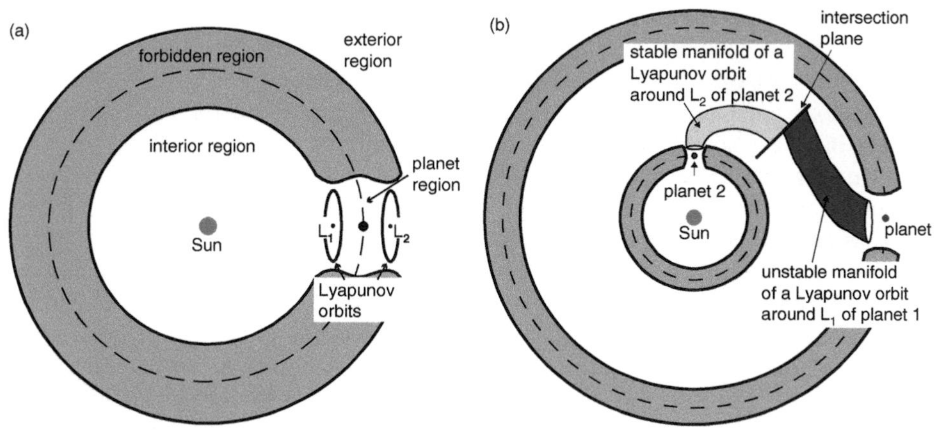

Fig. 5.1. Left: Projection of an energy surface onto position-space (schematic) for a value of the Jacobi integral for which the spacecraft is able to transit between the exterior and the interior region. Right: Sketch of the "patched 3-body approach" [26, 27]. The idea is to travel within certain invariant manifold "tubes", with possibly an impulsive correction maneuver at the intersection plane.

manifolds are cylinders that partition the three dimensional energy surface into two sets: (1) transit orbits, that locally pass between the *interior region* and the *planet region* in the case of an L_1-Lyapunov orbit or between the *exterior region* and the *planet region* in the case of L_2, and (2) non-transit orbits that stay in the exterior or interior region [30].

2. By "embedding" one PCR3BP into a second one, parts of the stable manifold of a Lyapunov orbit in one system may come close to the unstable manifold of a Lyapunov orbit in the other system (where, for a moment, it may help to imagine that the two systems do not move relative to each other), (Figure 5.1(b)). It may thus be possible for a spacecraft to "bridge the gap" between two pieces of trajectories in the vicinity of these manifolds by exerting an impulsive maneuver [26, 27].

One way to detect a close approach of two such invariant manifolds is to reduce the dimensionality of the problem. One computes the intersection of the two manifolds with a suitable intersection plane (Figure 5.1(b)) and determines points of close approach in this surface—for example by inspecting projections onto 2D-coordinate planes. This approach has in fact been used for a systematic construction of trajectories that follow prescribed itineraries around and between the Jovian moons [27]. In Section 5.6 of this chapter we go one step further and consider controlled problems via the incorporation of low thrust propulsion.

5.3 Set oriented numerics

Over the last decade *set oriented* numerical methods have been developed for the analysis of the global behavior of dynamical systems [6–10]. These numerical tools can, e.g., be used to approximate different types of invariant sets or invariant manifolds. They also allow to extract statistical information via the computation of natural invariant measures or almost invariant sets. In contrast to other numerical techniques these methods do not rely on the computation of *single long term* simulations but rather agglomerate the information obtained from *several short term* trajectories.

The methods are based on a multilevel subdivision procedure for the computation of certain invariant sets. This multilevel approach allows one to cover the object of interest—e.g., an invariant manifold or the support of an invariant measure—by several small subsets of state space. Since outer approximations are produced and long term simulations are avoided these methods are typically quite robust.

The numerical methods presented here are similar in spirit to the so-called *cell mapping approach* [22, 28]. However, there is a significant difference: the cell mapping approach relies on a partition of the *entire phase space* and thus the numerical effort depends crucially on its dimension. In contrast, in our set oriented apporach only a covering of the set of interest (e.g., the attractor) is constructed and so the computational effort essentially depends on the complexity of the underlying dynamics.

We would also like to mention that by now there exist several relevant extensions and adaptations of the set oriented approach as described here: the methods have been, e.g., combined with branch and bound techniques and methods from evolutionary optimization in order to solve global (multi-objective) optimization problems [13, 14, 33–35]. They also have been combined with shortest path algorithms from graph theory in order to compute the optimal value function of optimal control problems as well as globally stabilizing feedback laws [19, 25].

5.3.1 The multilevel subdivision algorithm

We now describe the multilevel subdivision algorithm for the computation of invariant sets which forms the basis for all methods described in this chapter. We consider a discrete time dynamical system

$$x_{j+1} = f(x_j), \quad j = 0, 1, 2, \dots, \tag{5.3}$$

where $f: \mathbb{R}^n \to \mathbb{R}^n$ is a diffeomorphism which may, e.g., be given by the time-T-map of some underlying vector field.

A subset $A \subset \mathbb{R}^n$ is called *invariant* if

$$f(A) = A.$$

Moreover, an invariant set A is an *attracting set* with *fundamental neighborhood* U if for every open set $V \supset A$ there is an $N \in \mathbb{N}$ such that $f^j(U) \subset V$ for all $j \geq N$. Observe that if A is invariant then the closure of A is invariant as well. Hence we restrict our attention to closed invariant sets A, and in this case we obtain

$$A = \bigcap_{j \in \mathbb{N}} f^j(U).$$

By definition all the points in the fundamental neighborhood U are attracted by A. For this reason the open set $\cup_{j\in\mathbb{N}} f^{-j}(U)$ is called the *basin of attraction* of A. If the basin of attraction of A is the entire $\mathbb{R}^n$ then A is called the *global attractor*. Note that the global attractor contains all invariant sets of the given dynamical system.

Definition 5.3.1. Let $Q \subset \mathbb{R}^n$ be a compact set. We define the *global attractor relative to Q* by

$$A_Q = \bigcap_{j \geq 0} f^j(Q). \tag{5.4}$$

The definition of A_Q implies that $A_Q \subset Q$ and that $f^{-1}(A_Q) \subset A_Q$, but not necessarily that $f(A_Q) \subset A_Q$. Furthermore, A_Q is compact since Q is compact. Finally, A_Q is a subset of the global attractor A, however, in general $A_Q \neq A \cap Q$.

5.3.1.1 Subdivision algorithm

The following algorithm provides a method for the approximation of relative global attractors. It generates a sequence $\mathcal{B}_0, \mathcal{B}_1, \dots$ of finite collections of compact subsets of $\mathbb{R}^n$ such that the diameter $\mathrm{diam}(\mathcal{B}_k) = \max_{B \in \mathcal{B}_k} \mathrm{diam}(B)$ converges to zero for $k \to \infty$.

Given an initial collection $\mathcal{B}_0$, we inductively obtain $\mathcal{B}_k$ from $\mathcal{B}_{k-1}$ for $k = 1, 2, \ldots$ in two steps:

1. *Subdivision*: Construct a new collection $\hat{\mathcal{B}}_k$ such that

$$\bigcup_{B \in \hat{\mathcal{B}}_k} B = \bigcup_{B \in \mathcal{B}_{k-1}} B \tag{5.5}$$

and

$$\operatorname{diam}(\hat{\mathcal{B}}_k) \leq \theta_k \operatorname{diam}(\mathcal{B}_{k-1}), \tag{5.6}$$

where $0 < \theta_{\min} \leq \theta_k \leq \theta_{\max} < 1$.

2. *Selection*: Define the new collection $\mathcal{B}_k$ by

$$\mathcal{B}_k = \left\{ B \in \hat{\mathcal{B}}_k : \exists \hat{B} \in \hat{\mathcal{B}}_k \text{ such that } f^{-1}(B) \cap \hat{B} \neq \emptyset \right\}. \tag{5.7}$$

Note that by construction $\operatorname{diam}(\mathcal{B}_k) \leq \theta_{\max}^k \operatorname{diam}(\mathcal{B}_0) \to 0$ for $k \to \infty$.

Example 5.3.1. We consider $f : \mathbb{R} \to \mathbb{R}$,

$$f(x) = \alpha x,$$

where $\alpha \in (0, \frac{1}{2})$ is a constant. Then the global attractor $A = \{0\}$ of f is a stable fixed point. We begin the subdivision procedure with $\mathcal{B}_0 = \{[-1, 1]\}$ and construct $\hat{\mathcal{B}}_k$ by bisection. In the first subdivision step we obtain

$$\mathcal{B}_1 = \hat{\mathcal{B}}_1 = \{[-1, 0], [0, 1]\}.$$

No interval is removed in the selection step, since each of them is mapped into itself. Now subdivision leads to

$$\hat{\mathcal{B}}_2 = \left\{ \left[-1, -\frac{1}{2}\right], \left[-\frac{1}{2}, 0\right], \left[0, \frac{1}{2}\right], \left[\frac{1}{2}, 1\right] \right\}.$$

Applying the selection rule (5.7), the two boundary intervals are removed, i.e.

$$\mathcal{B}_2 = \left\{ \left[-\frac{1}{2}, 0\right], \left[0, \frac{1}{2}\right] \right\}.$$

Proceeding this way, we obtain after k subdivision steps

$$\mathcal{B}_k = \left\{ \left[-\frac{1}{2^{k-1}}, 0\right], \left[0, \frac{1}{2^{k-1}}\right] \right\}.$$

We see that the union $\bigcup_{B \in \mathcal{B}_k} B$ is indeed approaching the global attractor $A = \{0\}$ for $k \to \infty$. The speed of convergence obviously depends on the contraction rate of the global attractor.

5.3.1.2 Convergence

The sequence of collections generated by the subdivision algorithm converges to the global attractor A_Q relative to Q. In addition to the following result one can also derive a statement about the speed of convergence in the case that the relative global attractor possesses a hyperbolic structure [8].

Proposition 5.3.1. *Let A_Q be the global attractor relative to the compact set Q, let $\mathcal{B}_0$ be a finite collection of closed subsets with $Q = \bigcup_{B\in\mathcal{B}_0} B$ and let $Q_k = \bigcup_{B\in\mathcal{B}_k} B$, $k = 0, 1, 2, \ldots$. Then*

$$\lim_{k\to\infty} h\left(A_Q, Q_k\right) = 0,$$

where $h(B, C)$ denotes the usual Hausdorff distance between two compact subsets $B, C \subset \mathbb{R}^n$ [8].

5.3.1.3 Implementation

We realize the closed subsets constituting the collections using generalized rectangles ("boxes") of the form

$$B(c, r) = \{y \in \mathbb{R}^n : |y_i - c_i| \leq r_i \text{ for } i = 1, \ldots, n\},$$

where $c, r \in \mathbb{R}^n$, $r_i > 0$ for $i = 1, \ldots, n$, are the center and the radius, respectively. In the k-th subdivision step we subdivide each rectangle $B(c, r)$ of the current collection by bisection with respect to the j-th coordinate, where j is varied cyclically, i.e., $j = ((k-1) \bmod n) + 1$. This division leads to two rectangles $B_-(c^-, \hat{r})$ and $B_+(c^+, \hat{r})$, where

$$\hat{r}_i = \begin{cases} r_i & \text{for } i \neq j \\ r_i/2 & \text{for } i = j \end{cases}, \quad c_i^\pm = \begin{cases} c_i & \text{for } i \neq j \\ c_i \pm r_i/2 & \text{for } i = j \end{cases}.$$

Starting with a single initial rectangle we perform the subdivision until a prescribed size σ of the diameter relative to the initial rectangle is reached.

The collections constructed in this way can easily be stored in a binary tree. In Figure 5.2 we show the representation of three subdivision steps in three dimensions ($n = 3$) together with the corresponding sets Q_k, $k = 0, 1, 2, 3$. Note that each collection and the corresponding covering Q_k are completely determined by the tree structure and the initial rectangle $B(c, r)$.

5.3.1.4 Realization of the intersection test

In the subdivision algorithm we have to decide whether for a given collection $\mathcal{B}_k$ the image of a set $B \in \mathcal{B}_k$ has a non-zero intersection with another set $B' \in \mathcal{B}_k$, i.e., whether $f(B) \cap B' = \emptyset$. In simple model problems such as our trivial Example 5.3.1 this decision can be made analytically. For more complex problems we have to use some kind of discretization. Motivated by similar approaches in the context of cell-mapping techniques [22], we choose a finite set T of test points (on a grid or distributed randomly) in each set $B \in \mathcal{B}_k$ and replace the condition $f(B) \cap B' = \emptyset$ by $f(T) \cap B' = \emptyset$.

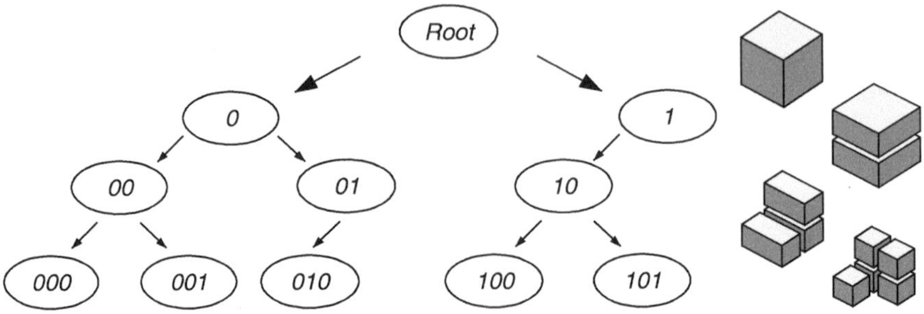

Fig. 5.2. Storage scheme for the collections and the corresponding coverings Q_k, $k = 0, 1, 2, 3$.

Example 5.3.2. Consider the Hénon map

$$f(x) = \begin{pmatrix} 1 - ax_1^2 + bx_2 \\ x_1 \end{pmatrix}. \tag{5.8}$$

for $b = 0.2$ and $a = 1.2$. Starting with the square $Q = [-2, 2]^2$, we display in Figure 5.3 the coverings of the global attractor relative to Q obtained by the algorithm after 8 and 12 subdivision steps.

The figures have been generated by the following GAIO script:

```
addpath(strcat(getenv('GAIODIR'),'/matlab'))
henon = Model('ohenon');
map = Integrator('Map');
map.model = henon;
henon.a = 1.2;
henon.b = 0.2;
```

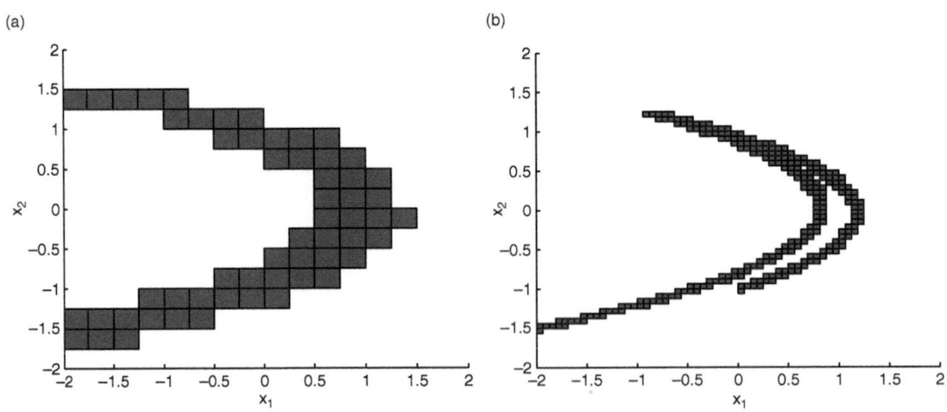

Fig. 5.3. Coverings of the global Hénon attractor after 8 (left) and 12 (right) subdivision steps.

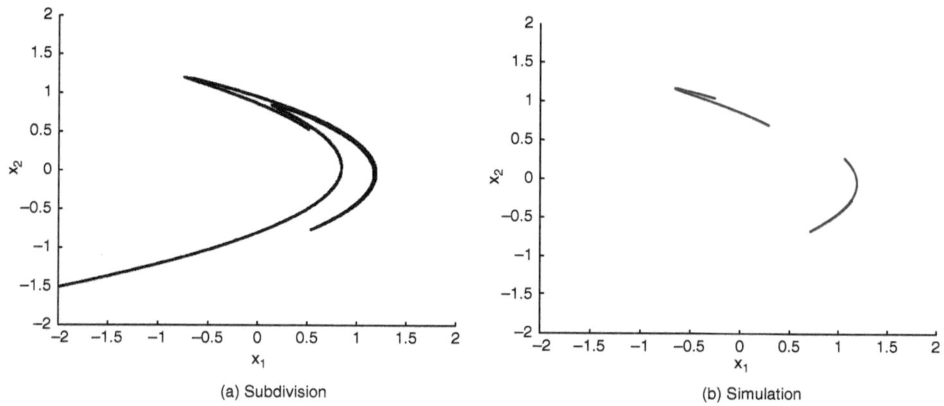

Fig. 5.4. (a) Approximation of the relative global attractor for the Hénon mapping after 18 subdivision steps; (b) attractor of the Hénon mapping computed by direct simulation.

```
edges = Points('Edges', 2, 100);
center = Points('Center', 2, 1);
tree = Tree([0 0], [2 2]);
tree.integrator = map;
tree.domain_points = edges;
tree.image_points = center;
rga(tree,8);
plotb(tree);
rga(tree,4);
clf; plotb(tree);
```

In Figure 5.4(a) we show the rectangles covering the relative global attractor after 18 subdivision steps. Note that a direct simulation would not yield a similar result. In Figure 5.4(b) we illustrate this fact by showing a trajectory, neglecting the transient behavior. This difference is due to the fact that the subdivision algorithm covers all invariant sets in Q, together with their unstable manifolds. In particular, the one-dimensional unstable manifolds of the two fixed points are approximated—but those cannot be computed by direct simulation.

5.4 Computing invariant manifolds

We now present a set oriented method for the computation of invariant manifolds. Although the method can in principle be applied to manifolds of arbitrary hyperbolic invariant sets we will restrict, for simplicity, to the case of a hyperbolic fixed point p.

We fix a (large) compact set $Q \subset \mathbb{R}^n$ containing p, in which we want to approximate part of the global unstable manifold $W^u(p)$ of p. To combine the subdivision algorithm with a continuation method, we realize the subdivision process using a nested sequence $\mathcal{P}_\ell$, $\ell \in \mathbb{N}$, of successively finer finite partitions of Q, i.e., for all $B \in \mathcal{P}_{\ell+1}$ there exist $B' \in \mathcal{P}_\ell$ such that $B \subset B'$ and $\operatorname{diam}(B) \leq \theta \operatorname{diam}(B')$ for some $0 < \theta < 1$.

Let $C \in \mathcal{P}_\ell$ be a neighborhood of the hyperbolic fixed point p such that the global attractor relative to C satisfies

$$A_C = W^u_{\mathrm{loc}}(p) \cap C.$$

Applying k steps of the subdivision algorithm to $\mathcal{B}_0 = \{C\}$, we obtain a covering $\mathcal{B}_k \subset \mathcal{P}_{\ell+k}$ of the local unstable manifold $W^u_{\mathrm{loc}}(p) \cap C$. By Proposition 3.1, this covering converges to $W^u_{\mathrm{loc}}(p) \cap C$ for $k \to \infty$. The following algorithm grows this initial covering until it covers a certain subset of the global unstable manifold of p.

5.4.1 Continuation algorithm

For a fixed k we define a sequence $\mathcal{C}_0^{(k)}, \mathcal{C}_1^{(k)}, \ldots$ of subsets $\mathcal{C}_j^{(k)} \subset \mathcal{P}_{\ell+k}$ by

1. *Initialization*:

$$\mathcal{C}_0^{(k)} = \mathcal{B}_k.$$

2. *Continuation*: For $j = 0, 1, 2, \ldots$ define

$$\mathcal{C}_{j+1}^{(k)} = \left\{ B \in \mathcal{P}_{\ell+k} : B \cap f(B') \neq \emptyset \text{ for some } B' \in \mathcal{C}_j^{(k)} \right\}.$$

Observe that the sets $C_j^{(k)} = \bigcup_{B \in \mathcal{C}_j^{(k)}} B$ form nested sequences in k, i.e., $C_j^{(0)} \supset C_j^{(1)} \supset \ldots$ for $j = 0, 1, 2, \ldots$.

5.4.2 Convergence result and error estimate

Set $W_0 = W^u_{\mathrm{loc}}(p) \cap C$ and define inductively for $j = 0, 1, 2, \ldots$

$$W_{j+1} = f(W_j) \cap Q.$$

Proposition 5.4.1. *[7]. The sets $C_j^{(k)}$ are coverings of W_j for all $j, k = 0, 1, \ldots$. Moreover, for fixed j, $C_j^{(k)}$ converges to W_j in Hausdorff distance if the number k of subdivision steps in the initialization goes to infinity.*

It can in general not be guaranteed that the continuation method leads to an approximation of the entire set $W^u(p) \cap Q$. The reason is that the unstable manifold of the hyperbolic fixed point p may "leave" Q but may as well "wind back" into it.

If we additionally assume the existence of a hyperbolic structure along the unstable manifold then we can establish results on the convergence behavior of the continuation method in a completely analogous way as in Ref. [8]. To this end assume that p is an element of an attractive hyperbolic set A. Then the unstable manifold of p is contained in A. We choose $Q = \overline{\bigcup_{x \in A} W^s_\eta(x)}$ for some sufficiently small $\eta > 0$, such that $A = A_Q$.

Let $\rho \geq 1$ be a constant such that for every compact neighborhood $\widetilde{Q} \subset Q$ of A_Q we have $h(A_Q, \widetilde{Q}) \leq \delta \Rightarrow \widetilde{Q} \subset U_{\rho\delta}(A_Q)$.

Proposition 5.4.2. *[23]. Assume that in the initialization step of the continuation method we have*

$$h(W_0, C_0^{(k)}) \leq \zeta \operatorname{diam} \mathcal{C}_0^{(k)}$$

for some constant $\zeta > 0$. If $C_j^{(k)} \subset W_\eta^s(W_j)$ for $j = 0, 1, 2, \ldots, J$, then

$$h(W_j, C_j^{(k)}) \leq \operatorname{diam} \mathcal{C}_j^{(k)} \max\left(\zeta, 1 + \beta + \beta^2 + \cdots + \beta^j \zeta\right) \tag{5.9}$$

for $j = 1, 2, \ldots, J$. Here $\beta = C\lambda\rho$ and C and λ are the characteristic constants of the hyperbolic set A.

The estimate (5.9) points up the fact that for a given initial level k and λ near 1—corresponding to a weak contraction transversal to the unstable manifold—the approximation error may increase dramatically with an increasing number of continuations steps (i.e., increasing j).

Example 5.4.1. As a numerical example we consider the (spatial) circular restricted three-body problem (Section 5.2) with $\mu = 3.040423398444176 \times 10^{-6}$ for the Sun/Earth system.

Motivated by the requirements of the mission design for the *NASA Genesis discovery mission* we aim for the computation of the unstable manifold of a certain unstable periodic orbit (a so-called *halo orbit*) in the vicinity of the L_1 Lagrange point. In light of Proposition 5.4.2, a naive application of the continuation method would—due to the Hamiltonian nature of the system—not lead to satisfactory results in this case. We therefore apply a modified version of this method [23]. Roughly speaking the idea is not to continue the current covering by considering *one* application of the map at *each* continuation step, but instead to perform only *one* continuation step while computing *several* iterates of the map.

More formally, we replace the second step in the continuation method by:

(ii) *Continuation*: For some $J > 0$ define

$$\mathcal{C}_J^{(k)} = \left\{B \in \mathcal{P}_{\ell+k} : \exists\, 0 \leq j \leq J : B \cap f^j(B') \neq \emptyset \text{ for some } B' \in \mathcal{C}_0^{(k)}\right\}.$$

The convergence statement in Proposition 5.4.1 is adapted to this method in a straightforward manner. One can also show that—as intended—the Hausdorff-distance between compact parts of the unstable manifold and the computed covering is of the order of the diameter of the partition, see Ref. [23] for details. However, and this is the price one has to pay, one no longer considers short term trajectories here and therefore accumulates methodological and round-off errors when computing the iterates f^j.

A second advantage of the modified continuation method is that whenever the given dynamical system stems from a flow ϕ^t one can get rid of the necessity to consider a time-T-map and instead replace the continuation step by

(ii) *Continuation*: For some $T > 0$ define

$$\mathcal{C}_T^{(k)} = \left\{ B \in \mathcal{P}_{\ell+k} : \exists\, 0 \le t \le T : B \cap \phi^t(B') \neq \emptyset \text{ for some } B' \in \mathcal{C}_0^{(k)} \right\}.$$

This facilitates the usage of integrators with adaptive step-size control and finally made the computations feasible for this example. Figure 5.5 shows the result of the computation, where we set $T = 7$ and used an embedded Runge-Kutta scheme of order 8 as implemented in the code DOPRI853[2] [21] with error tolerances set to 10^{-9}. See again [23] for more details on this computation. A movie illustrating a flight along this manifold is available at http://www-math.upb.de/~agdellnitz/Software/halo.html.

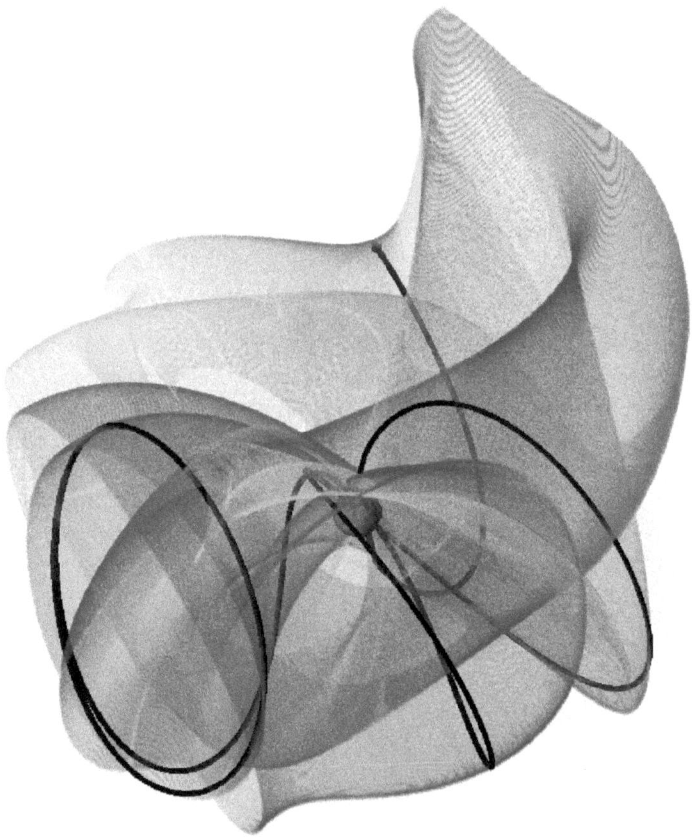

Fig. 5.5. Covering of part of the global unstable manifold of an unstable periodic orbit in the circular restricted three-body problem (projection onto configuration space). The blue body depicts the Earth, the black trajectory is a sample orbit which leaves the periodic orbit in the direction of the Earth. The coloring indicates the temporal distance from the periodic orbit. (see Color plate 1)

[2] `http://www.unige.ch/~hairer/software.html.`

5.5 Detecting connecting orbits

One of the key ideas in applying dynamical systems techniques in space mission design is to use *connecting orbits* between different invariant sets (e.g., periodic orbits or invariant tori) as energy efficient trajectories for the spacecraft.

In this section we show how the set oriented continuation algorithm for the computation of invariant manifolds described in the previous section can be used in order to detect connecting orbits. For simplicity, we restrict our considerations to the situation where connecting orbits between different steady state solutions have to be detected. We consider a parameter dependent ordinary differential equation

$$\dot{x} = f(x, \lambda), \tag{5.10}$$

where $f\colon \mathbb{R}^d \times \Lambda \to \mathbb{R}^d$ is a smooth vector field and $\Lambda \subset \mathbb{R}$ is an interval. Denote by x_λ and y_λ, $\lambda \in \Lambda$, two one-parameter families of hyperbolic steady state solutions of (5.10)—allowing that $x_\lambda = y_\lambda$. We are interested in the detection of a connecting orbit between two steady states $x_{\bar{\lambda}}$, $y_{\bar{\lambda}}$ while the system parameter λ is varied. In order to ensure that, in principle, connecting orbits can generically occur we assume that

$$\dim(W^u(x_\lambda)) + \dim(W^s(y_\lambda)) = d \quad \text{for all } \lambda \in \Lambda.$$

Here $W^u(x_\lambda)$ and $W^s(y_\lambda)$ denote the unstable resp. the stable manifold of the corresponding steady states.

We do not just aim for a rough guess of the parameter value $\bar{\lambda}$ but also for a guess of the connecting orbit itself. Using these data as initial values one may employ standard techniques on the computation of hetero-/homoclinic orbits [3, 17].

The discrete dynamical system which we are considering is the time-τ map of the flow of (5.10). We approximate this map using an explicit numerical integration scheme and denote by $\mathcal{U}_j^{(k)}(\lambda)$ and $\mathcal{S}_j^{(k)}(\lambda)$ the covering of the unstable resp. the stable manifold of x_λ resp. y_λ obtained by the continuation algorithm (Section 5.4) after k subdivision and j continuation steps. Let

$$\mathcal{I}_j^{(k)}(\lambda) = \mathcal{U}_j^{(k)}(\lambda) \cap \mathcal{S}_j^{(k)}(\lambda).$$

The idea of the following algorithm is to find intersections of the box coverings $\mathcal{U}_j^{(k)}(\lambda)$ and $\mathcal{S}_j^{(k)}(\lambda)$ for different values of λ and k. Using the hierarchical data structure introduced in Section 5.3 these intersections can efficiently be computed. In fact, both coverings are stored within the same tree and boxes belonging to different coverings are marked by different flags. The boxes which belong to both coverings are then easily identified as those that are marked by both flags.

Roughly speaking we are going to use the fact that if there exists a connecting orbit for $\lambda = \bar{\lambda}$, then the smaller the distance $|\lambda - \bar{\lambda}|$ the bigger we can choose the number of subdivisions k while still finding a non-empty intersection $\mathcal{I}_j^{(k)}(\lambda)$. That is, if we plot the maximal k for which a non-empty intersection is found versus λ, we expect to see a schematic picture as illustrated in Figure 5.6.

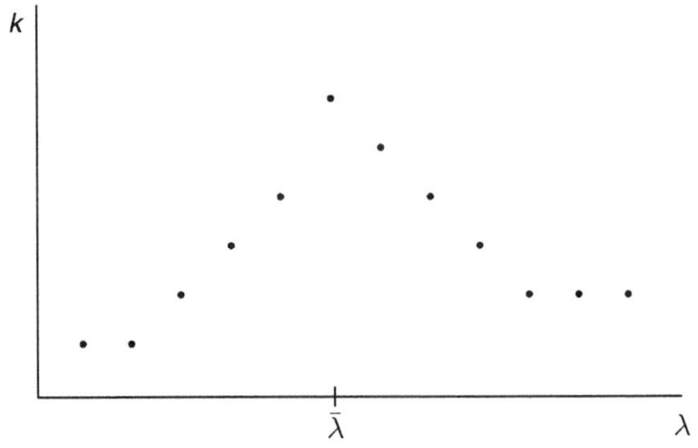

Fig. 5.6. Maximal number of subdivisions k for which an intersection of the coverings $\mathcal{U}_j^{(k)}(\lambda)$ and $\mathcal{S}_j^{(k)}(\lambda)$ has been found versus the parameter λ (schematic).

5.5.1 *The hat algorithm*

Let $\tilde{\Lambda} \subset \Lambda$ be a finite set of parameter values—e.g., a set of equidistant values of λ inside Λ. Taking $\tilde{\Lambda}$, $k_{\max} \in \mathbb{N}$ and $j_{\max} \in \mathbb{N}$ as inputs the following algorithm computes a function $m : \tilde{\Lambda} \to \mathbb{N}$ such that local maximizers of m are close to parameter values $\bar{\lambda}$ for which there may exist a connecting orbit between $x_{\bar{\lambda}}$ and $y_{\bar{\lambda}}$.

m = `hat(`$\tilde{\Lambda}$, `k`$_{\max}$, `j`$_{\max}$`)`
`for all` $\lambda \in \tilde{\Lambda}$
    k`:=0`
    `do`
        j`:=0`
        `do`
            `compute` $\mathcal{U}_j^{(k)}(\lambda)$ `and` $\mathcal{S}_j^{(k)}(\lambda)$
            j`:=`j`+1`
        `while` $\mathcal{I}_j^{(k)}(\lambda) = \emptyset$ `and` $k < j_{\max}$
        k`:=`k`+1`
    `while` $\mathcal{I}_j^{(k)}(\lambda) \neq \emptyset$ `and` $k < j_{\max}$
    $m(\lambda) = k$
`end`

In the case where $x_\lambda = y_\lambda$ one obviously has a non-empty intersection of $\mathcal{U}_j^{(k)}(\lambda)$ and $\mathcal{S}_j^{(k)}(\lambda)$ for all $\lambda \in \Lambda$. In this case one needs to modify the intersection test accordingly. In practice this is done by excluding parts of the box coverings which are inside some neighborhood of the steady state solution x_λ.

We now show that the hat algorithm can indeed be used for the detection of non-degenerate heteroclinic co-dimension one bifurcations [20]. Let ϕ^t denote the flow of the system (5.10). Then we define for an equilibrium p and $T \geq 0$

$$W_T^u(p) = \bigcup_{0 \leq t \leq T} \phi^t(W_{\mathrm{loc}}^u(p)), \qquad W_T^s(p) = \bigcup_{-T \leq t \leq 0} \phi^t(W_{\mathrm{loc}}^s(p)),$$

where $W^{u,s}_{\text{loc}}(p)$ are local (un)stable manifolds of p. If there exists an orbit connecting x_λ and y_λ for $\lambda = \bar{\lambda}$ then there is a $T \geq 0$ such that

$$W^u_T(x_{\bar{\lambda}}) \cap W^s_T(y_{\bar{\lambda}}) \neq \emptyset.$$

(In practice, the minimal T with this property is unknown and this is the reason why one should choose a rather large value for the parameter j_{max} in the input of the hat algorithm.) Accordingly the intersection $\mathcal{J}^{(k)}_j(\bar{\lambda})$ will be non-empty for all box coverings $\mathcal{U}^{(k)}_j(\bar{\lambda})$ and $\mathcal{S}^{(k)}_j(\bar{\lambda})$ satisfying

$$W^u_T(x_{\bar{\lambda}}) \subset \bigcup_{B \in \mathcal{U}^{(k)}_j(\bar{\lambda})} B \quad \text{and} \quad W^s_T(x_{\bar{\lambda}}) \subset \bigcup_{B \in \mathcal{S}^{(k)}_j(\bar{\lambda})} B.$$

In fact since there exists a connecting orbit for $\lambda = \bar{\lambda}$, $\mathcal{J}^{(k)}_j(\bar{\lambda})$ will be non-empty for all k if j is big enough. But also the converse is true:

Proposition 5.5.1. *[12]. If for some $\bar{\lambda} \in \tilde{\Lambda}$ and $j \in \mathbb{N}$ the intersection $\mathcal{J}^{(k)}_j(\bar{\lambda})$ is non-empty for all k, then there exists an orbit of (5.10) connecting $x_{\bar{\lambda}}$ and $y_{\bar{\lambda}}$.*

For the statement of the following result it is convenient to introduce a specific choice for the set $\tilde{\Lambda}$. If $\Lambda = [a, b]$ then we define for $n \in \mathbb{N}$

$$h = \frac{b-a}{n} \quad \textit{and} \quad \tilde{\Lambda}_h = \{a + ih : i = 0, 1, \ldots, n\}.$$

Proposition 5.5.2. *Suppose that for some $\bar{\lambda} \in \Lambda$ the system (5.10) undergoes a non-degenerate heteroclinic co-dimension one bifurcation with respect to the steady state solutions x_λ and y_λ. Then for each integer $k_{max} > 0$ there are $h > 0$ and $j_{max} > 0$ such that those $\tilde{\lambda} \in \tilde{\Lambda}_h$ for which $|\tilde{\lambda} - \bar{\lambda}|$ is minimal satisfy $m(\tilde{\lambda}) = k_{max}$. These values are in particular local maximizers of $m : \tilde{\Lambda} \to \mathbb{N}$. (Here m denotes the function computed by the hat algorithm.)*

Proof. Suppose that j_{max} is chosen in such a way that $\mathcal{J}^{(k)}_j(\bar{\lambda})$ is non-empty for all k and $j = j_{max} - 1$. Then, by construction of the hat algorithm, $m(\bar{\lambda}) = k_{max}$. Since $W^u_T(x_\lambda)$ and $W^s_T(y_\lambda)$ depend continuously on λ we can conclude that there is an $\eta > 0$ such that $m(\lambda) = k_{max}$ for all $\lambda \in (\bar{\lambda} - \eta, \bar{\lambda} + \eta)$. Now choose $h > 0$ so small that $\tilde{\Lambda}_h \cap (\bar{\lambda} - \eta, \bar{\lambda} + \eta) \neq \emptyset$. □

5.5.2 Numerical examples

Consider the system [4]

$$\begin{aligned} \dot{x} &= y \\ \dot{y} &= -\lambda y + x^3 + x^2 - 3x + 1. \end{aligned} \tag{5.11}$$

Our aim is to detect an orbit within the region $[-3, 2] \times [-3, 3]$ which connects the equilibrium $x_\lambda = (-1 - \sqrt{2}, 0)$ to the equilibrium $y_\lambda = (1, 0)$. We use the time-0.2-map of

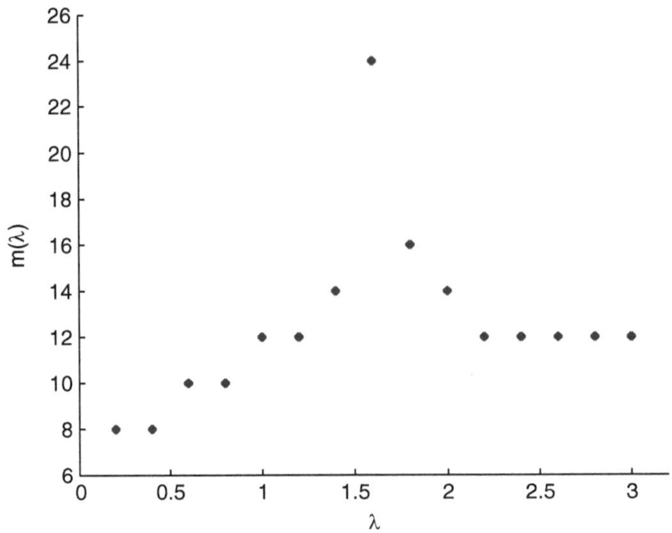

Fig. 5.7. The function m for system (5.11).

the corresponding flow and choose $\tilde{\Lambda} = \{0.2, 0.4, 0.6, \ldots, 2.8, 3.0\}$, as well as $k_{max} = 26$ and $j_{max} = 20$ as inputs for the hat algorithm. Figure 5.7 shows the graph of the resulting function m. It indicates that for $\lambda \approx 1.6$ there indeed exists a connecting orbit.

In order to illustrate the behavior of the hat algorithm we additionally show the box coverings obtained (a) for a fixed parameter value $\lambda = 1.6$ and for different numbers of subdivisions k (Figure 5.8) and (b) for a fixed depth $k = 14$ and different parameter values λ (Figure 5.9).

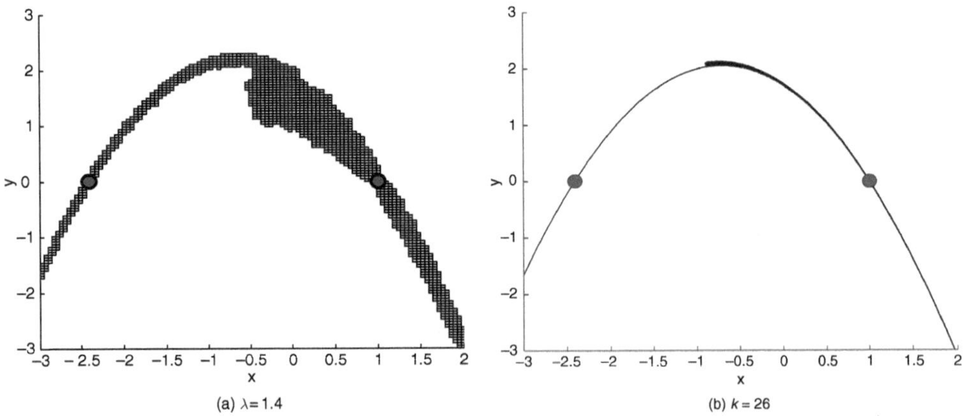

Fig. 5.8. The coverings $\mathcal{U}^{(k)}(1.6)$ and $\mathcal{S}^{(k)}(1.6)$ for different k. The two equilibria are marked by dots.

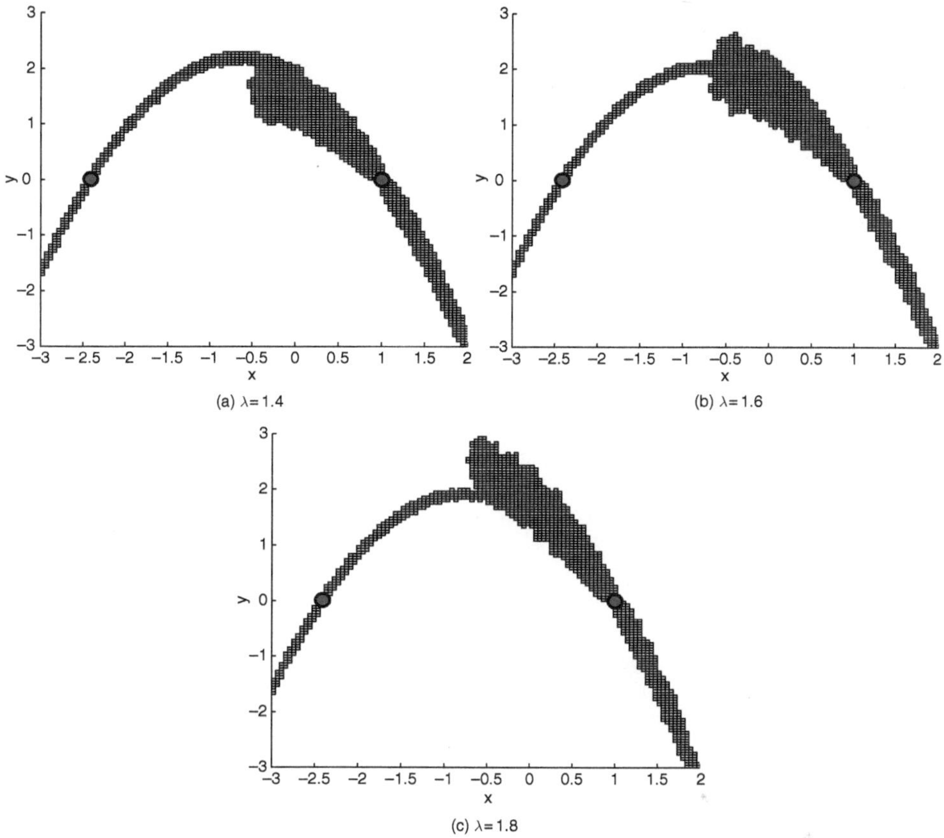

Fig. 5.9. The coverings $\mathcal{U}^{(14)}(\lambda)$ and $\mathcal{S}^{(14)}(\lambda)$ in dependence on λ.

5.5.3 The Lorenz system

As a second example we consider the Lorenz system

$$\dot{x} = \sigma(y - x)$$
$$\dot{y} = \rho x - y - xz$$
$$\dot{z} = xy - \beta z$$

with parameter values $\sigma = 10$ and $\beta = \frac{8}{3}$. It is well known that there exists a homoclinic orbit for the origin near $\rho = 13.93$ [36]. We consider the time-T map of the corresponding flow with $T = 0.2$ and apply the hat algorithm with $\tilde{\Lambda} = \{7, 8, 9, \ldots, 19, 20\}$, $k_{max} = 30$ and $j_{max} = 8$. Figure 5.10 shows the result of this computation as well as the number of boxes in the intersection of $\mathcal{U}^{(27)}(\rho)$ and $\mathcal{S}^{(27)}(\rho)$ for $\rho \in \{13, 13.5, 14, 14.5, 15\}$. Again we illustrate the computations by plotting several computed coverings, see Figures 5.11 and 5.12.

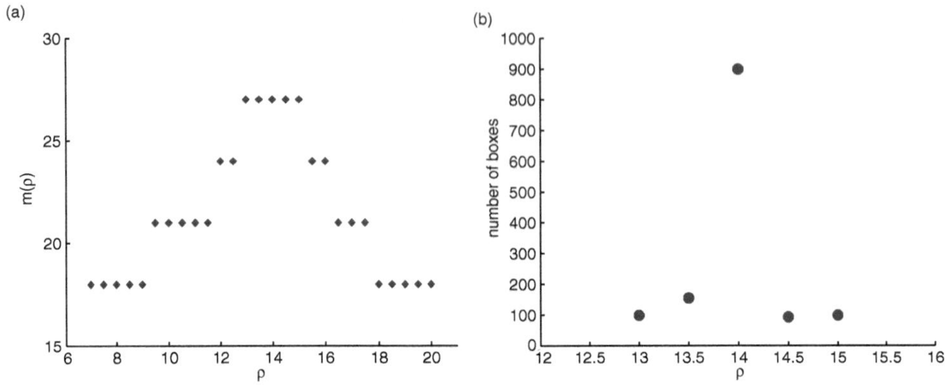

Fig. 5.10. Results for the Lorenz system. Left: The function m. Right: The number of boxes in $\mathcal{J}^{(27)}(\rho)$ as a function of ρ.

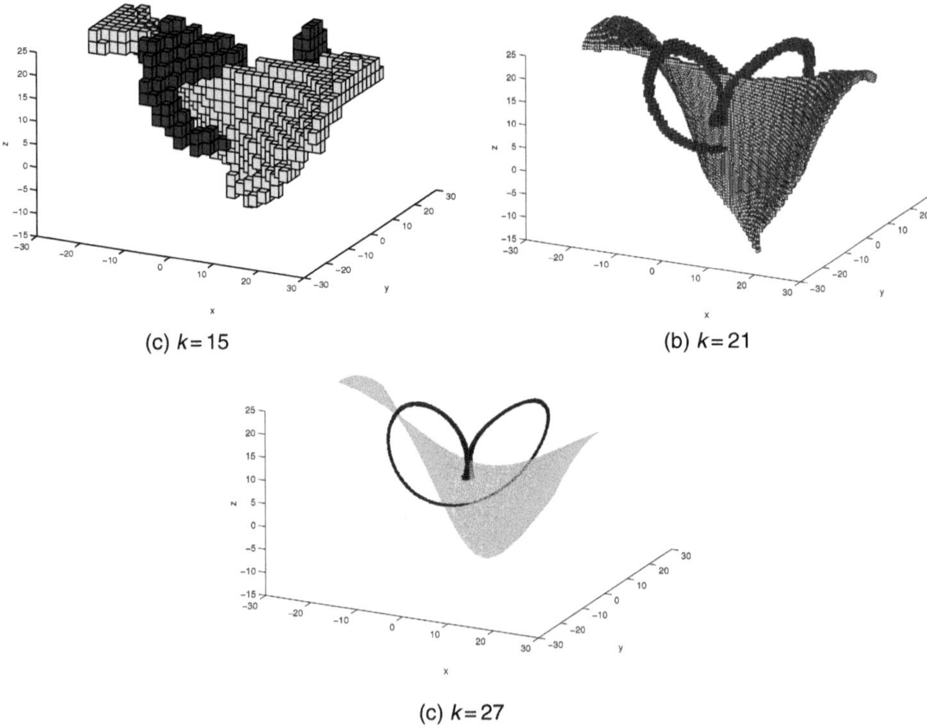

Fig. 5.11. Lorenz system: The coverings $\mathcal{U}_{(k)}(14)$ (blue) and $\mathcal{S}_{(k)}(14)$ (yellow) for different k. (see Color plate 2)

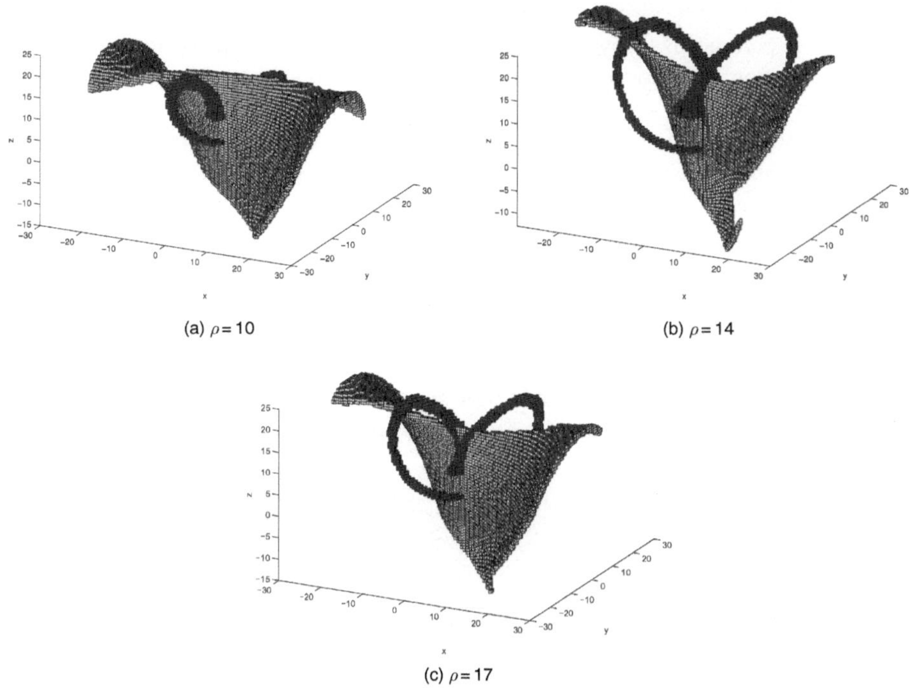

Fig. 5.12. Lorenz system: The coverings $\mathcal{U}^{(21)}(\rho)$ (blue) and $\mathcal{S}^{(21)}(\rho)$ (yellow) in dependence of ρ. (see Color plate 3)

5.6 Extension to controlled systems

The "patched 3-body approach" for the construction of energy efficient trajectories as sketched in Section 5.2 is tailored for spacecraft with impulsive thrusters. In this section, we propose an extension of this approach to the case of continuously controlled spacecraft (as realized by certain low thrust propulsion systems). Roughly speaking, the stable and unstable manifold tubes are replaced by certain (forward and backward) reachable sets in phase space. Using set oriented numerical tools it is possible to efficiently compute coverings of these sets as well as of the intersection of them with suitably chosen cross sections. We illustrate the approach by considering a low thrust mission to Venus.

5.6.1 A controlled three-body problem

In current mission concepts, like for the ESA interplanetary mission *BepiColombo* to Mercury or the *Smart I* mission, ion propulsion systems are being used that continuously exert a small force on the spacecraft ("low-thrust propulsion"). In order to model the motion of these spacecraft, we amend the planar circular restricted three-body problem (5.1) by a suitably defined control term. We restrict our considerations to the special case of a control force whose direction is defined by the spacecraft's velocity such that the control term is parametrized by a single real value u, determining the magnitude of the control acceleration. We do not take into account here that the mass of the spacecraft changes during its flight.

The velocity vector of the spacecraft has to be viewed with respect to the inertial coordinate system and not the rotating one. In view of this, one is lead to the following control system (Figure 5.13):

$$\ddot{x}+2\dot{x}^{\perp} = \nabla\Omega(x)+u\frac{\dot{x}+\omega x^{\perp}}{\|\dot{x}+\omega x^{\perp}\|}. \tag{5.12}$$

Here, $u = u(t) \in [u_{\min}, u_{\max}] \subset \mathbb{R}$ denotes the magnitude of the control force, $x = (x_1, x_2)$, $x^{\perp} = (-x_2, x_1)$ and ω is the common angular velocity of the primaries.

In a mission to Venus the spacecraft will get closer to the Sun, meaning that part of its potential energy with respect to the Sun will be transformed into kinetic energy. As a consequence, the spacecraft's velocity will have to be reduced during its flight such that it matches the one of Venus. Thus, in our concrete application the control values u will actually be negative.

5.6.2 Coupling controlled three-body problems

Obviously, every solution of (5.1) is also a solution of (5.12) for the control function $u \equiv 0$. We are going to exploit this fact in order to generalize the patched three-body approach as described in Section 5.2.2 to the case of controlled three-body problems. We are still going to use the L_1- and L_2-Lyapunov orbits as "gateways" for the transition between the interior, the planet, and the exterior regions. However, instead of computing the relevant invariant manifolds of these periodic orbits, we compute certain *reachable sets* [5], i.e., sets in phase space that can be accessed by the spacecraft when employing a certain control function.

5.6.2.1 Reachable sets

We denote by $\phi(t, z, u)$ the solution of the control system (5.12) for a given initial point $z = (x, \dot{x})$ in phase space and a given admissible control function $u \in \mathcal{U} = \{u : \mathbb{R} \to$

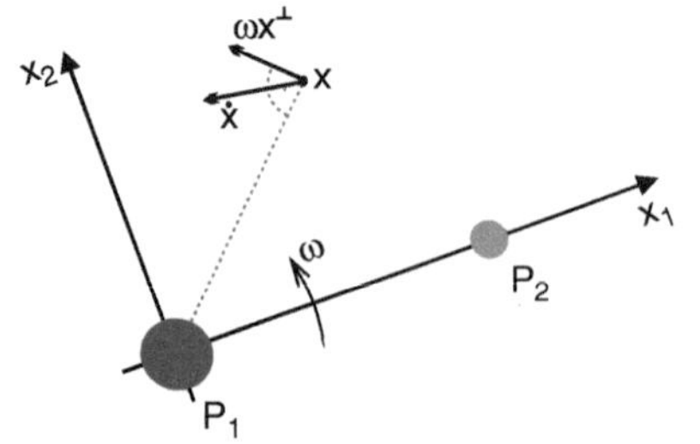

Fig. 5.13. The velocity of the spacecraft with respect to the inertial frame is given by $\dot{x}+\omega x^{\perp}$.

$[u_{\min}, u_{\max}]$, u admissible$\}$. The set of admissible control functions will be determined by the design of the thrusters. For example, it could consist of piecewise constant functions, where the minimal length of an interval on which the function is constant is determined by how fast the magnitude of the accelerating force can be changed within the thrusters.

For a set S in phase space and a given function $\tau : S \times \mathcal{U} \to \mathbb{R}$, we call $\mathcal{R}(S, \tau) = \{\phi(\tau(x, u), x, u) \mid u \in \mathcal{U}, x \in S\}$ the set which is (τ-) *reachable from S*.

5.6.2.2 Patched controlled three-body systems

We now extend the patched three-body approach to the context of a controlled system. Roughly speaking, the extension can be summarized as follows: For two suitable subsets $\mathcal{O}_1$ and $\mathcal{O}_2$ in phase space (typically two sets in the vicinity of an L_1–Lyapunov orbit of the outer planet and an L_2–Lyapunov orbit of the inner one, respectively) one computes associated reachable sets $\mathcal{R}_1 \subset \Sigma_1$ and $\mathcal{R}_2 \subset \Sigma_2$ within suitably chosen intersection planes Σ_1 and Σ_2 in each system. After a transformation of one of these reachable sets into the other rotating system, one determines their intersection. By construction, points in this intersection define trajectories that link the two "gateway sets" $\mathcal{O}_1$ and $\mathcal{O}_2$.

More precisely, the procedure is as follows:

1. Identify suitable sets $\mathcal{O}_1$ and $\mathcal{O}_2$ in the phase space of the two three-body problems, respectively. They should be chosen such that all points in $\mathcal{O}_1$ belong to trajectories that transit from the planet region into the interior region (of the outer planet) and those in $\mathcal{O}_2$ transit from the exterior region to the planet region (of the inner planet). Furthermore, in each of the two three-body problems, choose an intersection plane $\Sigma_i = \{\theta = \theta_i\}$, $i = 1, 2$, (where (r, θ) are polar coordinates for the position of the spacecraft and θ_i is a suitable angle, see also step 3). Typically the sets $\mathcal{O}_1$ and $\mathcal{O}_2$ will lie close to certain Lyapunov orbits.
2. For points $x_1 \in \mathcal{O}_1$ and $x_2 \in \mathcal{O}_2$ and an admissible control function u, let

$$\tau_1(x_1, u) = \inf\{t > 0 \mid \phi(t, x_1, u) \in \Sigma_1\} \quad \text{and}$$
$$\tau_2(x_2, u) = \sup\{t < 0 \mid \phi(t, x_2, u) \in \Sigma_2\}.$$

   For $i = 1, 2$, compute

$$\mathcal{R}(\mathcal{O}_i, \tau_i) = \{\phi(\tau_i(x, u), x, u) \mid x \in \mathcal{O}_i, u \in \mathcal{U}\} \subset \Sigma_i. \tag{5.13}$$

3. In order to transform one of the reachable sets $\mathcal{R}(\mathcal{O}_1, \tau_1)$ or $\mathcal{R}(\mathcal{O}_2, \tau_2)$ into the other rotating frame, let $\theta(t)$ be the phase angle between the two planets as seen in the rotating frame of the inner planet. Choose a time t_0 such that $\theta(t_0) = \theta_1 - \theta_2$. (Alternatively, based on a prescribed time t_0, one could choose the section angles θ_1 and θ_2 in step 1 such that $\theta(t_0) = \theta_1 - \theta_2$.) Using t_0, transform $\mathcal{R}(\mathcal{O}_1, \tau_1)$ into the rotating frame of the inner planet, yielding the set $\hat{\mathcal{R}}(\mathcal{O}_1, \tau_1) \subset \Sigma_2$. Note that here we exploit the fact that both systems are autonomous.
4. Compute the intersection (see below)

$$\hat{\mathcal{R}}(\mathcal{O}_1, \tau_1) \cap \mathcal{R}(\mathcal{O}_2, \tau_2) \subset \Sigma_2. \tag{5.14}$$

   By construction, for each point $x \in \hat{\mathcal{R}}(\mathcal{O}_1, \tau_1) \cap \mathcal{R}(\mathcal{O}_2, \tau_2)$, there exist admissible control functions u_1 and u_2 and times $t_1 = -\tau(\tilde{x}, u_1)$, $t_2 = -\tau(\bar{x}, u_2)$, such that $\phi(t_1, \tilde{x}, u_1) \in \mathcal{O}_1$

and $\phi(t_2, x, u_2) \in \mathcal{O}_2$, where $\tilde{x}$ are the coordinates of x with respect to the rotating frame of the outer planet at the phase angle $\theta(t_0)$ between the two planets. By construction of the sets $\mathcal{O}_1$ and $\mathcal{O}_2$ we have thus found a controlled trajectory that transits from the outer planet region into the inner planet region.

5.6.2.3 Implementation

Let $\mathcal{P}$ be a finite partition of some relevant bounded part of Σ_2. We compute coverings $\mathcal{P}_1, \mathcal{P}_2 \subset \mathcal{P}$ of $\hat{\mathcal{R}}(\mathcal{O}_1, \tau_1)$ and $\mathcal{R}(\mathcal{O}_2, \tau_2)$ by integrating a finite set of test points in $\mathcal{O}_1$ and $\mathcal{O}_2$, respectively. In the example computations, we restrict ourselves to constant control functions with values in a grid in $[u_{\min}, u_{\max}]$. For each of these values and each test point, we numerically integrate the control system (5.12) by an embedded Runge-Kutta scheme with adaptive stepsize control as implemented in the code DOP853 by Hairer and Wanner [21]. After each integration step, we check whether the computed trajectory has crossed the intersection plane Σ_1 resp. Σ_2 and, if this is the case, start Newton's method in order to obtain a point in the intersection plane. In the case of $\hat{\mathcal{R}}(\mathcal{O}_1, \tau_1)$, we transform this point into the other system and, in both cases, add the corresponding partition element to the collection $\mathcal{P}_1$ resp. $\mathcal{P}_2$.

Analogously to the approach in Section 5.5, we then extract all partition elements P from $\mathcal{P}$ which belong to both coverings. In each element, we furthermore store the minimal Δv that is necessary to reach this set from either $\mathcal{O}_1$ or $\mathcal{O}_2$. Whenever the intersection $\mathcal{P}_1 \cap \mathcal{P}_2$ consists of more than one partition element, this enables us to choose trajectories with a minimal Δv (with respect to the chosen parameters).

5.6.2.4 Application for a mission to venus

We now apply the approach described in the previous paragraphs to the design of a mission to Venus. In 2005, the European Space Agency has launched *VenusExpress*[3], a mission to Venus that sends a *MarsExpress*-like spacecraft into an elliptical orbit around Venus via a Hohmann transfer. Transfer time from Earth is around 150 days, while the required Δv amounts to roughly 1500 m/s [1, 18]. The interplanetary low-thrust orbit that we are going to construct now corresponds to a flight time of roughly 1.4 years, applying a Δv of approximately 3300 m/s. Since typical low-thrust propulsion systems (as in the ESA mission *Smart I* and the planned cornerstone mission *BepiColombo* for example) have a specific impulse which is approximately an order of magnitude larger than the one of chemical engines, these figures amount to a dramatic decrease in the amount of on-board fuel: at the expense of roughly the threefold flight time the weight of the fuel can be reduced to at least 1/3 of what is used for *VenusExpress*.

5.6.2.5 Computational details

We are now going to comment on the specific details of the computation for the Earth–Venus transfer trajectory. We are considering the two three-body systems

[3] http://sci.esa.int/science-e/www/area/index.cfm?fareaid=64

Sun–Earth–Spacecraft and Sun–Venus–Spacecraft with μ-values of $\mu_{SE} = 3.04041307864 \times 10^{-6}$ and $\mu_{SV} = 2.44770642702 \times 10^{-6}$, respectively.

1. For the construction of the 'gateway set' $\mathcal{O}_1$ we consider the L_1-Lyapunov orbit $\mathcal{L}_1$ associated with the value $C_1 = 3.0005$ of the Jacobi integral in the Sun–Earth system. This value results from experimenting with several different values and eventually bears further optimzation potential. We compute the intersection A_1 of its interior local unstable manifold (i.e., the piece of its local unstable manifold that extends into the interior region) with the section $\Sigma = \{x_1 = 0.98\}$ in the given energy surface $\{C = C_1\}$. Let $\bar{A}_1$ denote the points that are enclosed by the closed curve A_1 in this two-dimensional surface. We set

$$\mathcal{O}_1 = \mathcal{L}_1 \cup (\bar{A}_1 \backslash A_1).$$

Analogoulsy, we compute A_2, $\bar{A}_2$ and $\mathcal{O}_2$ in the Sun–Venus system, using again a value of $C_2 = 3.0005$ for the Jacobi integral. As intersection planes we choose $\Sigma_1 = \Sigma_2 = \{\theta = \frac{\pi}{4}\}$, since this turned out to yield the good compromise between transfer time and Δv.

2. We have been using constant control functions only, employing 800 mN as an upper bound for the maximal thrust. This bound is in accordance with the capabilities of the thrusters that are planned to be used in connection with the *BepiColombo* mission. Here we assumed a mass of 4000 kg for the spacecraft.

3./4. Figure 5.14 shows coverings of the sets $\hat{\mathcal{R}}(\mathcal{O}_1, \tau_1)$ (red) and $\mathcal{R}(\mathcal{O}_2, \tau_2)$ (blue), as well as a covering of their intersection (yellow), projected onto the $(x_1, \dot{x}_1)$-plane. The associated optimal trajectory (i.e., the one with a minimal combined Δv for both

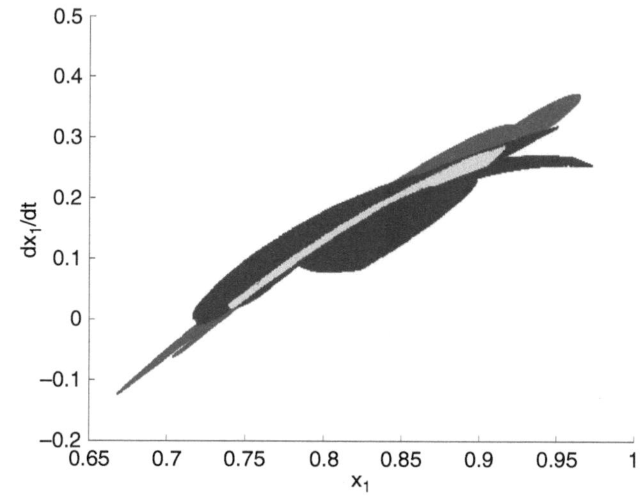

Fig. 5.14. Intersection of two reachable sets in a common intersection plane. Red: reachable set of the gateway set of Earth, blue: reachable set of the gateway set of Venus; yellow: their intersection. Shown is a projection of the covering in three space onto the $(x_1, \dot{x}_1)$-plane (normalized units). (see Color plate 4)

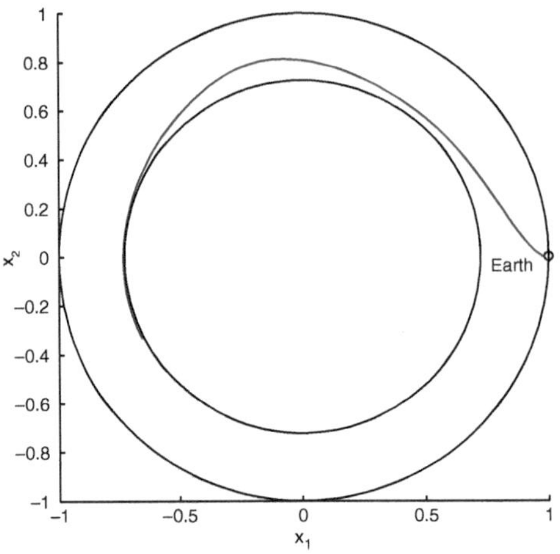

(a) Rotating frame of Earth, normalized units

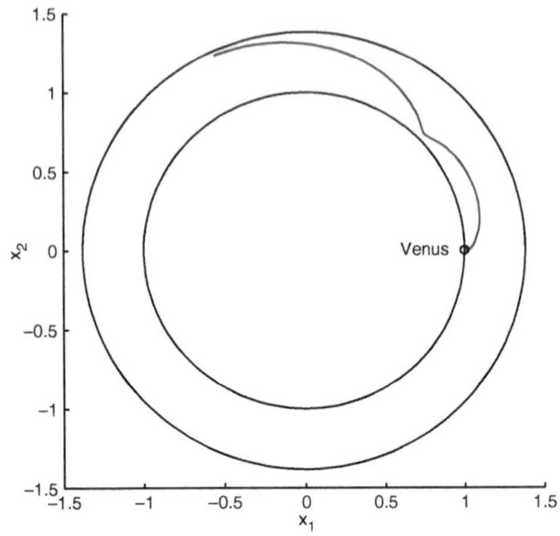

(b) Rotating frame of Venus, normalized units

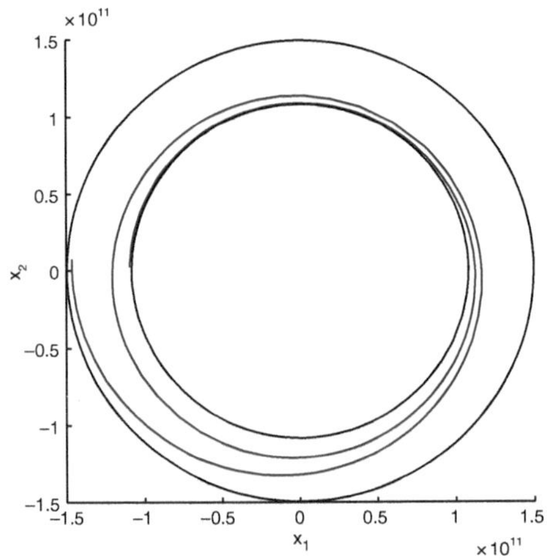

(c) Inertial frame, units in *km*

Fig. 5.15. Approximate interplanetary trajectory, joining the gateway sets $\mathcal{O}_1$ (near the Sun–Earth L_1) and $\mathcal{O}_2$ (near the Sun–Venus L_2).

pieces of the trajectory) is shown in Figure 5.15 as seen in the inertial frame as well as in both rotating frames. It requires a (constant) control force of $u_1 = -651$ mN in the first phase (i.e., while travelling from $\mathcal{O}_1$ to Σ_1) and of $u_2 = -96$ mN in the second phase. The corresponding flight times are $|\tau_1| = 0.51$ and $|\tau_2| = 0.92$ years, amounting to a total Δv of approximately 3300 m/s. We note that there still exists a discontinuity in the computed trajectory when switching from the first to the second phase. This is due to the fact that the two pieces of the trajectory are only forced to end in the same box in the intersection plane. However, the radii of the boxes are rather small, namely roughly 10 000 km in position space and $\approx$35 m/s in the velocity coordinates. This is why we expect the computed trajectory to be a very good initial guess for a standard local solver (e.g., a collocation or multiple shooting approach, see Refs.[15, 16, 37, 39]) for a suitably formulated optimal control problem. In fact, we used the computed trajectory as an initial guess for the solution of a four-body model by a recently developed variational approach to the computation of optimal open loop controls [24].

5.6.2.6 The complete journey

In Ref. [11] we complement the above example computation by computing transfer trajectories between the gateway sets $\mathcal{O}_1, \mathcal{O}_2$ and the corresponding planets. We end up with a flight time of roughly 1.8 years and a corresponding Δv of slightly less than 4000 m/s for the complete journey from Earth to Venus.

5.7 Conclusion

Set oriented methods provide a robust framework for the numerical solution of problems in space mission design. Due to the fact that outer approximations of the objects of interest are computed, these methods in particular enable a reliable detection of certain energy efficient trajectories like connecting orbits under the gravitational dynamics or in the low-thrust setting. A particular inherent advantage of this approach is the ability to systematically take into account the propagation of errors as well as the effects of uncertainty—a feature which only has been touched upon in current work and which will be explored in future investigations.

As mentioned above, the numerical effort of our approach crucially depends on the dimension of the object that is being approximated (rather than on the dimension of the underlying phase space). Correspondingly, while it is rather straightforward to treat the case of the full three-dimensional configuration space in Section 5.6, it will be a challenging task to incorporate a fully actuated (i.e., 3D) control, since the dimension of the reachable sets which are being computed will typically be increased.

References

1. Venus express mission definition report. (2001). *European Space Agency, ESA-SCI*, **6**.
2. Abraham, R. and Marsden, J.E. (1978). *Foundations of Mechanics*. Second Edition, Addison-Wesley.
3. Beyn, W.-J. (1990). The numerical computation of connecting orbits in dynamical systems. *IMA Journal of Numerical Analysis*, **9**, pp. 379–405.
4. Beyn, W.-J. (1990). The numerical computation of connecting orbits in dynamical systems. *IMA J. Numer. Anal.*, **10**(3), pp. 379–405.
5. Colonius, F. and Kliemann, W. (2000). The dynamics of control. Systems & Control: Foundations & Applications. Birkhäuser Boston Inc., Boston, MA, 2000.
6. Dellnitz, M., Froyland, G. and Junge, O. (2001). The algorithms behind GAIO-set oriented numerical methods for dynamical systems. In *Ergodic Theory, Analysis, and Efficient Simulation of Dynamical Systems*, pages 145–174, 805–807. Springer, Berlin.

7. Dellnitz, M. and Hohmann, A. (1996). The computation of unstable manifolds using subdivision and continuation. In H.W. Broer, S.A. van Gils, I. Hoveijn, and F. Takens, editors, *Nonlinear Dynamical Systems and Chaos*, pages 449–459. Birkhäuser, *PNLDE* 19.
8. Dellnitz, M. and Hohmann, A. (1997). A subdivision algorithm for the computation of unstable manifolds and global attractors. *Numerische Mathematik*, **75**, pp. 293–317.
9. Dellnitz, M. and Junge, O. (1999). On the approximation of complicated dynamical behavior. *SIAM Journal on Numerical Analysis*, **36**(2), pp. 491–515.
10. Dellnitz, M. and Junge, O. (2002). Set oriented numerical methods for dynamical systems. *In Handbook of Dynamical Systems*, **2**, pp. 221–264. North Holland, Amsterdam.
11. Dellnitz, M., Junge, O., Post, M. and Thiere, B. (2006). On target for Venus—set oriented computation of energy efficient low thrust trajectories. Celestial Mechanics and Dynamical Astronomy, to appear.
12. Dellnitz, M., Junge, O., Post, M. and Thiere, B. (2001). The numerical detection of connecting orbits. *Discrete Contin. Dyn. Syst. Ser. B*, **1**(1), pp. 125–135.
13. Dellnitz, M., Schütze, O. and Hestermeyer, T. (2005). Covering Pareto sets by multilevel subdivision techniques. *JOTA*, **24**(1), pp. 113–136.
14. Dellnitz, M., Schütze, O. and Sertl, St. (2002). Finding zeros by multilevel subdivision techniques. *IMA J. Numer. Anal.*, **22**(2), pp. 167–185.
15. Deuflhard, P. and Bornemann, F. (2002). *Scientific Computing with Ordinary Differential Equations, Texts in Applied Mathematics*, **42**, Springer-Verlag, New York.
16. Deuflhard, P., Pesch, H.-J. and Rentrop, P. (1976). A modified continuation method for the numerical solution of nonlinear two-point boundary value problems by shooting techniques. *Numer. Math.*, **26**(3), pp. 327–343.
17. Doedel, E.J. and Friedman, M.J. (1989). Numerical computation of heteroclinic orbits. *Journal of Computational and Applied Mathematics*, **26**, pp. 155–170.
18. Fabrega, J., Schirmann, T., Schmidt, R. and McCoy, D. (2003). Venus express: The first European mission to Venus. *International Astronautical Congress*, IAC-03-Q.2.06, pp. 1–11.
19. Grüne, L. and Junge, O. (2005). A set oriented approach to optimal feedback stabilization. *Systems Control Lett.*, **54**(2), pp. 169–180.
20. Guckenheimer, J. and Holmes, Ph. (1983). *Nonlinear Oscillations, Dynamical Systems, and Bifurcations of Vector Fields*. Springer.
21. Hairer, E., Nørsett, S.P. and Wanner, G. (1993). *Solving Ordinary Differential Equations. I, Springer Series in Computational Mathematics*, **8**, Springer-Verlag, Berlin, second edition, Nonstiff problems.
22. Hsu, H. (1992). Global analysis by cell mapping. *Int. J. Bif. Chaos*, **2**, pp. 727–771.
23. Junge, O. (1999). *Mengenorientierte Methoden zur numerischen Analyse dynamischer Systeme*. PhD thesis, University of Paderborn.
24. Junge, O., Marsden, J.E. and Ober-Blöbaum, S. (2005). Discrete mechanics and optimal control. *Proceedings of the 16th IFAC World Congress (electronic)*.
25. Junge, O. and Osinga, H.M. (2004). A set oriented approach to global optimal control. *ESAIM Control Optim. Calc. Var.*, **10**(2), pp. 259–270 (electronic).
26. Koon, W.S., Lo, M.W., Marsden, J.E. and Ross, S.D. (2000). Shoot the moon. *AAS/AIAA Astrodynamics Specialist Conference, Florida*, **105**, pp. 1017–1030.
27. Koon, W.S., Lo, M.W., Marsden, J.E. and Ross, S.D. (2002). Constructing a low energy transfer between jovian moons. *Contemporary Mathematics*, **292**, pp. 129–145.
28. Kreuzer, E. (1987). *Numerische Untersuchung nichtlinearer dynamischer Systeme*. Springer.
29. Lo, M.W., Williams, B.G., Bollman, W.E., Han, D., Hahn, Y., Bell, J.L., Hirst, E.A., Corwin, R.A., Hong, P.E., Howell, K.C., Barden, B. and Wilson, R. (2001). Genesis mission design. *Journal of Astronautical Sciences*, **49**, pp. 169–184.
30. McGehee, R.P. (1969). *Some homoclinic orbits for the restricted 3-body problem*. PhD thesis, University of Wisconsin.
31. Meyer, K.R. and Hall, R. (1992). *Hamiltonian Mechanics and the n-body Problem*. Springer-Verlag, Applied Mathematical Sciences.
32. Lo, M.W., Williams, B.G., Bollman, W.E., Han, D.S., Hahn, Y.S., Bell, J.L., Hirst, E.A., Corwin, R.A., Hong, P.E., Howell, K.C., Barden, B. and Wilson, R. (2001). Genesis mission design. *J. Astron. Sci.*, **49**, pp. 169–184.

33. Schütze, O. (2003). A new data structure for the nondominance problem in multi-objective optimization. In *Evolutionary Multi-Criterion Optimization*, (Carlos M. Fonseca, Peter J. Fleming, Eckart Zitzler, Kalyanmoy Deb, and Lothar Thiele, eds.), *Lecture Notes in Computer Science*, **2632**, pp. 509–518. Springer.
34. Schütze, O., Mostaghim, S., Dellnitz, M. and Teich, J. (2003). Covering Pareto sets by multilevel evolutionary subdivision techniques. In *Evolutionary Multi-Criterion Optimization*, (Carlos M. Fonseca, Peter J. Fleming, Eckart Zitzler, Kalyanmoy Deb, and Lothar Thiele, eds.), *Lecture Notes in Computer Science*, **2632**, pp. 118–132. Springer.
35. Sertl, S. and Dellnitz, M. (2006). Global optimization using a dynamical systems approach. *J. Glob. Opt.*, in press.
36. Spreuer, H. and Adams, E. (1993). On the existence and the verified determination of homoclinic and heteroclinic orbits of the origin for the Lorenz equations. In *Validation Numerics*, (R. Albrecht, G. Alefeld, and H.J. Stetter, eds.), pp. 233–246. Springer.
37. Stoer, J. and Bulirsch, R. (2002). *Introduction to numerical analysis*, **12**, Springer-Verlag, New York, 2002.
38. Szebehely, V. (1967). *Theory of Orbits—The Restricted Problem of Three Bodies*. Academic Press.
39. von Stryk, O. (1993). Numerical solution of optimal control problems by direct collocation. In *Optimal Control—Calculus of Variations, Optimal Control Theory and Numerical Methods* (R. Bulirsch, A. Miele, J. Stoer, and K.-H. Well, ed.), *Internat. Ser. Numer. Math.*, **111**, pp. 129–143. Birkhäuser, Basel.

6 Space Trajectory Optimization and L^1-Optimal Control Problems

I. MICHAEL ROSS

Department of Mechanical and Astronautical Engineering, Naval Postgraduate School, Monterey, CA 93943

Contents

6.1 Introduction

The engineering feasibility of a space mission is overwhelmingly dictated by the amount of propellant required to accomplish it. This requirement stems from the simple notion that if propellant consumption was not a prime driver, then amazing things are possible such as geostationary spacecraft in low-Earth, non-Keplerian orbits. The need for lowering the fuel consumption is so great for a space mission that optimal open-loop guidance is flown during the critical endo-atmospheric segment of launch (even for the manned Space-Shuttle) in preference to non-optimal feedback guidance. In fact, the Holy Grail of ascent guidance can be simply described as fuel-optimal feedback control [1].

The cost of fuel in space is exponentially larger than its terrestrial cost because space economics is currently driven by space transportation costs rather than the chemical composition of fuel. Recently, this simple point became more mundane when the U.S. Government was charged more than twice the peace-time market-value of gasoline due to the increased cost of transportation in a war zone [2]. That is, the cost of fuel is not just intrinsic; it is also driven by a routine of operations or the lack of it thereof. Given that space operations (access to space) are not yet routine, fuel in space continues to be extraordinarily expensive thereby dictating the feasibility of any proposed architecture.

It is worth noting that since current launch costs continue to be high, the economics of refueling an aging spacecraft need to be offset by the possibility of launching a cheaper, advanced spacecraft. As a result of this economic fact, multiple spacecraft of undetermined number need to be refueled simply to break even [3]. Thus, in the absence of an economically viable strategy for refueling, minimum-fuel maneuvers will continue to dominate the design, guidance, control, and operations of a space system.

In principle, formulating the problem of designing minimum fuel trajectories is quite simple: the rocket equation provides the necessary physics, and the problem can be formulated either as a Mayer problem (maximizing the final mass) or as an equivalent Lagrange problem [4]. In these well-documented formulations, the mass-flow-rate equation is part of the dynamical system and one needs to explicitly account for the type of fuel used in terms of the specific impulse of the propellant. Including the coupling of the propulsion system with the mechanical system makes such a problem formulation undesirable during a preliminary phase of mission analysis as it is difficult to independently evaluate the merits of a trajectory or guidance algorithm that is intimately connected to a particular engine or propellant characteristic. Thus, mission analysts frequently use the normalizing concept of the characteristic velocity [4] that is sometimes simply referred to as the total "delta-V" requirement even when impulsive maneuvers are not employed. The most obvious way to compute these delta-Vs is to take Euclidean norms. In this chapter, we show that these Euclidean norms are part of a class of L^1 cost functions and not the popular quadratic costs. As noted in Ref. [5], this point is frequently misunderstood in the literature resulting in the design of poor guidance and control algorithms that incur fuel penalties as high as 50%. On the other hand, L^1 cost functions based on absolute values have been widely considered going back as far as the 1960s; see, for example, Ref. [6]. In the language introduced in this chapter, these early L^1 cost functions can be described as l^1-variants of the L^1 norm while the correct Euclidean-based cost functions are the l^2-variants of the L^1 norm.

In an effort to clarify the above points, this chapter begins with first principles. By considering various thruster configurations and the physics of the propulsion system, we motivate a definition of l^p-variants of the L^1 norm. That is, by considering the way the engines are mounted onto the spacecraft body we naturally arrive at l^p versions of the L^1 norm of the thrust. These class of L^1 norms of the thrust directly measure fuel consumption. By extending this definition to thrust acceleration, the resulting mathematics shows a proper way to decouple the propulsion system's performance from that of the trajectory so that a correct analysis can be carried out. Although these physics-based formulations are somewhat formal, it creates apparent problems in theory and computation because the cost function is nonsmooth (i.e., the integrand is non-differentiable). Rather than employ formal nonsmooth analysis, [7, 8] we develop an alternative approach that transforms the nonsmooth problems to problems with smooth functions while maintaining the nonsmooth geometric structure. The price we pay for this approach is an increase in the number of variables and constraints. Such transformation techniques are quite rampant in analysis; that is, the exchange of an undesirable effect to a desirable one by paying an affordable price. A well-known example of this barter in spacecraft dynamics is the parameterization of $SO(3)$: a 4-vector "quaternion" in S^3 is frequently preferred over a singularity-prone employment of three Eulerian angles.

In order to demonstrate the merits of solving the apparently more difficult nonsmooth L^1 optimal control problem, we use a double-integrator example to highlight the issues, and motivate the practical importance of a Sobolev space perspective for optimal control. Case studies for the nonlinear problem of orbit transfer demonstrate the theory and computation of solving practical problems. Lest it be misconstrued that practical problems are essentially smooth, or that the nonsmooth effects can be smoothed, we briefly digress to illustrate points to the contrary. To this end, consider a modern electric-propulsion system. When the electric power to the engine, P_e, is zero, the thrust force, T, is zero. Thus, $(P_e, T) = (0, 0)$ is a feasible point in the power-thrust space; see Figure 6.1. As P_e is continuously increased, T remains zero until P_e achieves a threshold value, $P_{e,0}$. At $P_e = P_{e,0}$, the engine generates a thrust of $T = T_0 > 0$ as shown in Figure 6.1. This is the minimum non-zero value of thrust the engine can generate. Thus, the feasible values of thrust for a practical electric engine is given by the union of two disjoint sets,

$$T \in [T_0,\ T_{\max}(P_{\max})] \cup \{0\}, \tag{6.1}$$

where $T_{\max}(P_{\max})$ is the power-dependent maximum available thrust, and P_{max} is the maximum available power which may be less than the engine power capacity, $P_{e,max}$, due to housekeeping power requirements, available solar energy and a host of other real-world factors. Note that $T_{max}(P_{max}) \leq T_{sup}$ where T_{sup} is the maximum possible thrust. Thus, the practical control variable for such engines is electrical power and not thrust. In this case, the thrust force becomes part of the controlled vector field in the dynamical equations governing the spacecraft motion. Consequently, the real-world problem data is truly nonsmooth. Smoothing the data (e.g., by curve fitting) generates infeasible values of thrust [10] at worst and non-optimal controls at best—both of which are truly undesirable as already noted. Clearly, in accounting for the stringent fuel requirements of practical space missions, nonsmooth phenomena are inescapable. Thus, contrary to conventional wisdom, the more practical the problem, the more the required mathematics.

Throughout this chapter, we use the words propellant and fuel interchangeably since the differences between them are relatively irrelevant for the discussions that follow.

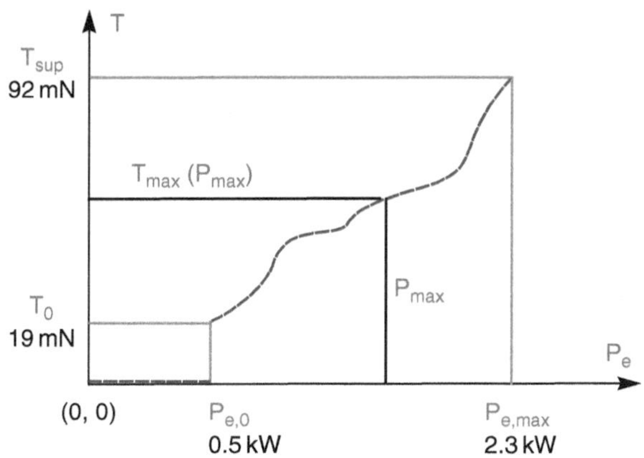

Fig. 6.1. Feasible region for a practical electric powered space propulsion system; indicated numerical values are for the NSTAR engine [9].

6.2 Geometry and the mass flow equations

Suppose that we have a single thruster that steers the spacecraft by gimbaling (see Figure 6.2 (a)). Let (x, y, z) be orthogonal body-fixed axes and $\mathbf{T} = (T_x, T_y, T_z) \in \mathbb{R}^3$ be the thrust force acting on a spacecraft. Then, the rocket equation is given by,

$$\dot{m} = -\frac{\sqrt{T_x^2 + T_y^2 + T_z^2}}{v_e} = -\frac{\|\mathbf{T}\|_2}{v_e} \tag{6.2}$$

where v_e is the exhaust speed, $\dot{m}$ is the mass-flow rate, and

$$\|\mathbf{T}\|_p := (|T_x|^p + |T_y|^p + |T_z|^p)^{1/p}$$

is the l^p-norm [11] of the thrust vector. If thrusting is achieved by six (ungimbaled) identical engines (see Figure 6.2 (b)) rigidly mounted to the body axes, then we have,

$$\dot{m} = -\frac{|T_x| + |T_y| + |T_z|}{v_e} = -\frac{\|\mathbf{T}\|_1}{v_e}. \tag{6.3}$$

If we have one main engine to perform the guidance while vernier engines are used to steer the thrust vector (as in launch vehicles, for example; see also Figure 6.2 (c)), we can write,

$$\dot{m} \simeq -\frac{\|\mathbf{T}\|_\infty}{v_e}, \tag{6.4}$$

where the approximation implies that we are ignoring the fuel consumption arising from the use of the vernier engines. Thus, the rocket equation can be unified as,

$$\dot{m} = -\frac{\|\mathbf{T}\|_p}{v_e} \qquad p = 1,\ 2 \text{ or } \infty, \tag{6.5}$$

where we have ignored the fact that this equation is an approximation for $p = \infty$. Note that p is now a design option (i.e., gimbaled single engine or multiple ungimbaled engines).

The unified rocket equation holds in other situations as well. For example, in cases when it is inconvenient to use spacecraft body axes, Eq. (6.2) can be used if (T_x, T_y, T_z) are any orthogonal components of $\mathbf{T}$. In such cases, steering must be interpreted to be

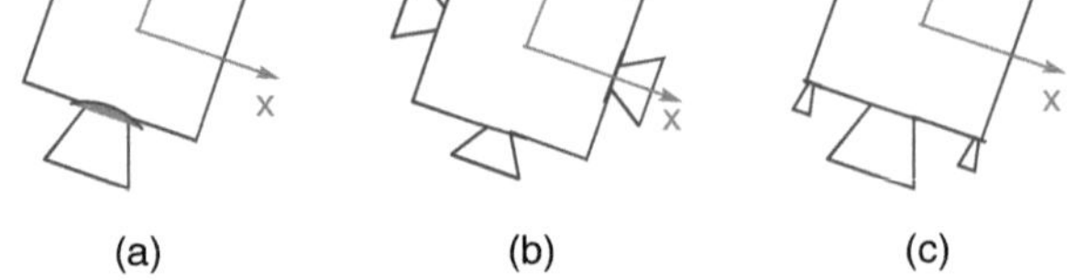

Fig. 6.2. Space vehicle thruster configurations: (a) l^2, (b) l^1, and (c) l^∞ mass flow rates; additional thrusters not shown.

provided by attitude control (with a transfer function of unity). Similarly, Eq. (6.3) can be used even when the axes are neither orthogonal nor body-fixed. The versatility of such formulations has been used quite extensively elsewhere [10, 12–15]. Finally, note that Eq. (6.5) applies whether or not the thrust region is continuous, discrete (e.g., on-off thrusters), or even disjoint as in Eq. (6.1).

In regarding T as control variable, we note that physics bounds its control authority; hence, we have $\mathbf{T} \in \mathbb{U} \subset \mathbb{R}^3$ where $\mathbb{U}$ is the control space, a compact set. Suppose that $\mathbf{T}$ can be varied continuously (i.e., $\mathbb{U}$ is a continuous set). In the l^2 mass-flow-rate configuration, a bound on the thrust implies a bound on the l^2 norm; hence the control space for this configuration is a Euclidean ball, indicated as $\mathbb{U}^2$ in Figure 6.3. On the other hand, in the l^1 mass-flow-rate configuration, bounds on the thrust generated by each thruster implies a bound on the l^∞-norm of $\mathbf{T}$. Thus, for identical engines, the control space for the l^1-configuration is the "l^∞ ball," a solid cube, denoted as $\mathbb{U}^1$, in Figure 6.3 (cutaway view).

It is instructive to look at the mass-flow rate as a region in $\mathbb{R}^3$ by associating to $|\dot{m}|$ the same direction as the net thrust force. Thus the set,

$$\mathbb{F}^p := \left\{\mathbf{F}^p \in \mathbb{R}^3 : \mathbf{F}^p = \|\mathbf{T}\|_p \left(\mathbf{T}/\|\mathbf{T}\|_2\right),\ \mathbf{T} \in \mathbb{U}^p\right\} \tag{6.6}$$

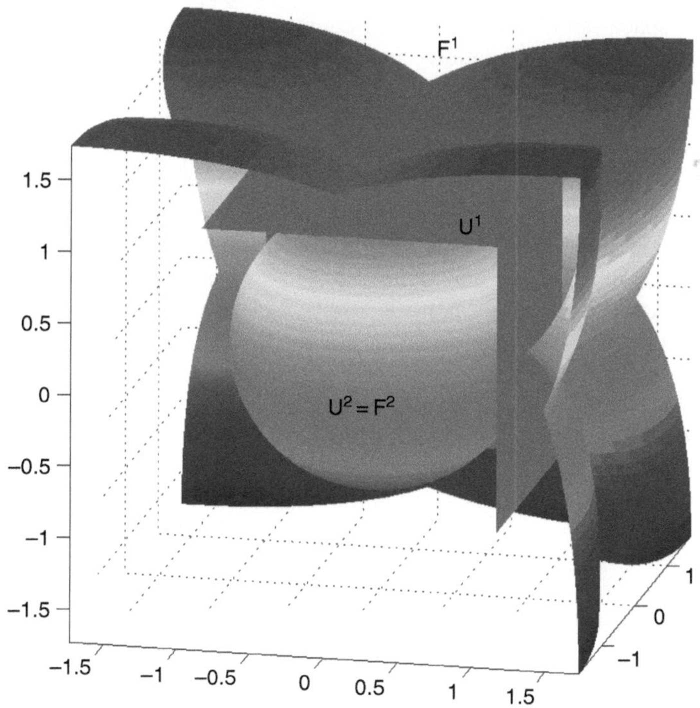

Fig. 6.3. Cutaway views of the geometries of the control space and their corresponding mass-flow rates. (see Color plate 5)

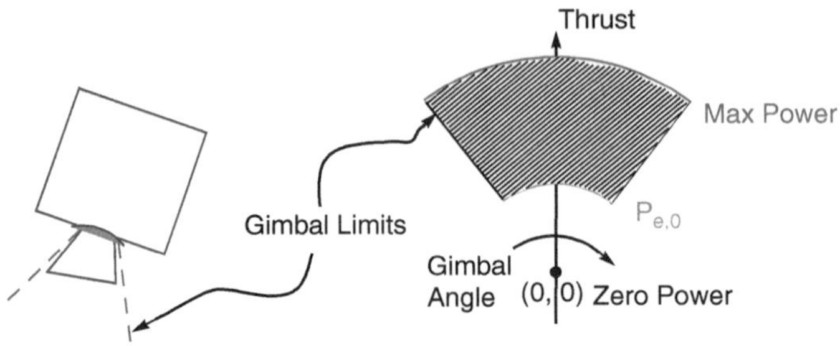

Fig. 6.4. A two-dimensional illustration of a practical control space, $\mathbb{U}$, in low-thrust trajectory optimization.

associates every net thrust firing, $\mathbf{T}$, a vector $\mathbf{F}^p$ whose Euclidean norm is the absolute value of the mass-flow-rate scaled by $-1/v_e$ (Eq. (6.5)). Clearly, we have, $\mathbb{F}^2 = \mathbb{U}^2$. On the other hand, $\mathbb{F}^1 \neq \mathbb{U}^1$. A cutaway view of the space $\mathbb{F}^1$ is the petal-shaped region shown in Figure 6.3. The mismatch between the geometries of the mass-flow-rate and the control space can generate some apparently peculiar control programs and fuel consumptions. Although not articulated in terms of geometric mismatches, it was Bilimoria and Wie [16] who first showed that a mismatch between the inertia ellipsoid (a sphere in their example), and the control space (an l^∞ ball) generates counter-intuitive time-optimal maneuvers in the sense that the rigid-body rotations are almost always not about the eigenaxis. This phenomenon was re-discovered in Ref. [17].

In practical applications, the control space, $\mathbb{U}$, can be quite different from the sets discussed above, and these characteristics can lead to quite interesting controllers. For example, in the l^2 mass-flow-rate configuration (see Figure 6.2 (a)), if the engine gimbals are limited and the propulsion is electric, then $\mathbb{U}$ is a non-convex disjoint set as illustrated in Figure 6.4 (see also Eq. (6.1)). Thus, solving practical problems requires a more careful modeling of the control space, and quite often, $\mathbb{U}$ has a complex geometric structure arising from systems' engineering considerations such as the placement of the thrusters, cant angles, and so on. Our intent here is not to document these issues but to simply note that as a result of the structure of $\mathbb{U}$, practical optimal trajectories [10, 15] can differ substantially from textbook cases [4, 18].

Although our focus here is largely thrusting space vehicles, we note that all of the preceding notions apply to air vehicles as well. This is because, for air vehicles, the mass-flow equations are the same except that one uses $c = 1/v_e$ as the thrust-specific fuel consumption parameter.

6.3 Cost functions and Lebesgue norms

Propellant consumption is simply the change in mass of the spacecraft. If v_e is a constant, then from Eq. (6.5) we have

$$m(t_0) - m(t_f) = -\int_{t_0}^{t_f} \dot{m}\, \mathrm{d}t = \frac{1}{v_e}\int_{t_0}^{t_f} \|\mathbf{T}(t)\|_p\, \mathrm{d}t, \tag{6.7}$$

where $\mathbf{T}(\cdot) : [t_0, t_f] \to \mathbf{T} \in \mathbb{R}^3$ is the thrust vector function of time. Thus, we can say that the L^1-norm of the scalar function, $[t_0, t_f] \mapsto \|\mathbf{T}\|_p \in \mathbb{R}$, is a measure of the fuel consumption, and is, in fact, equal to the propellant consumption with a proportionality factor, $1/v_e$. If v_e is not a constant, then of course $1/v_e$ must be inside the integral in Eq. (6.7) and takes the role of a weight function. Thus, in performing minimum-fuel analysis independent of the propulsion system, it is obvious from Eq. (6.7), that the proper family of cost functions is indexed by p and can be defined as,

$$J[\mathbf{T}(\cdot)] := \int_{t_0}^{t_f} \|\mathbf{T}(t)\|_p \, dt, \tag{6.8}$$

where J is the functional, $\mathbf{T}(\cdot) \mapsto \mathbb{R}$. In solving optimal control problems, it is useful to be cognizant of the space, $\mathcal{U}$, of admissible controls so that the problem formulation can be changed to search for controls in a more desirable space should the solution in a particular formulation turn out to be less than desirable. As Pontryagin et al. [19] note, $\mathcal{U}$ is frequently taken to be the (incomplete) space of piecewise continuous bounded functions for engineering applications but expanded to the space of measurable bounded functions for rigorous mathematical proofs. Deferring the implications of this observation, we simply note that $\mathbf{T}(\cdot) \in \mathcal{U}$, so that the functional J in Eq. (6.8) is understood to mean, $J : \mathcal{U} \to \mathbb{R}$. In subsequent sections, we will evaluate J from a larger space, $\mathcal{X} \times \mathcal{U} \times \mathbb{R}^n$, where $\mathcal{X}$ is the function space corresponding to the state variable so that the functional J is understood to mean, $J : \mathcal{X} \times \mathcal{U} \times \mathbb{R}^n \to \mathbb{R}$. It will be apparent later that the proper space for $\mathcal{X}$ (and $\mathcal{U}$) is a Sobolev space [20], as it forms the most natural space for both theoretical [21] and computational considerations [22].

6.3.1 *Quadratic cost is not $p = 2$*

By a minor abuse of notation, we denote by J_2 the cost function for $p = 2$; thus, by setting $p = 2$ in Eq. (6.8) we have,

$$(J_2)^2 = \left(\int_{t_0}^{t_f} \|\mathbf{T}(t)\|_2 \, dt \right)^2. \tag{6.9}$$

Similarly, let J_Q denote the standard quadratic cost function; then, we have,

$$\begin{aligned} J_Q &:= \int_{t_0}^{t_f} (T_x^2(t) + T_y^2(t) + T_z^2(t)) \, dt \\ &= \int_{t_0}^{t_f} \|\mathbf{T}(t)\|_2^2 \, dt \\ &\neq \left(\int_{t_0}^{t_f} \|\mathbf{T}(t)\|_2 \, dt \right)^2. \end{aligned}$$

Thus,

$$(J_2)^2 \neq J_Q.$$

The importance of this observation is that integration does not commute with the operation of taking powers. Thus, the oft-used argument that minimizing a quantity is the same as minimizing its square applies to J_2^2, which measures fuel consumption, but minimizing J_2 is not the same as minimizing J_Q. In physical terms, this is equivalent to noting that $v_e^2 \left(m(t_0) - m(t_f)\right)^2 \neq J_Q$; see Eq. (6.7).

6.3.2 *Fuel expenditures are measured by L^1 norms*

For a scalar-valued function, $f: \mathbb{R} \supseteq \Omega \to \mathbb{R}$, the L^p-norm of f, denoted by $\|f\|_{L^p}$ $(< \infty)$ is defined by [11],

$$\|f\|_{L^p} := \left(\int_{\Omega} |f(t)|^p \, \mathrm{d}t \right)^{1/p}, \tag{6.10}$$

where $|\cdot|$ denotes the absolute value. For vector-valued functions, $\mathbf{f} : \mathbb{R} \supseteq \Omega \to \mathbb{R}^n$, $n > 1$, the L^p-norm, $\|\mathbf{f}\|_{L^p}$ is frequently defined to be derived from Eq. (6.10) with $|\cdot|$ replaced by the Euclidean norm in $\mathbb{R}^n$. Thus, for example, if $n = 2$ so that $\mathbf{f}(t) = (f_1(t), f_2(t))$, then, by this definition of a norm, $\|\mathbf{f}\|_{L^p}$ is given by,

$$\|\mathbf{f}\|_{L^p} := \left(\int_{\Omega} \left(\sqrt{f_1^2(t) + f_2^2(t)} \right)^p \mathrm{d}t \right)^{1/p}.$$

Applying this definition for the function, $\mathbf{T}(\cdot)$, we get,

$$\|\mathbf{T}(\cdot)\|_{L^1} = \int_{t_0}^{t_f} \sqrt{T_x^2(t) + T_y^2(t) + T_z^2(t)} \mathrm{d}t$$

$$\|\mathbf{T}(\cdot)\|_{L^2}^2 = \int_{t_0}^{t_f} (T_x^2(t) + T_y^2(t) + T_z^2(t)) \mathrm{d}t.$$

Clearly, $J_Q = \|\mathbf{T}(\cdot)\|_{L^2}^2$, the L^2-norm of $\mathbf{T}(\cdot)$ and as shown in the previous subsection does not measure fuel. On the other hand, $\|\mathbf{T}(\cdot)\|_{L^1}$, does indeed measure fuel consumption and follows from Eq. (6.8) with $p = 2$.

Since finite-dimensional norms are equivalent, we can also define the L^p-norm of a vector-valued function, $\mathbf{f}$, in Eq. (6.10) with $|\cdot|$ replaced by the l^1 norm in $\mathbb{R}^n$. Thus, for $\mathbf{f}(t) = (f_1(t), f_2(t))$, we can define, $\|\mathbf{f}\|_{L^p}$, as

$$\|\mathbf{f}\|_{L^p} := \left(\int_{\Omega} \left(|f_1(t)| + |f_2(t)| \right)^p \mathrm{d}t \right)^{1/p}.$$

Using this definition, we get,

$$\|\mathbf{T}(\cdot)\|_{L^1} = \int_{t_0}^{t_f} \left(|T_x(t)| + |T_y(t)| + |T_z(t)| \right) \mathrm{d}t$$

$$\|\mathbf{T}(\cdot)\|_{L^2}^2 = \int_{t_0}^{t_f} \left(|T_x(t)| + |T_y(t)| + |T_z(t)| \right)^2 \mathrm{d}t.$$

From Eq. (6.8), by substituting $p = 1$, it follows that $\|\mathbf{T}(\cdot)\|_{L^1}$ is indeed a measure of the fuel consumption while $\|\mathbf{T}(\cdot)\|_{L^2}$ once again fails the test.

6.3.3 L^1 cost and l^p geometry

In addition to performing minimum-fuel analysis independent of the the propulsion system, one sometimes prefers to ignore the change in mass, particularly if the burn time is small and/or the specific impulse is high. In this case, the control may be taken as the thrust acceleration, $\mathbf{u} = \mathbf{T}/m$. By using the same arguments leading to Eq. (6.8), we can now state a fundamental result: the cost functions for minimum fuel control are a family of L^1-norms of the control function, $t \mapsto \mathbf{u}$. Specifically, the minimum fuel cost (see Eq. (6.8)) is the L^1-norm of the l^p-norm function $[t_0, t_f] \mapsto \|\mathbf{u}\|_p \in \mathbb{R}$

$$J[\mathbf{u}(\cdot)] = \int_{t_0}^{t_f} \|\mathbf{u}(t)\|_p \, \mathrm{d}t, \tag{6.11}$$

where we may use $\mathbf{u}$ to be either the thrust or the acceleration with the latter form of the control accompanied by the caveat mentioned above. Among others, one possible reason why this "l^p-variant" of the L^1-norm is not used as a cost function is that the running cost, i.e., the integrand in Eq. (6.11), is not differentiable with respect to the parameter $\mathbf{u}$. Deferring the details of the implications of this non-differentiability, we note that the Pontryagin version [19] of the Minimum (Maximum) Principle does not require differentiability of the integrand with respect to the control parameter; only differentiability with respect to the states is required. Nonetheless, it is worth noting that new versions of the Minimum Principle [8, 21, 23] do not even require differentiability with respect to the states: thanks to the era of nonsmooth analysis pioneered by Clarke, Sussmann and others [8, 23–25].

6.3.4 Penalty for not using the L^1 cost

The penalty in propellant consumption for designing trajectories not based on the proper family of L^1 cost functions can be summarized by the following fundamental fact.

Proposition 6.3.1. *Given two optimal control problems, F and G, that only differ in the optimality criteria, the F-cost of the G-optimal solution can never improve the F-cost of the F-optimal solution. For minimization problems, we have,*

$$J_F[\mathbf{x}_F(\cdot), \mathbf{u}_F(\cdot)] \leq J_F[\mathbf{x}_G(\cdot), \mathbf{u}_G(\cdot)].$$

The proof of this proposition is elementary; see Ref. [5].

If we now let the functional J_F be the L^1 cost and J_G be any other cost functional (such as a quadratic cost), it is clear that the system trajectory for Problem G cannot yield better fuel performance than the L^1 cost.

In addition to penalties in fuel consumption, additional penalties may arise in the design of the control system itself. For example, the thrust force (or acceleration) appears linearly in a Newtonian dynamical system: this is a direct consequence of Newton's Laws of motion and not a simplification from linearization. In minimizing such control-affine systems, barring the possibility of singular arcs, the L^1-optimal controller has a bang-off-bang structure. On the other hand, quadratic-cost-optimal controllers are continuous controllers. Continuous thrusting is frequently not desirable for spacecraft guidance and control since these controllers typically create undesirable effects on the payload. For example, thrusting increases the microgravity environment on the space station or induces undesirable effects on precision pointing payloads. Hence it is preferable to do much of the science during the "off periods". Thus, it is important to be cognizant of not creating new systems-engineering problems that were non-existent prior to active control considerations. The double integrator example in the next section illustrates all the main points including a quantification of the fuel penalty incurred in not using the L^1 cost.

6.3.5 A note on global optimality

Obviously, zero propellant is the absolute lowest possible cost. This fact can be mathematically stated as,

$$\inf_{\mathbf{u}(\cdot)\in\mathcal{U}} \left(\int_{t_0}^{t_f} \|\mathbf{u}(t)\|_p \, dt \right) = 0,$$

where $p = 1, 2$ or ∞ as before. Thus, if the L^1 cost is zero, it is apparent that we have a globally fuel-optimal solution. In other words, there is no need to prove necessary or sufficient conditions for global optimality if the L^1 cost is zero. Such globally optimal solutions are extremely useful in the design of spacecraft formations, and are further discussed in Refs. [12–14]. An interesting consequence of the existence of such solutions is that there may be several global minima. A simple approach to finding these solutions is to design cost functionals, $J_G[\mathbf{x}(\cdot), \mathbf{u}(\cdot)]$, that are not necessarily the L^1 cost, but are such that

$$\int_{t_0}^{t_f} \|\mathbf{u}_G^*(t)\|_p \, dt = 0,$$

where $\mathbf{u}_G^*$ is the control solution to some Problem G. The advantage of such problem formulations from both a theoretical and computational perspective is that the optimal system trajectories can be different from one problem formulation to another while yielding the same zero fuel cost. Thus, for example, if we were to solve a quadratic cost problem (as Problem G) and the system trajectory generated a solution such that the control was zero, then it is also a zero propellant solution. Since there is no guarantee that the state trajectory converges (theoretically and computationally) to the same trajectory as the L^1 solution, this seemingly undesirable property can be exploited to seek alternative global minimums. Such a strategy is used in Refs. [12–14] to design various spacecraft formations.

6.4 Double integrator example

The second-order control system, $\ddot{x} = u$, is widely studied [26] due to the simple reason that it is a quintessential Newtonian system: any information gleaned from a study of double-integrators has broad implications. In this spirit, we formulate an L^1 optimal control problem as,

$$\mathbf{x}^T := [x, v] \quad \mathbf{u} := [u] \quad \mathbb{U} := \{u : |u| \leq 6\}$$

$$(L^1P) \begin{cases} \text{Minimize} & J_1[\mathbf{x}(\cdot), \mathbf{u}(\cdot)] = \int_0^1 |u(t)| \, dt \\ \text{Subject to} & \dot{x} = v \\ & \dot{v} = u \\ & (x_0, v_0) = (0, 0) \\ & (x_f, v_f) = (1, 0) \end{cases}$$

Although the absolute value function, $u \mapsto |u|$, is not differentiable, the Pontryagin version of the Minimum Principle is still applicable as noted earlier. It is straightforward to show that the solution to Problem L^1P is given by,

$$u_1(t) = \begin{cases} 6 & t \in \Delta_1 \\ 0 & t \in \Delta_2 \\ -6 & t \in \Delta_3 \end{cases}$$

$$x_1(t) = \begin{cases} 3t^2 & t \in \Delta_1 \\ 3\Delta(2t - \Delta) & t \in \Delta_2 \\ 6(t + \Delta - \Delta^2) - 3(1 + t^2) & t \in \Delta_3 \end{cases}$$

$$v_1(t) = \begin{cases} 6t & t \in \Delta_1 \\ 6\Delta & t \in \Delta_2 \\ 6(1 - t) & t \in \Delta_3 \end{cases}$$

$$\lambda_{x_1}(t) = \frac{2}{2\Delta - 1}$$

$$\lambda_{v_1}(t) = \frac{1 - 2t}{2\Delta - 1},$$

where Δ_i, $i = 1, 2, 3$ are three subintervals of $[0, 1]$ defined by,

$$\Delta_1 = [0, \Delta], \quad \Delta_2 = [\Delta, 1 - \Delta], \quad \Delta_3 = [1 - \Delta, 1]$$

and

$$\Delta = \frac{1}{2} - \sqrt{\frac{1}{12}} \simeq 0.211.$$

In addition, we have,

$$J_1[\mathbf{x}_1(\cdot), \mathbf{u}_1(\cdot)] = \int_0^1 |u_1(t)| \, dt = 12\Delta \simeq 2.536. \tag{6.12}$$

Now suppose that we change the cost function in Problem L^1P to a quadratic cost while keeping everything else identical; then, we can write,

$$
(\text{LQP}) \begin{cases} \text{Minimize} & J_Q[\mathbf{x}(\cdot), \mathbf{u}(\cdot)] = \int_0^1 u^2(t)\,\mathrm{d}t \\ \text{Subject to} & \dot{x} = v \\ & \dot{v} = u \\ & (x_0, v_0) = (0, 0) \\ & (x_f, v_f) = (1, 0) \end{cases}
$$

The optimal solution is given by,

$$
\begin{aligned}
u_Q(t) &= 6 - 12t \\
x_Q(t) &= t^2(3 - 2t) \\
v_Q(t) &= 6t(1 - t) \\
\lambda_{x_Q}(t) &= -12 \\
\lambda_{v_Q}(t) &= 12t - 6
\end{aligned}
$$

and

$$
J_Q[\mathbf{x}_Q(\cdot), \mathbf{u}_Q(\cdot)] = \int_0^1 u_Q^2(t)\,\mathrm{d}t = 12 \tag{6.13}
$$

That $\max_{t\in[0,1]} |u_Q(t)| = 6$ explains why the control space in Problem L^1P was bounded accordingly.

6.4.1 LQP v/s L^1P

In comparing the performance of the two controllers, it is quite a simple matter to evaluate the L^1-cost of the quadratic control as,

$$
J_1[\mathbf{x}_Q(\cdot), \mathbf{u}_Q(\cdot)] = \int_0^1 |u_Q(t)|\,\mathrm{d}t = 3.0 \tag{6.14}
$$

Comparing this result with Eq. (6.12), we find that the quadratic controller is 18.3% more expensive (in fuel) than the L^1-optimal controller; obviously, a substantial margin. Further differences between the controllers are more evident in Figure 6.5. In comparing the two controllers, it is quite obvious that the L^1-controller is more desirable than the quadratic controller due to all the reasons outlined in Section 6.3.4. Quantitatively we note that we have a preferred zero-control action for approximately 58% of the time interval.

Despite the large differences between the two optimal controls, Figure 6.6 appears to indicate that there is little difference between the state trajectories. This apparently small difference comes about because plots such as Figure 6.6 do not adequately capture the true distance between two functions in the correct topology. The proper space to view

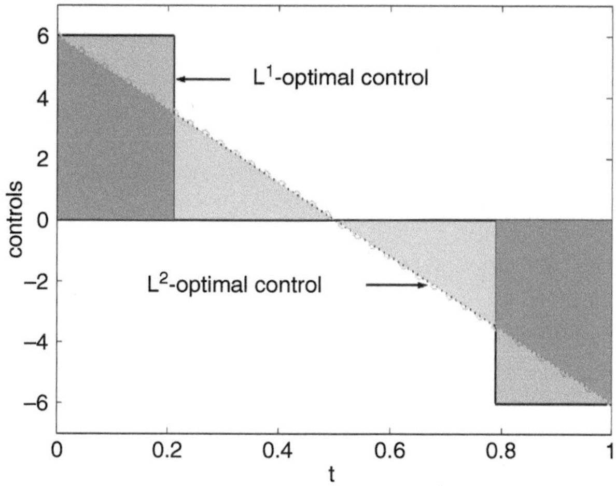

Fig. 6.5. Control plots for the quadratic and L^1-optimal control problems.

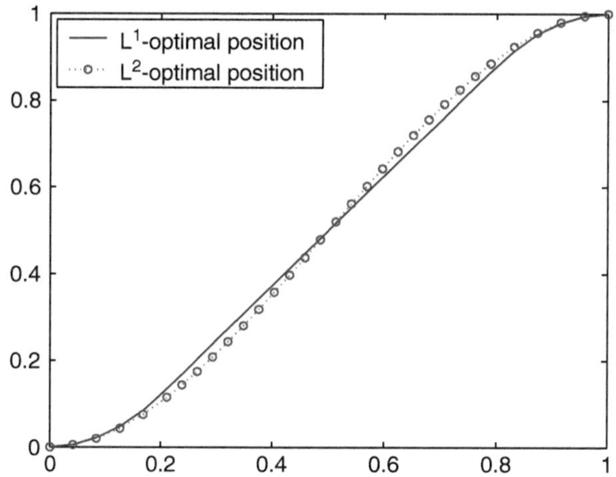

Fig. 6.6. Position plots for the quadratic and L^1-optimal control problems.

functions in control theory is a Sobolev space [20–22]. This space, denoted as, $W^{m,p}$, consists of all functions, $f : \mathbb{R} \supseteq \Omega \to \mathbb{R}$ whose jth-derivative is in L^p (see Eq. (6.10)) for all $0 \le j \le m$. The Sobolev norm of f is defined as,

$$\|f\|_{W^{m,p}} := \sum_{j=0}^{m} \left\| f^{(j)} \right\|_{L^p}. \tag{6.15}$$

Thus, in observing the plots in Figure 6.6 as being close to one another, we are implicitly viewing them in some L^p-norm. For example,

$$\left\|x_1(\cdot) - x_Q(\cdot)\right\|_{L^\infty} := ess \sup_{t\in[0,1]} \left|x_1(t) - x_Q(t)\right| = \max_{t\in[0,1]} \left|x_1(t) - x_Q(t)\right| \simeq 0.03.$$

When we observe the same functions in a Sobolev norm, say, $W^{1,\infty}$, then we have,

$$\left\|x_1(\cdot) - x_Q(\cdot)\right\|_{W^{1,\infty}} = \max_{t\in[0,1]} \left|x_1(t) - x_Q(t)\right| + \max_{t\in[0,1]} \left|\dot{x}_1(t) - \dot{x}_Q(t)\right| \simeq 0.30,$$

where we have replaced *ess* sup by max as before. Thus, the functions plotted in Figure 6.6 are ten times further apart in the Sobolev norm when compared to the corresponding Lebesgue norm. Since $\dot{x} = v$, the velocity plot shown in Figure 6.7 is more representative of the distance between the position functions.

The above arguments are essentially primal in flavor. A dual space perspective provides a more complete picture as covector spaces are fundamental to optimization. In this perspective [27], the position plot (Figure 6.8) must be jointly considered with the position costates. This view is quite justified by the large separation between the costates as evident in Figure 6.8. Thus, costates serve very important purposes in computational optimal control theory and are further illustrated in Section 6.7.

It is worth noting that if the control space, $\mathbb{U} = \{u : |u| \leq 6\}$, is changed to $\mathbb{U} = \mathbb{R}$, the solution to Problem LQP remains unaltered while a solution to Problem L^1P does not exist. In order to contemplate a solution to Problem L^1P for $\mathbb{U} = \mathbb{R}$, the space of admissible controls must be expanded from Lebesgue measurable functions (the assumption in the Pontryagin version of the Minimum Principle) to the space of generalized functions [28].

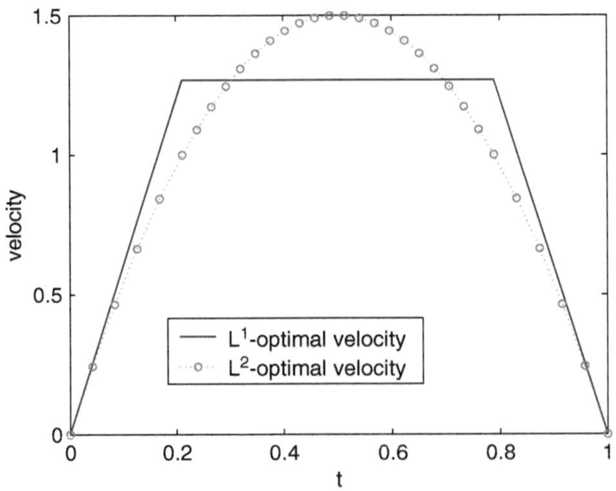

Fig. 6.7. Velocity plots for the quadratic and L^1-optimal control problems.

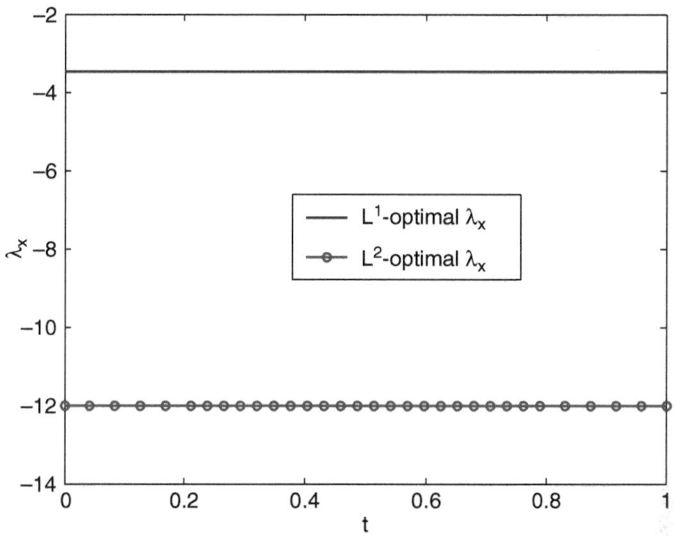

Fig. 6.8. Dual position trajectories for the quadratic and L^1-optimal control problems.

Circumventing these technicalities by using a continuation method of Lawden [29], it is straightforward to show [5] that the optimal control is given by

$$u_\delta(t) = \delta(t) - \delta(1-t)$$

where δ is the Dirac delta function. The states are then given by,

$$x_\delta(t) = t$$

$$v_\delta(t) = \begin{cases} [0, 1] & t = 0 \\ 1 & t \in (0, 1), \\ [0, 1] & t = 1 \end{cases}$$

where $v_\delta(t)$ is expressed in a set-valued form consistent with nonsmooth calculus; [8] see also Ref. [30] for a practical demonstration of nonsmooth concepts. In this spirit, the L^1-cost of impulse control is given by,

$$J_1[\mathbf{x}_\delta(\cdot), \mathbf{u}_\delta(\cdot)] = 2 \tag{6.16}$$

Thus, the quadratic controller (see Eq. (6.14)) is 50% more expensive than the impulse controller. Note however that the impulse cost is only a mathematical phenomenon whereas the cost obtained by solving the $L1P$ is indeed the true cost of fuel. Furthermore, these differences in cost have nothing to do with "gravity-" or "drag-loss," terminology that is quite common in orbital mechanics to describe other phenomena.

6.5 Issues in solving nonlinear L^1-optimal control problems

While the previous sections illuminated the core principles in formulating the nonsmooth L^1 problems and the penalties incurred in solving "simpler" smooth problems, the approach used in Section 6.4 is not portable to solving astrodynamical systems because closed-form solutions to even simple optimal control problems are unknown. In order to frame the key issues in computing L^1-optimal controls for general dynamical systems, we summarize the problem statement as,

$$(B)\begin{cases}\text{Minimize} & J[\mathbf{x}(\cdot),\mathbf{u}(\cdot),t_0,t_f]=\int_{t_0}^{t_f}\|\mathbf{u}(t)\|_p\,\mathrm{d}t\\ \text{Subject to} & \dot{\mathbf{x}}(t)=\mathbf{f}(\mathbf{x}(t),\mathbf{u}(t))\\ & \mathbf{u}(t)\in\mathbb{U}\\ & (\mathbf{x}_0,\mathbf{x}_f,t_0,t_f)\in\mathbb{E}\end{cases},$$

where $\mathbb{E}\subset\mathbb{R}^{N_x}\times\mathbb{R}^{N_x}\times\mathbb{R}\times\mathbb{R}$ is some given endpoint set and $\mathbb{U}$ is a compact set as before. State constraints of the form, $\mathbf{x}(t)\in\mathbb{X}$ can also be added to the problem, but we focus on Problem B as formulated above to only limit the scope of the discussion; the ideas extend to these situations as well. The functional J is the map, $\mathcal{X}\times\mathcal{U}\times\mathbb{R}\times\mathbb{R}\mapsto\mathbb{R}$. As indicated earlier, although we typically take $\mathcal{X}=W^{1,1}$ for theoretical purposes, we limit $\mathcal{X}$ to the space $W^{1,\infty}$ for computation. Summarizing the result of Section 6.3.5, we have,

Proposition 6.5.2. *Any tuple,* $(\mathbf{x}^*(\cdot),\mathbf{u}^*(\cdot),t_0^*,t_f^*)$, *for which* $J[\mathbf{x}^*(\cdot),\mathbf{u}^*(\cdot),t_0^*,t_f^*]=0$ *is a globally optimal solution to Problem B.*

There are essentially three methods for solving optimal control problems [31], all of which require a careful analysis of the Hamiltonian Minimization Condition [23] (HMC).

6.5.1 The Hamiltonian minimization condition

At each instant of time, t, the HMC is a nonsmooth static optimization problem,

$$(\mathrm{HMC})\begin{cases}\underset{\mathbf{u}}{\text{Minimize}} & H(\boldsymbol{\lambda},\mathbf{x},\mathbf{u})=\|\mathbf{u}\|_p+\boldsymbol{\lambda}^T\mathbf{f}(\mathbf{x},\mathbf{u})\\ \text{Subject to} & \mathbf{u}\in\mathbb{U}\end{cases},$$

where H is the *control* Hamiltonian [23]. In the framework of the Minimum Principle, $\boldsymbol{\lambda}\in\mathbb{R}^{N_x}$ is the costate where $t\mapsto\boldsymbol{\lambda}$ satisfies the adjoint equation while in Bellman's dynamic programming framework, $\boldsymbol{\lambda}=\partial\varphi/\partial\mathbf{x}$ where, $\varphi:\mathbb{R}\times\mathbb{R}^{N_x}\rightarrow\mathbb{R}$, is a function that satisfies the Hamilton–Jacobi–Bellman (HJB) partial differential equation [8],

$$\mathcal{H}(\varphi_{\mathbf{x}}(t,\mathbf{x}),\mathbf{x})+\varphi_t(t,\mathbf{x})=0, \tag{6.17}$$

where $\mathcal{H} : \mathbb{R}^{N_x} \times \mathbb{R}^{N_x} \to \mathbb{R}$ is the *lower* Hamiltonian [8] defined as,

$$\mathcal{H}(\boldsymbol{\lambda}, \mathbf{x}) := \min_{\mathbf{u} \in \mathbb{U}} H(\boldsymbol{\lambda}, \mathbf{x}, \mathbf{u}). \tag{6.18}$$

In recognizing that Problem HMC is fundamental to solving optimal control problems, we discuss some key issues pertaining to this problem.

6.5.2 Issues in solving Problem HMC

In Section 6.4, the control variable was one-dimensional ($N_u = 1$). This facilitated solving Problem HMC simply by inspection without resorting to nonsmooth calculus [8]. To solve problems in higher dimensional spaces, we need a more systematic procedure. Rather than resort to formal nonsmooth analysis, a procedure that is tenable to both analysis and computation is to convert the nonsmooth HMC to an equivalent problem where the functions describing the objective function and the constraint set are smooth. Such conversion techniques, well-known in nonlinear programming, can be achieved by exchanging the complication of nonsmoothness in a lower dimensional space to a simpler problem in higher dimensions. As noted in Section 6.1, similar trades are rampant in engineering analysis. In order to focus our attention to the conversion issue, we demonstrate this procedure for the HMC by limiting the scope of the problem to the case when $\mathbf{f}$ is differentiable with respect to $\mathbf{u}$, and $\mathbb{U}$ is given in terms of inequalities as follows,

$$\mathbb{U} := \left\{\mathbf{u} \in \mathbb{R}^{N_u} : \mathbf{h}^L \le \mathbf{h}(\mathbf{u}) \le \mathbf{h}^U\right\},$$

where $\mathbf{h} : \mathbb{R}^{N_u} \to \mathbb{R}^{N_h}$ is a differentiable function and $\mathbf{h}^L, \mathbf{h}^U \in \mathbb{R}^{N_h}$ are the lower and upper bounds on the values of the function $\mathbf{h}$ respectively. Much of the analysis to follow easily extends to more complex situations (see for example, Section 6.8 of this chapter and Ref. [32]), but our intent here is not an enumeration of these situations but to demonstrate a methodology. Hence, we choose to illustrate the concepts for one of the most prevalent cases in engineering applications.

As noted previously, the function, $\mathbf{u} \mapsto H(\boldsymbol{\lambda}, \mathbf{x}, \mathbf{u})$, is nonsmooth because $\|\mathbf{u}\|_p$ is nonsmooth. For example, when $p = 2$ and $N_u = 3$,

$$\|\mathbf{u}\|_2 = \sqrt{u_1^2 + u_2^2 + u_3^2}.$$

The function, $\mathbf{u} \mapsto \|\mathbf{u}\|_2$, is not differentiable at the origin $(0, 0, 0)$. This is illustrated in Figure 6.9 for $\mathbf{u} \in \mathbb{R}^2$. As Betts [33, 34] notes, this single point can cause major problems in computation. The singular point cannot be ignored even in a theoretical framework as it is the most desirable point: as evident in Section 6.4, it occurs for about 58% of the time interval in the solution to Problem L^1P. That is, one point in the solution to Problem HMC can easily get smeared over a substantial time interval. In mathematical terms, this is simply an effect of the chain rule in evaluating the derivative of $t \mapsto \|\mathbf{u}\|_2$ by way of the gradient of $\mathbf{u} \mapsto \|\mathbf{u}\|_2$. Noting that the function $\mathbf{u} \mapsto \|\mathbf{u}\|_2^2$ is differentiable,

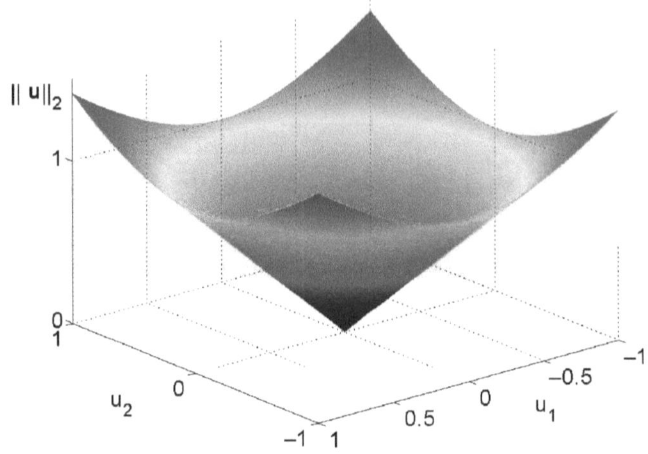

Fig. 6.9. Illustrating the nonsmooth structure of $\mathbf{u} \mapsto \|\mathbf{u}\|_2$. (see Color plate 6)

the nonsmooth HMC for $p = 2$ can be converted to a smooth one by introducing a pseudo-control variable $u_4 := \|\mathbf{u}\|_2$. That is, Problem HMC for $p = 2$ (and $N_u = 3$) can be transformed to a smooth nonlinear programming (Nl^2P) problem in an augmented control variable, $\mathbf{u}_a \in \mathbb{R}^{N_u+1}$, as,

$$\mathbf{u}_a^T := [\mathbf{u}^T, u_4] \equiv [u_1, u_2, u_3, u_4]$$

$$(\text{N}l^2\text{P}) \begin{cases} \underset{\mathbf{u}_a}{\text{Minimize}} & H(\boldsymbol{\lambda}, \mathbf{x}, \mathbf{u}_a) = u_4 + \boldsymbol{\lambda}^T \mathbf{f}(\mathbf{x}, \mathbf{u}) \\ \text{Subject to} & \mathbf{u} \in \mathbb{U} \\ & \|\mathbf{u}\|_2^2 - u_4^2 = 0 \\ & u_4 \geq 0 \end{cases},$$

where we have retained the use of the symbol H for the new Hamiltonian by a minor abuse of notation. Since the original problem was nonsmooth, the inequality, $u_4 \geq 0$, essentially retains the nonsmooth geometric structure of the problem although the function used in the inequality is now differentiable. Thus, the standard Karush–Kuhn–Tucker (KKT) conditions for Problem Nl^2P can be applied. The minimum-fuel orbit transfer example discussed in Section 6.7 further discusses the KKT conditions in conjunction with the larger problem of actually solving the optimal control problem.

The situation for $p = 1$ is similar, except that it requires the introduction of many more control variables. This is because the function,

$$\mathbf{u} \mapsto \|\mathbf{u}\|_1 = |u_1| + |u_2| + |u_3|$$

is non-differentiable at the origin, $(0, 0, 0)$, as well as all other points where $u_i = 0$, $i = 1, 2, 3$ (see Figure 6.10). By introducing variables, $v_i \geq 0$, $w_i \geq 0$, $i = 1, 2, 3$, the

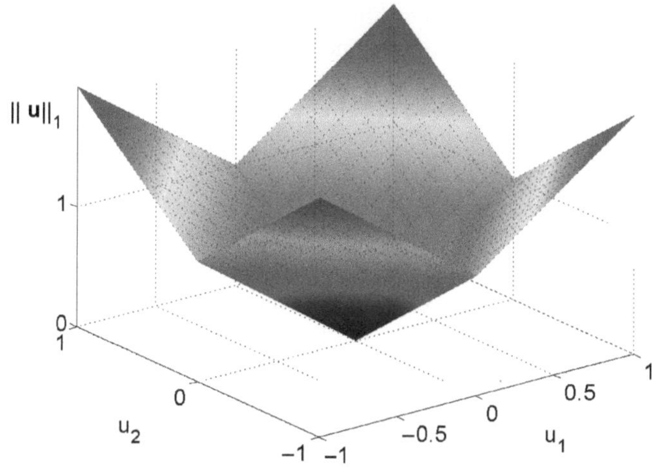

Fig. 6.10. Illustrating the nonsmooth structure of $\mathbf{u} \mapsto \|\mathbf{u}\|_1$. (see Color plate 7)

nonsmooth HMC problem for $p = 1$ can be transformed to a smooth nonlinear programming (Nl^1P) problem in an augmented control variable, $\mathbf{u}_a \in \mathbb{R}^{2N_u}$, as,

$$\mathbf{u}_a := [\mathbf{v}, \mathbf{w}]$$

$$(\mathrm{N}l^1\mathrm{P}) \begin{cases} \underset{\mathbf{u}_a}{\text{Minimize}} & H(\boldsymbol{\lambda}, \mathbf{x}, \mathbf{u}_a) = \mathbf{1}^T\mathbf{u}_a + \boldsymbol{\lambda}^T\widetilde{\mathbf{f}}(\mathbf{x}, \mathbf{u}_a) \\ \text{Subject to} & \widetilde{\mathbf{h}}^L \le \widetilde{\mathbf{h}}(\mathbf{u}_a) \le \widetilde{\mathbf{h}}^U \\ & \mathbf{v} \ge \mathbf{0} \\ & \mathbf{w} \ge \mathbf{0} \end{cases},$$

where $\mathbf{1}$ is an $\mathbb{R}^{2N_u}$-vector of ones and tildes over the symbols implies transformed functions and variables when $\mathbf{u}$ is transformed to $\mathbf{u}_a$. For example, when $\mathbf{f}$ is control-affine,

$$\mathbf{f}(\mathbf{x}, \mathbf{u}) = \mathbf{a}(\mathbf{x}) + \mathbf{B}(\mathbf{x})\mathbf{u},$$

where $\mathbf{a} : \mathbb{R}^{N_x} \to \mathbb{R}^{N_x}$ and $\mathbf{B} : \mathbb{R}^{N_x} \to \mathbb{R}^{N_x \times N_u}$, then $\widetilde{\mathbf{f}}$ is given by,

$$\widetilde{\mathbf{f}}(\mathbf{x}, [\mathbf{v}, \mathbf{w}]) = \mathbf{a}(\mathbf{x}) + \mathbf{B}(\mathbf{x})[\mathbf{v} - \mathbf{w}].$$

Once Problem HMC is converted to an NLP with smooth functions, the KKT conditions then describe the necessary conditions for a putative optimal controller.

6.5.3 HMC on HJB

A cursory examination of Problems Nl^1P and Nl^2P reveal that it may be quite difficult to obtain a closed-form solution. An examination of the KKT conditions for these problems strengthen this observation which has far-reaching consequences.

In the absence of a closed-form solution to Problem HMC, an explicit expression for the map, $(\boldsymbol{\lambda}, \mathbf{x}) \mapsto \mathbf{u}$, cannot be obtained. This means that the lower Hamiltonian (Eq. (6.18)) cannot be constructed explicitly. That is, it would be impossible to even write down explicitly the HJB partial differential equation. This elementary observation almost immediately eliminates the HJB as a practical means for solving problems beyond academic ones.

In cases where the controls can be eliminated, the HJB suffers from at least two additional well-known problems [8, 21, 26] that merit recounting. As is the case for a large number of problems, a differentiable solution to the HJB does not exist for the L^1-optimal control problem; however, if the notion of differentiability is expanded along the lines of nonsmooth analysis, then, according to the celebrated result of Crandall and Lions [8, 21], the Bellman value function is a unique viscosity solution to the HJB. This theoretical breakthrough has not yet translated to practical problem solving, as even smooth partial differential equations continue to be challenging problems.

The third problem associated with Eq. (6.17) is the absence of good computational techniques for solving partial differential equations involving more than three independent variables. Even for a coplanar orbit transfer problem (discussed further in Section 6.7), $N_x = 4$. For practical three-dimensional space models, $N_x = 6$; hence, the number of independent variables in φ is seven. Given that the vast majority of computational techniques for solving partial differential equations is limited to two independent variables, it is clear that solving the HJB for a practical problem is far from feasible.

It is worth noting at this stage that even if the HJB can be solved numerically, it loses one of its major attractions: the ability to generate feedback solutions in closed form. This is simply because, a numerical solution to the HJB implies a table lookup data for feedback control or an approximation at best for a closed-form solution by way of a surface fit for Bellman's value function. Thus, although the Hamilton–Jacobi framework is quite elegant, the absence of a viable methodology that overcomes the major technical hurdles to solve a generic problem limits its utility to low dimensional academic problems; hence, we eliminate this approach from further consideration.

6.5.4 HMC on the Minimum Principle

Unlike the HJB framework, the Minimum Principle does not require an explicit solution to Problem HMC. This first step immediately trumps the HJB from a solvability perspective; however, an application of the Minimum Principle results in a nonlinear, differential-algebraic boundary value problem (BVP). Given that even linear differential BVPs do not have closed-form solutions, finding analytical solutions to optimal controls does appear to be quite daunting. This task is quite formidable even from a numerical perspective as the Hamiltonian BVP has a fundamental sensitivity problem that results from its symplectic structure [26]. As discussed by Bryson and Ho [26], when a shooting-type method is applied to solve the Hamiltonian BVP, the sensitivity of the initial conditions with respect to the final conditions is so large that the values of the intervening variables often exceeds the numerical range of a computer. While multiple-shooting algorithms alleviate this particular issue, the vast number of other problems associated with shooting methods as detailed by Betts [33, 34] makes them fundamentally unsuitable for computing optimal controls.

From a modern perspective [22, 46], a BVP is essentially a problem of solving a generalized equation of the form, $0 \in \mathcal{F}(x)$, where $\mathcal{F}$ is a set-valued map. By resisting the temptation to use shooting methods, generalized equations can be solved more robustly by a combination of operator methods that retain the structure of $\mathcal{F}$ and nonlinear programming techniques [24]. Details of this approach are well documented by Betts [34] and Hager [22].

6.6 Solving nonlinear L^1-optimal control problems

As a result of the observations of the preceding paragraphs, solving optimal control problems, L^1 or otherwise, are widely perceived as difficult problems. Over the last decade, as a result of major advancements in approximation theory and optimization techniques, solving optimal control problems, particularly smooth problems, are no longer considered to be difficult. This is evident by the broad class of complex optimal control problems that have been solved with relative ease [1, 12–14, 30, 33–36]. This approach is essentially a modification and modernization of Euler's abandoned idea in solving calculus-of-variations problems combined with Lagrange's multiplier theory [37]. An early version of this approach is due to Bernoulli. This neo-Bernoulli–Euler–Lagrange approach, is encapsulated as the Covector Mapping Principle (CMP) and represents a triad of ideas for solving optimal problems [22, 31, 37]. When infused with modern computational power, the CMP facilitates real-time computation of optimal controls [38–40] thus enabling a neo-classical approach to feedback guidance and control.

6.6.1 A brief history of the covector mapping principle

According to Mordukhovich [25], Euler discovered the Euler–Lagrange equations by discretizing the fundamental problem of the calculus of variations, and then passing to the limit. Upon receiving Lagrange's letter containing "... a beautiful and revolutionary idea ... Euler dropped his own method, espoused that of Lagrange, and renamed the subject the calculus of variations" [41]. Thus, the invention of direct methods [42] of the 1960s are, conceptually, Euler's abandoned idea before the limiting process [31]. Of course, modern direct methods [33, 34] typically discretize the problem using higher-order methods although Eulerian approximations continue to be widely used. A key point in modernizing Euler's original idea is the absence of the limiting process by solving the problem on a digital computer for a sufficiently fine grid, much the same way as we solve initial value problems by a Runge-Kutta method for some non-zero step size. We therefore expect the discrete solution to satisfy point wise the Euler–Lagrange equations. That this expectation does not necessarily bear fruit is one of the many reasons why indirect methods were popular (until the early 1990s) despite their well-known problems in solving symplectic (Hamiltonian) boundary-value problems [26, 31, 34].

Unlike the Euler-Lagrange approach with its distinct primal "flavor," had Euler combined his discretization approach with Lagrange's multiplier theory, the history of the calculus of variations might have taken on an early dual flavor. This combination did not take place until 200 years later, after the discovery of Minimum Principle [25]. What is remarkable about this combination is that the discrete problem does not generally satisfy the discrete Minimum Principle without an additional assumption of convexity. Since no convexity assumptions are required for the validity of the continuous-time Minimum Principle, discrete solutions are viewed with great suspicion. Significant fodder for this suspicion is provided by higher-order methods. That is, rather than improve the quality of the solution, a higher-order discretization can lead to a completely disastrous solution [22, 35, 43]. These experiments of the late 1990s paved a way for a deeper understanding of optimal control theory by connecting the first principles to approximation theory and computation in Sobolev spaces [22, 35]. In other words, convergence of the approximation takes center stage for both theory and practice [25, 35].

6.6.2 Convergence and the covector mapping principle

The emerging issues in the neo-Bernoulli–Euler–Lagrange approach can be effectively visualized by the diagram shown in Figure 6.11. Here, Problem B is not necessarily limited to the problem discussed in Section 6.5 although our major focus continues to be the L^1-optimal control problem. The bottom of Figure 6.11 represents a generalization of Euler's initial idea of discretizing Problem B to Problem B^N where N denotes the number of discrete points. These are the classical direct methods. If convergence can be proved, then passing to the limit, $N \to \infty$, solves the original continuous problem in the limit (see the bottom convergence arrow in Figure 6.11). A convergence theorem is also a practical necessity since it ensures us that we can obtain solutions to an arbitrary precision (within the limits of digital precision). Note that Euler assumed convergence

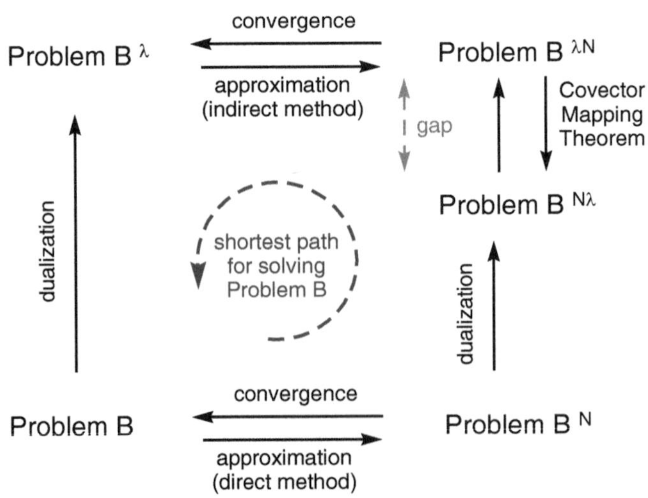

Fig. 6.11. The Covector Mapping Principle.

which can be shown to be valid for the simplest problem (general problem during Euler's days) of the calculus of variations (for Euler discretizations) but are generally invalid in the context of the Minimum Principle as indicated earlier.

When Problem B is a modern optimal control problem, Problem B^N is a nonlinear programming (NLP) problem if $\mathbb{U}$ is a continuous set; in general, it is a mixed-variable programming problem [22, 32]. Hence, Problem $B^{N\lambda}$ refers to the set of necessary conditions obtained by applying the Karush–Kuhn–Tucker (KKT) theorem. On the other hand, Problem $B^{\lambda N}$ refers to the discretization of the continuous differential-algebraic BVP obtained by applying the Minimum Principle. As indicated earlier, Problems $B^{\lambda N}$ and $B^{N\lambda}$ do not necessarily generate the same solution. That is, dualization and discretization are not commutative operations. Recent research by Hager [35] and Betts et al. [44] provide additional fodder to this concept. While Hager has shown that convergent Runge-Kutta methods fail to converge Betts et al. have shown that non-convergent methods converge. What has become clear is that a new theoretical framework is quite essential to understand seemingly contradictory results.

6.6.3 Linking theory, practice and computation

For the solution of the approximate problem (B^N) to be indistinguishable in a practical sense from some unknown exact theoretical solution to Problem B, we need to solve Problem B^N for a "sufficiently large grid". Thanks to the exponential convergence property of pseudospectral methods [45], these grids can be remarkably small. When combined with sparse efficient methods for solving NLPs, solutions to Problem B^N can be rendered virtually indistinguishable from theoretical solutions to Problem B if convergent methods (in the sense of discretization [46]) are adopted. Convergence of the discretization is sharply distinguished from the convergence of the NLP. A proper design of a computational method requires convergence analysis. The new ideas on convergence require set-valued analysis [22] and connections to the symplectic structure of Hamiltonian systems [31]. The absence of these connections lead to difficult problems or disastrous results even with convergent NLPs [35, 43]. Exploiting the global convergence properties of modern NLP algorithms [47] with relaxation techniques in discretization [31] implies that optimal control problems can be solved routinely.

The statements of the preceding paragraph are deeply theoretical since modern computational methods facilitate a practical demonstration of "epsilons, deltas, limits and sequences," the hallmark of functional analysis. Thus the practice of optimal control today is more firmly rooted and integrated with theory than ever before. This point is better understood by way of Figure 6.11; here, Problem B^λ is the Hamiltonian BVP discussed earlier and Problem $B^{\lambda N}$ represents the approximation (recall that any numerical method requiring a digital computer is an approximation). The reason certain well-known discretization methods (like a class of Runge-Kutta methods [35, 43]) fail for optimal control problems is that dualization and discretization are non-commutative operations indicated by the commutation gap shown in Figure 6.11. A zero gap does not guarantee convergence while convergence does not guarantee zero gap (except in the limit). In principle, this gap can be closed for finite N if there exist an order-preserving map between the duals [35, 46, 48]. Such maps have been obtained (i.e., Covector Mapping Theorems) for a special class of symplectic Runge-Kutta [35] methods and modifications [48] to pseudospectral methods [49]. Thus the Covector Mapping Principle essentially encapsulates the approximation issues that started with the work of Bernoulli, Euler and Lagrange [37]. It is thus apparent that the oft mentioned difficulties in solving optimal control problems can be completely circumvented today by modernizing and extending Euler's original ideas as depicted in Figure 6.11. This essentially implies that a robust general procedure that is tenable for solving practical problems is a practical combination of functional analysis with approximation theory. Indeed, in recent years, a broad class of complex optimal control problems have been solved under this framework with relative ease [1, 12–14, 30, 33–35, 48, 49]. Additional details on these ensemble of topics along with extensive references are discussed in Refs. [31] and [37].

6.6.4 Feedback guidance and control

Suppose that Problem B can be solved in real time. This means that for any $(t_0, \mathbf{x}_0)$, we can solve the optimal control problem in negligible time. Then, replacing the initial conditions by current conditions, $(t, \mathbf{x})$, it is apparent that we have a feedback map, $(t, \mathbf{x}) \mapsto \mathbf{u}$. In other words, real-time computation implies feedback control. Theoretically, real-time computation implies zero computation time; in practice, the real issue is the measurable effect, if any, of a non-zero computation time. Stated differently, a key issue in feedback control is the required minimum computational speed for feedback implementation rather than the imposition of the theoretical real-time computation of optimal controls. If we had perfect models and a deterministic system, feedback would be unnecessary provided the perfect model was used in the computation of the control. In other words, the higher the fidelity of the models used in the computation of control, the less the demand on real-time computation. Further, the need for computational speed is less if the time constant of the system is larger. Thus, if the system time constant is large and reasonably high fidelity models are chosen for the computation of control, implementing feedback controls by way of online optimization is not a difficult problem. These are precisely the conditions for orbit control: the time constant of a low Earth orbit (LEO) is the orbital period of about 90 minutes and nonlinear models of relatively high accuracy are available. Hence, if recomputed optimal thrusting programs were to be available every minute for LEO spacecraft, then it is possible to implement a sampled-data feedback control with 90 samples per orbit. As demonstrated in the next section and elsewhere [5, 10, 15], minimum-fuel orbit transfer problems can be solved on Pentium 4 computers in under 30 seconds (thus implying the possibility of 180 samples for LEO). Faster computational speeds are easily possible [38] with optimized code and/or by removing the overhead associated with the operating system (Windows) and the problem solving environment (MATLAB). For example, in Ref. [38], the optimal solution to a flexible robot arm was obtained in 0.03 seconds (thus making avail the possibility of a 30 Hz sampling frequency). Applications of such feedback solutions to other problems are extensively discussed elsewhere [38–40]. Thus, optimal feedback orbit control via real-time optimization is a clear modern-day reality.

6.7 L^1-Formulation of the minimum-fuel orbit transfer problem

We will now illustrate some of the ideas described in Sections 6.5 and 6.6 by solving a new formulation of the minimum-fuel orbit transfer problem. As noted earlier, the minimum-fuel orbit transfer problem is a central problem in orbit control. This problem can be easily formulated by posing it as a problem of maximizing the final mass; however, in this formulation, the astrodynamics of the problem is coupled to the propulsion system of the spacecraft by way of the specific impulse of the propellant (Eq. (6.7)). As noted in Section 6.1, it is frequently desirable to decouple the propulsion system performance from the astrodynamics of the problem by comparing the cost of a maneuver in terms of the characteristic velocity, i.e., the velocity change attributable to a generic propulsion system. This translates to using the l^2-variant of the L^1-cost. The following coplanar orbit transfer problem defines this formulation:

$$\mathbf{x}^T := [r, \theta, v_r, v_t] \quad \mathbf{u}^T := [u_r, u_t] \quad \mathbf{u} \in \mathbb{U}$$

$$\mathbb{U} := \left\{ \mathbf{u} \in \mathbb{R}^2 : \ \|\mathbf{u}\|_2 \le u_{max} \right\}$$

$$(\mathrm{O}) \begin{cases} \text{Minimize} & J[\mathbf{x}(\cdot), \mathbf{u}(\cdot), t_f] = \displaystyle\int_{t_0}^{t_f} \|\mathbf{u}(t)\|_2 \, \mathrm{d}t \\ \text{Subject to} & \dot{r} = v_r \\ & \dot{\theta} = \dfrac{v_t}{r} \\ & \dot{v}_r = \dfrac{v_t^2}{r} - \dfrac{1}{r^2} + u_r \\ & \dot{v}_t = -\dfrac{v_r v_t}{r} + u_t \\ & \mathbf{e}_0(t_0, \mathbf{x}_0) = \mathbf{0} \\ & \mathbf{e}_f(\mathbf{x}_f) = \mathbf{0} \end{cases}$$

The functions for the endpoint conditions,

$$\mathbf{e}_0(t, \mathbf{x}) := \begin{pmatrix} t \\ a_0[(v_r^2 + v_t^2)r - 2] + r \\ r[1 + e_0 \cos(\theta - \omega_0)] - (v_t r)^2 \\ v_r[1 + e_0 \cos(\theta - \omega_0)] - e v_t \sin(\theta - \omega_0) \end{pmatrix}$$

$$\mathbf{e}_f(\mathbf{x}) := \begin{pmatrix} a_f[(v_r^2 + v_t^2)r - 2] + r \\ r[1 + e_f \cos(\theta - \omega_f)] - (v_t r)^2 \\ v_r[1 + e_f \cos(\theta - \omega_f)] - e v_t \sin(\theta - \omega_f) \end{pmatrix}$$

describe the initial and final manifolds in Problem O in terms of the initial and final orbits respectively, where (a_0, e_0, ω_0) and (a_f, e_f, ω_f) are the standard orbital elements. Except for its resemblance to the dynamical model, this problem formulation is different in every respect when compared to the continuous-thrust problem posed by Moyer and Pinkham [50] and discussed in the texts by Bryson and Ho [26] and Bryson [18].

Let $\boldsymbol{\lambda}^T := [\lambda_r, \lambda_\theta, \lambda_{v_r}, \lambda_{v_t}]$ and $\mathbf{u}_a^T := [u_r, u_t, u]$, where $u = \|\mathbf{u}\|_2$; then, the Hamiltonian Minimization Condition (see Problem Nl^2P discussed earlier), simplifies to,

$$\begin{aligned} \underset{\mathbf{u}_a}{\text{Minimize}} \quad & H(\boldsymbol{\lambda}, \mathbf{x}, \mathbf{u}_a) = u + \lambda_{v_r} u_r + \lambda_{v_t} u_t + H_0(\boldsymbol{\lambda}, \mathbf{x}) \\ \text{Subject to} \quad & u_r^2 + u_t^2 - u^2 = 0 \\ & 0 \le u \le u_{max} \end{aligned}$$

where $H_0(\boldsymbol{\lambda}, \mathbf{x})$ denotes terms in the Hamiltonian that do not depend upon the controls. Note that the control space is a nonconvex cone in $\mathbb{R}^3$ (see Figure 6.9). The KKT conditions for this problem can be obtained by forming the Lagrangian of the Hamiltonian ($\overline{H}$),

$$\overline{H}(\boldsymbol{\mu}, \boldsymbol{\lambda}, \mathbf{x}, \mathbf{u}_a) = u + \lambda_{v_r} u_r + \lambda_{v_t} u_t + H_0(\boldsymbol{\lambda}, \mathbf{x}) + \mu_1 u + \mu_2 (u_r^2 + u_t^2 - u^2),$$

where μ_1 and μ_2 are the KKT (Lagrange) multipliers associated with the control constraints with μ_1 satisfying the complementarity condition,

$$\begin{array}{ccc} u = 0 & & \mu_1 \le 0 \\ 0 < u < u_{max} & \Leftrightarrow & \mu_1 = 0 \\ u = u_{max} & & \mu_1 \ge 0, \end{array} \tag{6.19}$$

while μ_2 is unrestricted in sign. Thus, the function, $t \mapsto \mu_1$, supplies the switching information. The vanishing of the gradient of the Lagrangian of the Hamiltonian, $\partial \overline{H} / \partial \mathbf{u}_a$, provides three additional necessary conditions,

$$\lambda_{v_r} + 2\mu_2 u_r = 0 \tag{6.20}$$

$$\lambda_{v_t} + 2\mu_2 u_t = 0 \tag{6.21}$$

$$1 + \mu_1 - 2\mu_2 u = 0. \tag{6.22}$$

From Eqs. (6.19) and (6.22) it follows that

$$u = 0 \Rightarrow \mu_1 = -1.$$

This result is quite interesting. If the optimal control program is not identically equal to zero (i.e., zero cost), the function $t \mapsto \mu_1$ must jump at the points where $t \mapsto u = \|\mathbf{u}\|_2$ goes from zero to some non-zero value either via singular (i.e., $\mathbf{u} \in int\, \mathbb{U}$) or bang-bang (i.e., $\mathbf{u} \in bdry\, \mathbb{U}$) thrusting. That this phenomenon does indeed occur is shown in Figure 6.12 for a sample solution corresponding to the following case:

$$a_0 = 1, \; a_f = 2, \; e_0 = 0.1, \; e_f = 0.2, \; \omega_0 = 1, \; \omega_f = 2, \; u_{max} = 0.05.$$

The plot shown in Figure 6.12 was *not* obtained by solving the "difficult" Hamiltonian BVP (i.e., an "indirect method" indicated in Figure 6.11), rather, it was obtained quite readily by an application of the CMP to the Legendre pseudospectral method [48]. In fact, Problem O was easily solved by way of the software package DIDO [51]. DIDO is a minimalist's approach to solving optimal control problems: only the problem formulation is required, and in a form that is almost identical to writing it on a piece of paper and

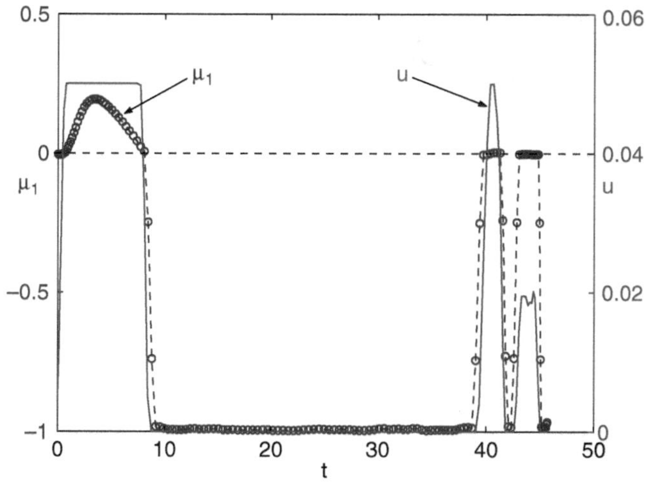

Fig. 6.12. Demonstrating the Hamiltonian Minimization Condition for Problem O; note the singular control and the vanishing of the switching function.

pencil. The latter is facilitated by the use of object-oriented programming readily available within the MATLAB problem solving environment.

A number of features are noteworthy in Figure 6.12. Observe the excellent correlation between the switching function, $t \mapsto \mu_1$, and the control trajectory, $t \mapsto u$, in conformance with the equations resulting from the Hamiltonian Minimization Condition (i.e., Eq. (6.19)). The last burn appears to be a singular arc (with u taking values near 0.02) as evident by $\mu_1 = 0$ (within numerical precision). The second burn appears to be a touch point case with μ_1 near zero but its slight uptick drives u towards its maximum value of 0.05.

The optimal trajectory along with the vectoring program is shown in Figure 6.13. Strictly speaking we do not know if the computed trajectory is optimal; however, we can conclude that it is at least an extremal by verifying the necessary conditions for optimality. Thus, one of the indicators of optimality is the agreement of the switching function with the control program shown in Figure 6.12. Many other indicators of optimality can be derived by an application of the Minimum Principle. For the purposes of brevity, we do not discuss them here; extensive examples of such ideas are presented elsewhere [13–15, 31, 36, 48, 51].

6.8 A simple extension to distributed space systems

A distributed space system (DSS) is a multi-agent control system that has long been recognized [52, 53] as a key technology area to enhance the scope of both military [52] and civilian [53] space applications. While much of the challenges are in distributing the functionality of a remote sensing problem, the difficulties in the design, control and

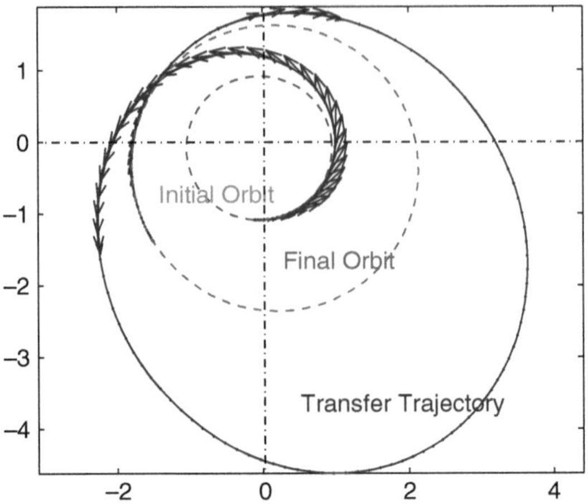

Fig. 6.13. A benchmark optimal low-thrust orbit transfer.

operations of the DSS arises chiefly as a result of managing the complexity of multiple agents. From the perspective of examining each agent separately, the problem is indeed quite formidable; however, a systems' approach to a DSS dramatically reduces some of the major problems in multi-agent control in much the same way as matrix analysis simplifies solving a system of linear equations by taking the view that a collection of linear equations is essentially one matrix equation. In order to appreciate how this perspective dramatically simplifies multi-agent control, consider a collection of $N_s \in \mathbb{N}$ spacecraft that constitute a DSS. Let $\mathbf{x}^i(t) \in \mathbb{R}^{N_{x^i}}$, $\mathbf{u}^i(t) \in \mathbb{R}^{N_{u^i}}$ denote the state and control vectors of the ith spacecraft at time t. Then, the fuel consumption for any one spacecraft, i, is given by,

$$J_s[\mathbf{x}^i(\cdot), \mathbf{u}^i(\cdot), t_0, t_f] = \int_{t_0}^{t_f} \|\mathbf{u}^i(t)\|_p \, \mathrm{d}t \tag{6.23}$$

By defining the system state and control variables for the DSS as,

$$\mathbf{x} = (\mathbf{x}^1, \ldots, \mathbf{x}^{N_s}) \tag{6.24}$$

$$\mathbf{u} = (\mathbf{u}^1, \ldots, \mathbf{u}^{N_s}) \tag{6.25}$$

the total fuel consumption is quite simply given by,

$$J[\mathbf{x}(\cdot), \mathbf{u}(\cdot), t_0, t_f] = \sum_{i=1}^{N_s} J_s[\mathbf{x}^i(\cdot), \mathbf{u}^i(\cdot), t_0, t_f]. \tag{6.26}$$

Note that Eq. (6.26) is not an l^p variant of the L^1 norm of $\mathbf{u}(\cdot)$ except in the special case of its l^1 version. This is one of the many reasons why solving multi-agent problems becomes difficult when compared to agent-specific techniques. On the other hand, when viewed through the prism of an optimal control problem, Eq. (6.26) is yet another nonsmooth cost functional. In certain applications, it may be necessary to require that each spacecraft in the DSS consume the same amount of propellant. This requirement can be stipulated by the constraints,

$$J_s[\mathbf{x}^i(\cdot), \mathbf{u}^i(\cdot), t_0, t_f] = J_s[\mathbf{x}^k(\cdot), \mathbf{u}^k(\cdot), t_0, t_f] \quad \forall\, i, k. \tag{6.27}$$

In generalizing Eq. (6.27) we write,

$$J_{ik}^L \le J_s[\mathbf{x}^i(\cdot), \mathbf{u}^i(\cdot), t_0, t_f] - J_s[\mathbf{x}^k(\cdot), \mathbf{u}^k(\cdot), t_0, t_f] \le J_{ik}^U, \tag{6.28}$$

where $J_{ik}^L, J_{ik}^U \in \mathbb{R}$ are part of the DSS requirements; for example, J_{ik}^L and J_{ik}^U may be non-zero numbers that facilitate a formulation of soft requirements on equal-fuel or rough numbers that facilitate budget allocation to the individual spacecraft. In any case, Eq. (6.28) is essentially a special case of a more general "mixed" state-control path constraint defined by,

$$\mathbf{h}^L \le \mathbf{h}(\mathbf{x}(t), \mathbf{u}(t), \mathbf{p}) \le \mathbf{h}^U, \tag{6.29}$$

where $\mathbf{h} : \mathbb{R}^{N_x} \times \mathbb{R}^{N_u} \times \mathbb{R}^{N_p} \to \mathbb{R}^{N_h}$ and $\mathbf{h}^L, \mathbf{h}^U, \in \mathbb{R}^{N_h}$. The components of $\mathbf{h}, \mathbf{h}^L$ and $\mathbf{h}^U$ are given by Eq. (6.28) while the components of $\mathbf{p}$ are just t_0 and t_f. Such constraints are discussed in more detail in Refs. [12, 15] and [32]. It is clear that pure control constraints are naturally included in Eq. (6.29). Additional components of $\mathbf{h}$ come from topological considerations. For example, by using a generic metric (not necessarily Euclidean) to define distances between two spacecraft, the requirement that no two spacecraft collide can be written as,

$$d(\mathbf{x}^i(t), \mathbf{x}^j(t)) \ge b^{i,j} > 0 \qquad \forall\, t \text{ and } i \ne j$$

where $d(\mathbf{x}^i, \mathbf{x}^j) \in \mathbb{R}_+$ is the distance metric. Clearly, collision constraints fall within the framework of the construction of the function, $\mathbf{h}$ (and its lower and upper bounds). Many other DSS requirements can be included as components of $\mathbf{h}$; for example, a broad class of formations can be defined by the inequality,

$$c_l^{i,j} \le d(\mathbf{x}^i(t), \mathbf{x}^j(t)) \le c_u^{i,j} \qquad \forall\, t, i, j, \tag{6.30}$$

where $c_l^{i,j}$ and $c_u^{i,j}$ are formation design parameters that are specific to a given space mission [12, 14].

The construction of the system dynamics for a DSS is quite simple. Suppose that the dynamics of each spacecraft of the DSS is given by (see for example, Problem O discussed in Section 6.7),

$$\dot{\mathbf{x}}^i(t) = \mathbf{f}^i(\mathbf{x}^i(t), \mathbf{u}^i(t); \mathbf{p}^i) \quad i = 1 \ldots N_s, \tag{6.31}$$

where $\mathbf{f}^i : \mathbb{R}^{N_{xi}} \times \mathbb{R}^{N_{ui}} \times \mathbb{R}^{N_{pi}} \to \mathbb{R}^{N_{xi}}$ is a given function, $\mathbf{u}^i \in \mathbb{U}^i \subseteq \mathbb{R}^{N_{ui}}$ and $\mathbf{p}^i \in \mathbb{R}^{N_{pi}}$ is a vector of (constant) design parameters. By using Eqs. (6.24) and (6.25), the dynamics of the DSS may be represented quite succinctly as,

$$\dot{\mathbf{x}}(t) = \mathbf{f}(\mathbf{x}(t), \mathbf{u}(t); \mathbf{p}) \quad \mathbf{u} \in \mathbb{U}, \tag{6.32}$$

where $\mathbb{U} = \mathbb{U}^1 \times \cdots \times \mathbb{U}^{N_s}$. Typically, the functions $\mathbf{f}^i$ are all the same so that $\mathbf{f}$ is simply N_s copies of $\mathbf{f}^1$. This fact can be exploited for real-time computation [38].

The optimal control system framework also facilitates a simplification of DSS management (design and operations) by exploiting the couplings between the dynamics, path constraints and the endpoint set, $\mathbb{E}$ (see Section 6.5). To observe this, consider a simple requirement of the form,

$$\mathbf{x}^i(t_f) \in \mathbb{E}^i \subset \mathbb{E}, \ \forall\, i. \tag{6.33}$$

In the framework of Problem B, it is sufficient to stipulate all the constraints of Eq. (6.33) as a single constraint,

$$\mathbf{x}^i(t_f) \in \mathbb{E}^i \subset \mathbb{E}, \qquad \text{for } \ i = 1 \tag{6.34}$$

or any other index, i. This is because, the path constraints (Eq. (6.29)) will automatically enforce the remainder of the constraints in Eq. (6.33). As a matter of fact, if Eq. (6.33) is chosen over Eq. (6.34), the feasible set may be empty as a result of over-specification. Consequently, we may want to use Eq. (6.33) during design considerations to explore possible over-specifications but use Eq. (6.34) during flight operations. In the latter context, we may designate $i = 1$ as the leader, but it essentially reduces to semantics rather than a leader-follower architecture. In other words, in this framework, there is no leader or follower; rather a true system of multiple spacecraft, or a DSS. Note however that if there was a mission requirement to designate a particular spacecraft as a leader and the others as followers, it can be easily accomplished by picking out the particular index, i, representing the leader. Then, when the leader moves along some trajectory, $t \mapsto \mathbf{x}^i$, the distance constraints along with any additional path constraints, Eq. (6.29), dictate how the remainder of the spacecraft must follow certain trajectories to meet the configuration constraints. Thus, if any one spacecraft had an additional configuration constraint, it would automatically transfer in some fashion to the remainder of DSS by way of the couplings between the various equations.

Certain formation-type DSS missions are vaguely defined in terms of periodicity simply because the engineering requirements are vague [52]. A natural way to account for these requirements is to adapt Bohr's notion of almost periodic functions [54, 55]. That is, rather than impose strict periodicity, we specify,

$$\varepsilon_l^i \le \mathbf{x}^i(t_0) - \mathbf{x}^i(t_f) \le \varepsilon_u^i \quad \forall\, i \quad \text{or} \quad \text{for } \ i = 1, \tag{6.35}$$

where ε_l^i and ε_u^i are formation design parameters representing almost periodicity and Eq. (6.35) is to be taken within the context of Eqs. (6.33) and (6.34). Note that Eq. (6.35) is not the same as specifying standard boundary conditions because the values of $\mathbf{x}^i(t_0)$ and $\mathbf{x}^i(t_f)$ are unknown. In the same spirit, we can define relative periodicity by writing $\mathbf{x}^i(t_0) - \mathbf{x}^j(t_0) = \mathbf{x}^i(t_f) - \mathbf{x}^j(t_f)$ or relax the equality for almost relative periodicity. That is, the DSS collective can have an aperiodic configuration if its constituents have almost relative periodicity. This is one of the reasons why the proper way to view Eq. (6.35) is in terms of the endpoint map, $\mathbb{E}$.

What is clear from the preceding discussions is that by treating the DSS as yet another system in an optimal control framework, the design and control of a DSS can be fully

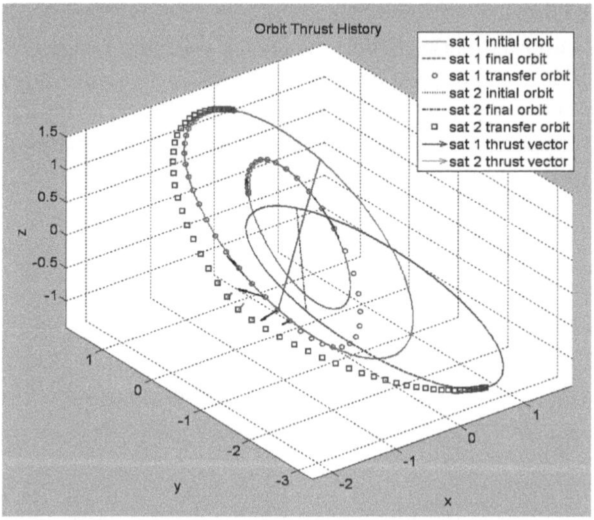

Fig. 6.14. A two-agent optimal space trajectory; from Ref. [15]. (see Color plate 8)

accounted for under the structure of Problem B discussed in Section 6.5. The missing details in Section 6.5 vis-à-vis the parameter $\mathbf{p}$ or the state constraints does not change the substance of the discussions as already alluded to earlier. Thus, by taking a systems' approach to the DSS formation problem, the mathematical problem can be framed under the constructs of Problem B. An application of this framework for formation design and control is discussed in Refs. [12, 13] and [14]. The same framework is used for configuration problems in Ref. [15]. A sample solution to a re-configuration problem is shown in Fig. 6.14. This plot [15] was obtained by casting the dynamics using the equinoctial element set for the state variables.

6.9 Conclusions

The L^1-optimal control problem forms a natural framework for formulating space trajectory optimization problems. Based on thruster configurations and the physics of the mass expulsion, several l^p variants of the L^1 norm of the thrust force can be articulated. Quadratic cost functions are inappropriate performance indices for space trajectory optimization problems. Nonsmooth issues dominate both theory and practice; in fact, practical problems are more likely to have non-convex, nonsmooth geometric structures. Transformation techniques can be applied to efficiently solve these problems. Real-time computation of the controls facilitates optimal feedback guidance and control. The same optimal control framework can be applied to design, control, and operate a distributed space system. These new possibilities are chiefly due to a confluence of two major tipping points that occurred in the late 1990s. The first advancement—and the most obvious one—was the widespread availability of extraordinary computing capabilities on ordinary computers. The second advancement was in the first-principles integration of optimal control theory with approximation theory under the unifying perspective of computation in Sobolev spaces. This perspective obviates the sensitivity issues arising from the symplectic structure of Hamiltonian systems. In addition, while requiring differentiability was once a reflection on the inadequacy of the available tools for analysis, it is no longer a major problem in either theory or computation.

Acknowledgments

I gratefully acknowledge the generous funding for this research provided by NASA-GRC and the Secretary of the Air Force. In particular, I would like to thank John P. Riehl (NASA) and Stephen W. Paris (Boeing) for their enthusiastic support of pseudospectral methods for solving mission planning problems for NASA.

References

1. Lu, P., Sun, H. and Tsai, B. (2003). Closed-loop endoatmospheric ascent guidance. *Journal of Guidance, Control and Dynamics*. **26**(2), pp. 283–294.
2. Teather, D. (2004). Halliburton cleared in fuel row. *The Guardian*, January 7.
3. Hibbard, R.L. (1996). Satellite on-orbit refueling: A cost-effect analysis. M.S. Thesis in Systems Technology, Naval Postgraduate School, Monterey, CA.
4. Marec, J.-P. (1979). *Optimal Space Trajectories*, Elsevier, New York.
5. Ross, I.M. (2004). How to find minimum-fuel controllers. *Proceedings of the AIAA Guidance, Navigation and Control Conference*, Providence, RI, August. AIAA Paper No. 2004-5346.
6. Neustadt, L.W. (1965). A general theory of minimum-fuel space trajectories. *SIAM Journal of Control*, Ser. A, **3**(2), pp. 317–356.
7. Clarke, F.H. (1990). *Optimization and Nonsmooth Analysis*, SIAM Publications, Philadelphia, PA.
8. Clarke, F.H., Ledyaev, Y.S., Stern, R.J. and Wolenski, P.R. (1998). *Nonsmooth Analysis and Control Theory*, Springer-Verlag, New York, NY.
9. Sovey, J.S., Rawlin, V.K. and Patterson, M.J. (2001). Ion propulsion development projects in U.S.: Space electric rocket test 1 to deep space 1. *Journal of Propulsion and Power*, **17**(3), pp. 517–526.
10. Croley, P.A. (2005). Reachable sets for multiple asteroid sample return missions. Astronautical Engineer's Degree Thesis, Naval Postgraduate School, Monterey, CA, June.
11. Kolmogorov, A.N. and Fomin, S.V. (1999). Elements of the Theory of Functions and Functional Analysis, **2**, Dover Publications, Mineola, NY.
12. Ross, I.M., King, J.T. and Fahroo, F. (2002). Designing optimal spacecraft formations. *Proceedings of the AIAA/AAS Astrodynamics Conference*, AIAA-2002-4635, Monterey, CA, August 5–8.
13. King, J.T. (2002). A Framework for designing optimal spacecraft formations. M.S. Thesis, Department of Aeronautical and Astronautical Engineering, Naval Postgraduate School, Monterey, CA, September.
14. Infeld, S.I., Josselyn, S.B., Murray, W. and Ross, I.M. (2004). Design and Control of Libration Point Spacecraft Formations. *Proceedings of the AIAA Guidance, Navigation and Control Conference*, Providence, RI, August. AIAA Paper No. 2004-4786.
15. Mendy, P.B. (2004). Multiple satellite trajectory optimization. Astronautical Engineer Thesis, Department of Mechanical and Astronautical Engineering, Naval Postgraduate School, Monterey, CA, December.
16. Bilimoria, K.D. and Wie, B. (1993). Time-optimal three-axis reorientation of a rigid spacecraft. *Journal of Guidance, Control, and Dynamics*, **16**(3), pp. 446–452.
17. Chyba, M. (2003). Underwater vehicles: A surprising non time-optimal path. *Proceedings of the 42nd IEEE Conference on Decision and Control*, Maui, Hawaii, December.
18. Bryson, A.E. (1999). Dynamic Optimization, Addison-Wesley Longman, Inc.

19. Pontryagin, L.S., Boltyanskii, V.G., Gamkrelidze, R.V. and Mischenko, E.F. (1962). The Mathematical Theory of Optimal Processes, Wiley-Interscience, New York, NY.
20. Adams, R.A. (1975). Sobolev Spaces, Academic Press, New York, NY.
21. Vinter, R.B. (2000). Optimal Control, Birkhäuser, Boston, MA.
22. Hager, W.W., Numerical analysis in optimal control, *International Series of Numererical Mathematics*, (K.-H. Hoffmann, I. Lasiecka, G. Leugering, J. Sprekels, and F. Troeltzsch, eds.), Birkhäuser, Basel, Switzerland, 2001, Vol. 139, pp. 83–93.
23. Sussmann, H.J. (2000). New theories of set-valued differentials and new versions of the maximum principle of optimal control theory. Nonlinear Control in the Year 2000, (A. Isidori, F. Lamnabhi-Lagarrigue, and Respondek, W., eds.), Springer-Verlag, London, pp. 487–526.
24. Rockafellar, R.T. (1993). Lagrange multipliers and optimality. *SIAM Review*, **35**, pp. 183–238.
25. Mordukhovich, B.S. (2005). *Variational Analysis and Generalized Differentiation, I: Basic Theory*, Vol. 330 of Grundlehren der Mathematischen Wissenschaften [Fundamental Principles of Mathematical Sciences] Series, Springer, Berlin.
26. Bryson, A.E. and Ho, Y.C. (1975). Applied Optimal Control, Hemisphere, New York.
27. Young, L.C. (1969). Lectures on the Calculus of Variations and Optimal Control Theory, Saunders, Philadelphia, PA.
28. Silva, G.N. and Vinter, R.B. (1997). Necessary conditions for optimal impulsive control problems. *SIAM Journal of Control and Optimization*, **35**(6), pp. 1829–1846, November.
29. Lawden, D.F. (1963). Optimal Trajectories for Space Navigation, Butterworths, London.
30. Ross, I.M. and Fahroo, F. (2004). Pseudospectral knotting methods for solving optimal control problems. *Journal of Guidance, Control and Dynamics*, **27**(3), pp. 397–405.
31. Ross, I.M. (2005) A roadmap for optimal control: The right way to commute. *Annals of the New York Academy of Sciences*, **1065**, pp. 210–231, December.
32. Ross, I.M. and D'Souza, C.D. (2005). Hybrid optimal control framework for mission planning. *Journal of Guidance, Control and Dynamics*, **28**(4), July–August, pp. 686–697.
33. Betts, J.T. (2001) Practical Methods for Optimal Control Using Nonlinear Programming, SIAM: Advances in Control and Design Series, Philadelphia, PA.
34. Betts, J.T. (1998). Survey of numerical methods for trajectory optimization. *Journal of Guidance, Control, and Dynamics*, **21**(2), pp. 193–207.
35. Hager, W.W. (2000). Runge-Kutta methods in optimal control and the transformed adjoint system. *Numerische Mathematik,* **87**, pp. 247–282.
36. Fahroo, F. and Ross, I.M. (2001). Costate estimation by a legendre pseudospectral method. *Journal of Guidance, Control and Dynamics*, **24**(2), pp. 270–277.
37. Ross, I.M. (2005). A historical introduction to the covector mapping principle. *AAS/AIAA Astrodynamics Specialist Conference*, Tahoe, NV, August 8–11, Paper AAS 05-332.
38. Ross, I.M. and Fahroo, F. (2006). Issues in the real-time computation of optimal control. *Mathematical and Computer Modelling*, An International Journal, **43**(9–10), pp. 1172–1188, May.
39. Sekhavat, P., Fleming A. and Ross, I.M. (2005). Time-Optimal Nonlinear Feedback Control for the NPSAT1 Spacecraft. *Proceedings of the 2005 IEEE/ASME International Conference on Advanced Intelligent Mechatronics*, AIM 2005, July 24–28, Monterey, CA.
40. Yan, H., Lee, D.J., Ross, I.M. and Alfriend, K.T. (2005). Real-time outer and inner loop optimal control using DIDO. *AAS/AIAA Astrodynamics Specialist Conference*, Tahoe, NV, August 8–11, Paper AAS 05-353.
41. Goldstine, H.H. (1981). A History of the Calculus of Variations from the 17th to the 19th Century, Springer-Verlag, New York, NY, p. 110.
42. Cullum, J. (1969). Discrete approximations to continuous optimal control problems. *SIAM Journal of Control*, **7**(1), February, pp. 32–49.
43. Dontchev, A.L., Hager, W.W. and Veliov, V.M. (2000). Second-order Runge-Kutta approximations in control constrained optimal control problems. *SIAM Journal of Numerical Analysis*, **38**(1), pp. 202–226.
44. Betts, J.T., Biehn, N. and Campbell, S.L. (2000). Convergence of nonconvergent IRK discretizations of optimal control problems with state inequality constraints. *SIAM Journal of Scientific Computation* **23**(6), pp. 1981–2007.
45. Trefethen, L.N. (2000). *Spectral Methods in MATLAB,* SIAM, Philadelphia, PA.
46. Ross, I.M. and Fahroo, F. (2002). A Perspective on Methods for Trajectory Optimization. *Proceedings of the AIAA/AAS Astrodynamics Conference*, Monterey, CA, August. AIAA Paper No. 2002-4727.

47. Boggs, P.T., Kearsley, A.J. and Tolle, J.W. (1999). A global convergence analysis of an algorithm for large-scale nonlinear optimization problems. *SIAM Journal of Optimization*, **9**(4), pp. 833–862.
48. Ross, I.M. and Fahroo, F. (2003). Legendre pseudospectral approximations of optimal control problems. *Lecture Notes in Control and Information Sciences*, **295**, Springer-Verlag, New York.
49. Elnagar, J., Kazemi, M.A. and Razzaghi, M. (1995). The pseudospectral legendre method for discretizing optimal control problems. *IEEE Transactions on Automatic Control*, **40**(10), pp. 1793–1796.
50. Moyer, H.G. and Pinkham, G. (1964). Several trajectory optimization techniques, Part II: Applications. Computing Methods in Optimization Problems, (A.V. Balakrishnan and L.W. Neustadt, eds.), New York, Academic Press.
51. Ross, I.M. (2004). User's manual for DIDO: A MATLAB application package for solving optimal control problems. Tomlab Optimization, Sweden, February.
52. New World Vistas (1995). Summary Volume, USAF Scientific Advisory Board, December.
53. Vincent, M.A. and Bender, P.L. (1987). The orbital mechanics of a space-borne gravitational wave experiment. Advances in the Astronautical Sciences, Astrodynamics, **65**, Part II, P. 1346. (J.K. Soldner, et al. ed.), Paper AAS 87-523. Full paper in AAS Microfiche Series, **55**.
54. Fischer, A. (1996). Structure of fourier exponents of almost periodic functions and periodicity of almost periodic functions. *Mathematica Bohemica*, **121**(3), pp. 249–262.
55. Corduneanu, C. (1968). Almost Periodic Functions, John Wiley, New York.

7 Orbital Mechanics of Propellantless Propulsion Systems

COLIN R. MCINNES[1] AND MATTHEW P. CARTMELL[2]
[1]*Department of Mechanical Engineering, University of Strathclyde*
[2]*Department of Mechanical Engineering, University of Glasgow*

Contents

7.1 Introduction

Conventional spacecraft are limited in their ability to deliver high-energy missions by a fundamental reliance on reaction mass. However, this constraint can in principle be overcome by a class of propellantless propulsion systems which either extract momentum from the environment (solar sails) or balance momentum through payload exchanges (tethers). This chapter provides a brief introduction to the physics of solar sail and tether propulsion systems, an analysis of some novel aspects of their orbital mechanics and a review of potential applications.

The classical rocket equation starkly illustrates the limitations of reaction propulsion, through an exponential scaling of initial mass m_1 with mission ΔV for some delivered mass m_2, such that $m_1 = m_2 \exp(\Delta V / g_o I_{sp})$, where I_{sp} is the specific impulse of the propulsion system ($g_o = 9.81\,\mathrm{ms}^{-2}$). Attempts to overcome this scaling law rely on improved propulsion technologies (higher specific impulse) or reducing payload mass through miniaturisation. While such approaches have been successful, as evidenced by recent flight tests of solar electric propulsion, the envelope of possible missions is still constrained by the reliance on a finite mass of propellant.

Solar sailing could overcome the limitations of the rocket equation by extracting momentum from the flux of photons which is continually emitted by the Sun. A large articulated reflector is used to reflect photons, changing their momentum, and so exerting a reaction force on the sail. For an ideal solar sail, the net force exerted by incident and reflected photons is normal to the sail surface. Therefore, by rotating the sail, the thrust vector can be directed in a hemisphere about the Sun-sail line. However, due to the reduction in projected sail area and the reduction in the component of photon momentum transferred normal to the sail, the magnitude of the thrust decreases rapidly as the sail normal is pitched away from the Sun-sail line. It can be shown that the thrust magnitude scales as $(\hat{\mathbf{r}}.\mathbf{n})^2$, where $\hat{\mathbf{r}}$ is the unit vector directed along the Sun-sail line and $\mathbf{n}$ is the unit vector normal to the sail surface. By rotating the sail, the solar sail can either gain or lose orbital angular momentum and so can spiral inwards towards the Sun or outwards away from the Sun. In addition, by pitching the sail such that a component of the thrust is directed out of the orbit plane, the solar sail orbit inclination can be 'cranked' up or down, allowing a rendezvous with any target body in the solar system. While solar sailing appears to enable new high-energy missions and exotic highly non-Keplerian orbits, challenges are posed to engineer a sail assembly with a low areal density which can be reliably deployed in-orbit.

Tethers also attempt to overcome the limitations of the rocket equation, by balancing the flow of momentum through the tether system. While future concepts for orbital towers offer the possibility of truly low cost access to space, nearer term concepts for momentum exchange tethers can enable the transfer of large payloads to and from low Earth orbit, without the use of propellant. In particular, staged tethers can allow the transfer of mass from low Earth orbit to a rotating tether orbiting the moon, which delivers the payload to the lunar surface. If mass is also transferred from the lunar surface back to low Earth orbit, then in principle the flow of momentum is balanced, allowing the tether transportation system to operate without the use of reaction mass.

In summary, both solar sails and momentum exhange tethers can, in principle, overcome the fundamental limitations imposed by reaction propulsion. For solar sails, new high-energy mission concepts are enabled, along with families of exotic highly non-Keplerian orbits described in this chapter. For momentum exchange tethers, large payloads can be transported at low cost by balancing the flow of momentum through the system, again described in this chapter. By stepping beyond the limitations of the rocket equation, propellantless propulsion has the potential to enable new and exciting possibilities for the future which are currently impossible for conventional reaction propulsion devices.

7.2 Solar sailing

7.2.1 Introduction

The concept of solar sailing can be traced to various authors, including the Russian pioneers Tsiolkovsky [1] and Tsander [2]. However, serious development had to wait until a major NASA/JPL study during the mid-1970s for a proposed rendezvous mission to comet Halley [3]. Although being finally approved for flight, the study sparked international interest in solar sailing for future mission applications. It is only in recent years that sustained efforts have been made to develop the component technologies (reflective thin films and deployable booms) and to integrate these into a practical solar sail assembly. Development programmes are underway at both NASA and ESA to develop solar sail

Fig. 7.1. NASA ground test of a 20×20 m solar sail (NASA/ATK).

technology through to flight status using ground testing and ultimately in-orbit demonstration missions (Figure 7.1). Should such demonstration missions prove successful, there are a wide range of potential mission applications which will then be enabled. The key missions on such a development roadmap are the Geosail mission, which uses a small solar sail to artificially precess a long elliptical orbit to maintain a space physics payload within the Earth's geomagnetic tail [4]; Geostorm, a space weather mission located sunward of the classical Sun–Earth L_1 point [5]; Polar Observer, a mission to station an imaging payload at an artificial equilibrium point high above L_1 [6]; Solar Polar Orbiter (SPO), a solar physics payload delivered to a close polar orbit about the Sun [7]; Interstellar Heliopause Probe (IHP), a small payload delivered to the heliopause at 200 AU in 25 years using a high performance solar sail [8]. In addition, solar sailing appears ideally suited to high-energy and/or long duration missions such as Mercury sample return [9], comet nucleus sample return and multiple small body rendezvous missions [10], as well as more exotic mission applications [11].

7.2.2 *Solar sail sizing*

The fundamental measure of performance of a solar sail is its characteristic acceleration, defined as the light pressure-induced acceleration experienced by the solar sail while oriented normal to the Sun at a heliocentric distance of 1 AU [12]. The characteristic acceleration is a function of both the efficiency of the solar sail design and the mass of the payload. At a distance of 1 AU the magnitude of the solar radiation pressure P exerted on a perfectly absorbing surface is $4.56 \times 10^{-6}\,\mathrm{Nm}^{-2}$. Therefore, allowing for the reflection of photons (factor of 2) and the finite efficiency of the sail η, the characteristic acceleration a_o is defined by

$$a_o = \frac{2\eta P}{\sigma}, \quad \sigma_T = \frac{m_T}{A}, \tag{7.1}$$

where σ_T is the total solar sail loading, with m_T the total mass of the solar sail and its payload and A the sail area. The sail efficiency η (typically ~ 0.85) is a function of both the optical properties of the sail film and the sail shape due to billowing and wrinkling. The total mass of the solar sail will now be partitioned into two components, the sail film and structural mass m_S and the payload mass m_P. Therefore, the characteristic acceleration of the solar sail may now be written as

$$a_o = \frac{2\eta P}{\sigma_S + (m_P/A)}, \quad \sigma_S = \frac{m_S}{A}, \tag{7.2}$$

where σ_S is the mass per unit area of the sail assembly. This so-called sail assembly loading is a key technology parameter and is a measure of the thickness of the sail film and the efficiency of the solar sail structural and mechanical design. Low performance solar sail concepts center on the use of commercially available 7.5 μm Kapton film with a projected a sail assembly loading of order $30\,\mathrm{gm}^{-2}$, which is adequate for near term technology demonstration missions. Other development work to fabricate ultra-thin sail films with a thickness of order 2 μm, and high stiffness, low mass booms is leading to a sail assembly loading of order $5\,\mathrm{gm}^{-2}$ [13], or even as low as $1\text{–}2\,\mathrm{gm}^{-2}$ for high performance spinning disk sails [14].

Now that the key solar sail design parameters have been defined, the process of sizing a solar sail will be considered. From Eq. (7.2) it can be seen that the solar sail payload mass may be written as

$$m_p = \left[\frac{2\eta P}{a_o} - \sigma_S\right] A. \tag{7.3}$$

Similarly, from Eq. (7.1) the total mass of the solar sail may be written as

$$m_T = \frac{2\eta P A}{a_o} \tag{7.4}$$

For a required characteristic acceleration, Eqs. (7.3) and (7.4) may now be used to size a solar sail while imposing constraints on the total mass of the solar sail to satisfy the capacity of the launch vehicle. A typical design space is shown in Figure 7.2 for a characteristic acceleration of $0.25\,\mathrm{mms}^{-2}$, which is representative of the level of performance required for science missions such as a Mercury sample return. Both the payload mass and the total launch mass are shown. It is clear that for a payload of order 500 kg and a sail assembly loading of order $5\,\mathrm{gm}^{-2}$, a large sail is required with a sail side longer than 100 m. This requirement clearly poses challenges for the reliable mechanical deployment of large, low mass structures and the fabrication, packing and deployment of thin reflective films.

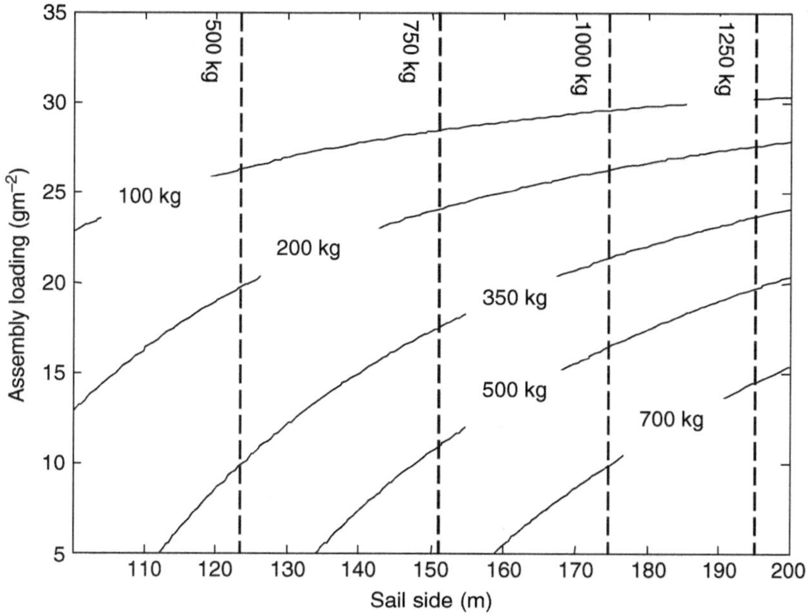

Fig. 7.2. Solar sail design space (solid line: payload mass, dashed line: total mass).

In addition to the design parameters discussed above, an additional parameter of interest can be defined. The payload mass fraction κ, defined by m_p/m_T can be obtained from Eqs. (7.3) and (7.4) as

$$\kappa = 1 - \frac{a_o \sigma_S}{2\eta P} \tag{7.5}$$

This is clearly another key parameter and is a measure of the efficiency of use of the solar sail. For a sail with a characteristic acceleration of $0.25\,\mathrm{mms^{-2}}$, which is again representative of the level of performance required for initial science missions, and a sail assembly loading of order $5\,\mathrm{gm^{-2}}$, the resulting payload mass fraction for the solar sail is 0.84, assuming an efficiency η of 0.85. It can be seen that improvements in sail assembly loading can either be used to increase the sail characteristic acceleration, and so reduce trip times, or can be used to improve the sail payload mass fraction, allowing a larger payload to be delivered for a fixed launch mass.

7.2.3 Solar sail performance

Since solar sails do not expel reaction mass, the conventional definition of specific impulse is inappropriate. This conventional definition relates the change in momentum of the spacecraft to the weight of propellant expelled. Since solar sails do not expel propellant they have, in principle, infinite specific impulse. However, for a finite mission duration, only a finite total impulse will be delivered by the solar sail. Infinite specific impulse is only available for infinite mission duration. In order to circumvent this difficulty with the conventional definition of specific impulse, an effective specific impulse [15] will be defined as the total impulse delivered per unit weight of propulsion system

$$I_{sp} = \frac{1}{W}\int_0^T F\mathrm{d}t, \tag{7.6}$$

where F is the thrust delivered for mission duration T and W is the weight of the propulsion system. For a solar sail of total mass m_T and characteristic acceleration a_o at solar distance R (AU), the sail thrust is given by

$$F = m_T a_o \cos^2\alpha \frac{1}{R^2}, \tag{7.7}$$

where α is the pitch angle of the sail normal relative to the Sun-sail line. The weight of the propulsion system will now be defined as the weight of the sail assembly so that for a sail assembly of mass m_S

$$W = m_s g_o. \tag{7.8}$$

The weight of the sail assembly can be then written in a more useful form as

$$m_S = m_T(1-\kappa), \tag{7.9}$$

where κ is the payload mass fraction of the solar sail, defined by Eq. (7.5). Therefore, the effective specific impulse of the solar sail can now be written as

$$I_{sp} = \frac{1}{1-\kappa}\frac{a_o}{g_o}T\left\langle\frac{\cos^2 a}{R^2}\right\rangle, \tag{7.10}$$

where <> indicates the mean value over mission duration T. It can be seen that the effective specific impulse of the solar sail increases linearly with mission duration and as the inverse square of mean solar distance. In addition, $I_{sp} \to \infty$ as $\kappa \to 1$ since the weight of the propulsion system vanishes in this limit.

For a solar sail to be effective it can be seen that it must have a large payload mass fraction, and be used for a long duration. Missions such as high-energy comet sample returns will therefore make significantly better use of solar sails than, for example, payload delivery to the Moon or Mars where the effective ΔV is low (unless multiple trips are considered). In addition, inner solar system missions where R is small, such as Mercury sample return, will also make effective use of solar sailing. Deep space missions can also be effective, but a close pass to the Sun is required. This can be seen from Eq. (7.10) and is also evident from trajectory optimisation studies where the sail passes close to the Sun before being accelerated to some cruise speed for deep space payload delivery [8]. Earth escape also makes efficient use of solar sailing, although escape times can be long [16].

The effective specific impulse for a range of solar sails is shown in Figure 7.3, where it is assumed that the mean pitch angle α is 35° to maximise the transverse thrust, and the payload mass fraction κ is 2/3. It is clear that even if the sail has a low characteristic acceleration, it can deliver an extremely large effective specific impulse if the mission duration is long. This is exactly the relationship which makes the mission applications of the artificial Lagrange points discussed later in Sections 7.4 and 7.5 so attractive.

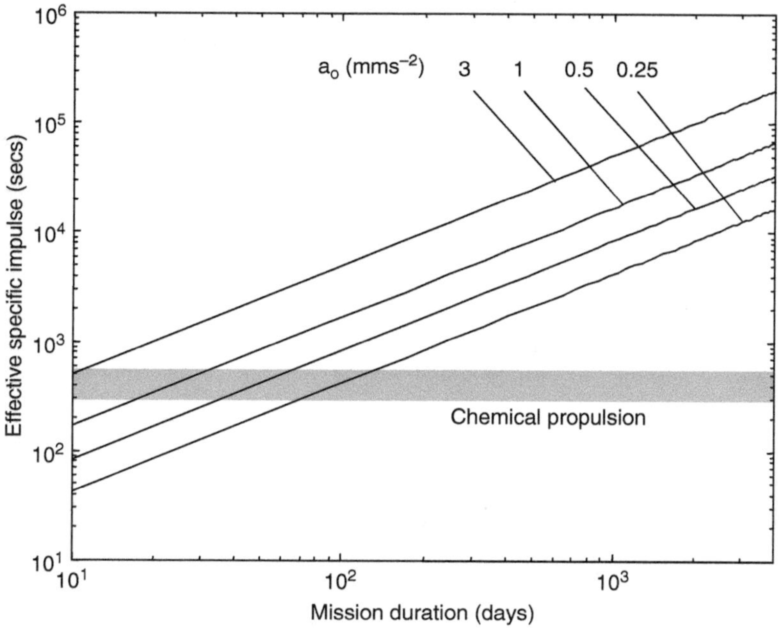

Fig. 7.3. Effective solar sail specific impulse.

7.3 Solar sail orbital mechanics

7.3.1 Introduction

The ratio of the solar radiation pressure force to the solar gravitational force exerted on the sail is defined by the sail lightness number β. Since both solar radiation pressure and solar gravity have an inverse square variation with solar distance, the sail lightness number is a constant for a given sail mass and area. It can be shown by considering the ratio of radiation pressure to gravitational forces that the sail lightness number is related to the total solar sail loading by $\sigma_T(gm^{-2}) = 1.53/\beta$, [12]. A high performance solar sail with a lightness number of 1 ($\sigma_T = 1.53\,\text{gm}^{-2}$) corresponds to a characteristic acceleration of $5.96\,\text{mms}^{-2}$. Such an advanced solar sail could exactly balance solar gravity, although near term solar sails are likely to have a characteristic acceleration of order $0.25\,\text{mms}^{-2}$, as noted earlier.

An ideal, plane solar sail will now be considered moving relative to an inertial frame of reference with origin at the Sun. The vector equation of motion of the solar sail is then defined by

$$\frac{d^2\mathbf{r}}{dt^2} + \frac{\mu}{r^2}\hat{\mathbf{r}} = \beta\frac{\mu}{r^2}\left(\hat{\mathbf{r}}.\mathbf{n}\right)^2\mathbf{n}, \tag{7.11}$$

where $\mathbf{r}$ is the position vector of the spacecraft with respect to the Sun, with $\hat{\mathbf{r}}$ the associated unit vector. The product of the gravitational constant and the mass of the Sun is defined by μ. For an ideal sail, the thrust vector is aligned along the sail unit normal $\mathbf{n}$, with the sail pitch angle α, defined as the angle between the sail normal and the radius vector such that $\cos\alpha = \hat{\mathbf{r}}.\mathbf{n}$.

7.3.2 Conic section orbits

When the sail normal is directed along the Sun-line, such that $\mathbf{n} = \hat{\mathbf{r}}$, families of conic section orbits are obtained with a modified gravitational parameter, $\tilde{\mu} = \mu(1-\beta)$, since again solar gravity and solar radiation pressure both vary as the inverse square of the solar distance. For a lightness number $\beta = 0$, a circular Keplerian orbit will be assumed. Then, with $\beta \neq 0$ such an orbit becomes elliptical for $0 \prec \beta \prec 1/2$. With a lightness number $\beta = 1/2$ there is a transition from an elliptical orbit to a hyperbolic orbit through a parabolic orbit, which defines the lightness number necessary for direct escape. When the lightness number increases such that $1/2 \prec \beta \prec 1$, a hyperbolic orbit is obtained. Then, when the lightness number is exactly unity, there is the interesting situation where solar gravity is exactly balanced by solar radiation pressure. This could enable rectilinear orbits or could allow the solar sail to levitate, stationary relative to the Sun. With extremely high performance solar sails exhibiting lightness numbers of greater than unity, the Sun now becomes placed at the opposite focus of an (inverted) hyperbolic orbit since the solar radiation pressure force exceeds the solar gravitational force acting on the solar sail [17].

7.3.3 Logarithmic spiral trajectories

When the solar sail thrust is orientated at a fixed, non-zero pitch angle to the Sun-sail line it can be shown that the solar sail can follow a logarithmic spiral trajectory [18–20]. The radial component of the sail thrust reduces the effective gravitational force on the sail, however the component of thrust in the transverse direction acts to increase (or decrease) the orbital angular momentum of the solar sail. For a logarithmic spiral trajectory, the local solar sail speed is always less than the local circular orbit speed. This means that coplanar transfer by logarithmic spiral, between two circular orbits, cannot be achieved without hyperbolic excess at launch to place the solar sail onto the logarithmic spiral, and then an impulse to circularise the orbit on arrival at the final circular orbit. These discontinuities in the boundary conditions pose problems in the practical application of logarithmic spirals to orbit transfers.

7.3.4 Minimum-time trajectories

For practical mission analysis purposes, optimization is required to minimize the transfer time between any two orbits. Since the sail attitude will be time varying, the boundary conditions required for the transfer may be met without the use of initial and final impulses, as required for the logarithmic spiral trajectory. First, the vector equation of motion will be re-cast as two, first order equations as

$$\dot{\mathbf{r}} = \mathbf{v} \tag{7.12a}$$

$$\dot{\mathbf{v}} = -\frac{\mu}{r^2}\hat{\mathbf{r}} + \beta\frac{\mu}{r^2}\left(\hat{\mathbf{r}}.\mathbf{n}\right)^2\mathbf{n} \tag{7.12b}$$

with boundary conditions imposed on the solar sail trajectory such that the solar sail state is defined by $(\mathbf{r}_o, \mathbf{v}_o)$ at initial time t_o and $(\mathbf{r}_f, \mathbf{v}_f)$ at final time t_f, where $t_f = t_o + T$. The goal is now to minimize the transfer time T subject to the constraints imposed by the boundary conditions of the transfer. To proceed with the optimization process the control Hamiltonian function for the problem is formed. To ensure optimization, the Hamiltonian must be maximized at all points along the trajectory through an appropriate choice of sail attitude **n**. The control Hamiltonian H is defined as

$$H = \mathbf{p}_r.\mathbf{v} - \frac{\mu}{r^2}\mathbf{p}_v.\hat{\mathbf{r}} + \beta\frac{\mu}{r^2}\left(\hat{\mathbf{r}}.\mathbf{n}\right)^2\mathbf{p}_v.\mathbf{n}, \tag{7.13}$$

where $\mathbf{p}_r$ and $\mathbf{p}_v$ are the co-states for position and velocity [21]. The velocity co-state is also referred to as the primer vector and defines the direction along which the solar sail thrust should be maximized. Note that unlike other low thrust transfer problems there is no co-state for the solar sail mass since it is clearly constant. The rate of change of the co-states is then obtained from the control Hamiltonian as

$$\dot{\mathbf{p}}_r = -\frac{\partial H}{\partial \mathbf{r}} \tag{7.14a}$$

$$\dot{\mathbf{p}}_v = -\frac{\partial H}{\partial \mathbf{v}} \tag{7.14b}$$

so that

$$\dot{\mathbf{p}}_r = \frac{\mu}{r^3}\mathbf{p}_v - \frac{3\mu}{r^5}\left(\mathbf{p}_r.\mathbf{r}\right)\mathbf{r} + 2\beta\frac{\mu}{r^3}\left(\hat{\mathbf{r}}.\mathbf{n}\right)\left(\mathbf{p}_v.\mathbf{n}\right)\left(\mathbf{n} + 2\left(\hat{\mathbf{r}}.\mathbf{n}\right)\hat{\mathbf{r}}\right) \tag{7.15a}$$

$$\dot{\mathbf{p}}_v = -\mathbf{p}_r. \tag{7.15b}$$

The sail attitude **n** which maximises the Hamiltonian can then be found from Eq. (7.13) as a function of the co-states $\mathbf{p}_r$ and $\mathbf{p}_v$. The equations of motion must therefore be integrated along with the co-state equations, and an iterative numerical algorithm used to determine the initial co-states that provide a minimum-time trajectory that satisfies the boundary conditions of the transfer. An example minimum-time trajectory from Earth to Mercury is shown in Figure 7.4. As an alternative to the indirect approach of optimal control theory, direct methods can be used [10]. These algorithms discretize the sail attitude time history and minimize the transfer time using methods such as sequential quadratic programming. While direct methods are not as accurate as indirect methods, due to the discretization of the sail attitude, they can be more robust and flexible. Novel evolutionary algorithms have also been recently applied to the solar sail trajectory optimization problem [22], as have minimum-time transfers to 1 year Earth synchronous circular orbits [23].

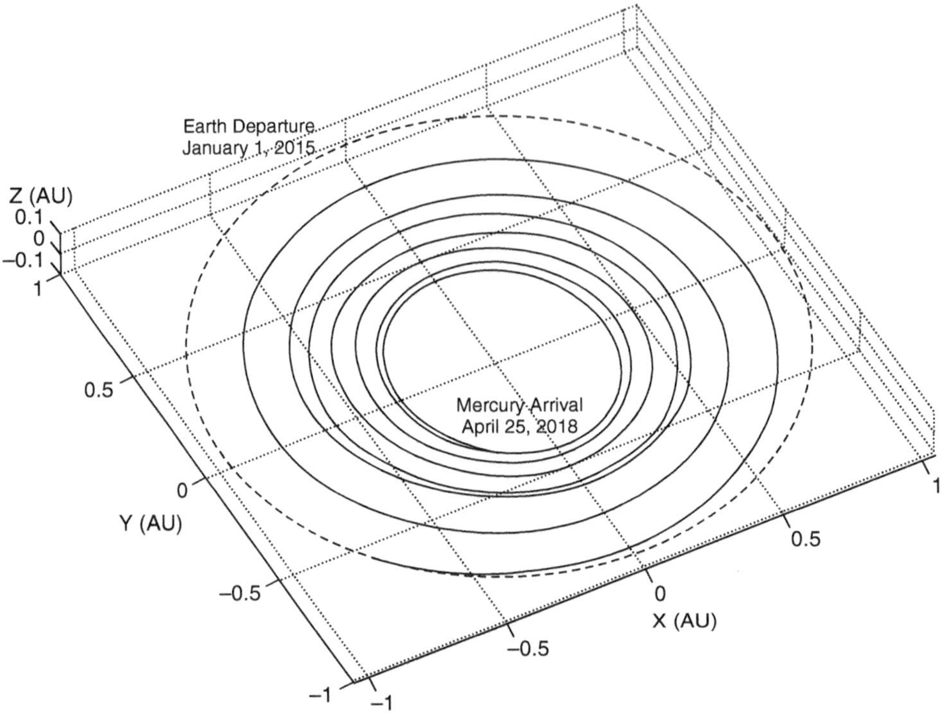

Fig. 7.4. Minimum-time spiral from Earth to Mercury ($a_0 = 0.25\,\mathrm{mms}^{-2}$).

7.4 Artificial three-body equilibria for solar sails

7.4.1 Non-Keplerian orbits

Due to the continually available thrust from solar radiation pressure, solar sails are capable of exotic, highly non-Keplerian orbits. Although some of these orbits require advanced, high performance solar sails, others are possible using relatively modest solar sails. The solar sail performance required for these orbits is a function of the local gravitational acceleration. Therefore, to displace an orbit high above the plane of the solar system requires an extremely high characteristic acceleration, while to generate an artificial Lagrange point near the Earth may only require a near-term solar sail. While these highly non-Keplerian orbits are not, in principle, forbidden for other forms of low-thrust propulsion, they can only be achieved for a limited duration, fixed by the propellant mass fraction of the spacecraft. However, a solar sail stationed at an artificial Lagrange point and requiring a low-characteristic acceleration can still deliver an extremely high effective specific impulse if the sail film is long-lived and so used for an extended duration, as discussed in Section 7.2.2.

Using an advanced solar sail it would be possible to choose its characteristic acceleration so that the solar radiation pressure force exactly balances the local solar gravitational force, corresponding to a lightness number β of 1. This is possible since, again, both of these forces have an inverse square variation with solar distance. The required characteristic acceleration for such a force balance is of order 6 mms^{-2}, corresponding to a mass per unit area of only 1.5 gm^{-2}, as discussed in Section 7.3.1. Such a high performance solar sail would enable solar physics missions which levitate above the solar poles, providing continuous observations, or indeed hover at any particular location in the solar system. Such a solar sail could also be used to displace circular Sun-centred orbits high above the plane of the solar system, with the orbit period chosen to be synchronous with the Earth or some other solar system body. Possible applications include stationing an infra-red telescope above the obscuring zodiacal dust within the ecliptic plane.

Using a more modest solar sail, the location of the Sun–Earth Lagrange points can be artificially displaced. For example, the interior L_1 point, 1.5 million km sunward of the Earth, is a favored location for solar physics missions. Since the solar sail adds an extra force to the dynamics of the problem the location of the L_1 point can be artificially displaced, closer to the Sun or even above the ecliptic plane. Since the local gravitational acceleration in the vicinity of L_1 is small (since centripetal force, solar and Earth gravity almost balance), only modest solar sails are required to provide a significant displacement of the classical L_1 point. For example, a solar sail with a characteristic acceleration of 0.25 mms^{-2} can double the distance of the L_1 point form the Earth. Such a new sunward equilibrium location appears useful for providing early warning of disruptive solar plasma storms before they reach Earth, an indeed formed the basis for the Geostorm mission concept, discussed later in Section 7.5.1. A solar sail with double the performance can be permanently stationed high above (or below) the classical L_1 point so that it appears above the Arctic (or Antarctic) regions of the Earth, again to be discussed later in Section 7.5.2.

7.4.2 Artificial three-body equilibria: ideal solar sail

This section will investigate the possibility of artificial Lagrange points for near-term, low performance solar sails [24–27]. Equilibrium solutions will be obtained for an ideal solar sail and then the problem will be re-visited with a more realistic partially reflecting solar sail. Apart from reducing the magnitude of the radiation pressure force exerted on the solar sail, the finite absorption of a realistic sail means that the radiation pressure force vector is no longer directed normal to the sail surface. Due to this effect, it will be shown that the volume of space available for artificial Lagrange points is extremely sensitive to the solar sail reflectivity.

Equilibrium solutions for an idealized, perfectly reflecting solar sail will now be derived. The ideal sail will be considered in a frame of reference co-rotating with two primary masses m_1 (Sun) and m_2 (Earth) at constant angular velocity $\boldsymbol{\omega}$, as shown in Figure 7.5. The sail attitude is again defined by a unit vector $\mathbf{n}$ normal to the sail surface, fixed in the rotating frame of reference. The units of the problem will be chosen such that the gravitational constant, the distance between the two primary masses and their sum are all taken to be unity.

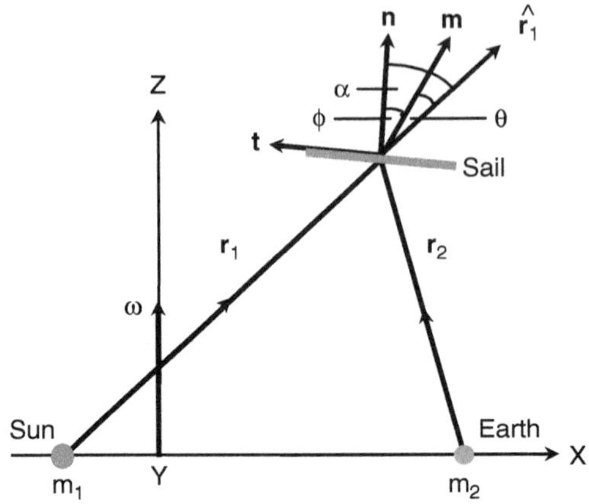

Fig. 7.5. Sun–Earth restricted three-body problem with a partially reflecting solar sail.

The vector equation of motion for a solar sail in this rotating frame of reference may then be written as

$$\frac{d^2\mathbf{r}}{dt^2} + 2\boldsymbol{\omega} \times \frac{d\mathbf{r}}{dt} + \nabla U = \mathbf{a} \tag{7.16}$$

with the three-body gravitational potential U and the solar radiation pressure acceleration $\mathbf{a}$ defined by

$$U = -\left[\frac{1}{2}(x^2+y^2) + \frac{1-\mu}{r_1} + \frac{\mu}{r_2}\right], \tag{7.17a}$$

$$\mathbf{a} = \beta\frac{1-\mu}{r_1^2}(\hat{\mathbf{r}}_1.\mathbf{n})^2\,\mathbf{n}, \tag{7.17b}$$

where $\mu = m_1/(m_1+m_2)$ is the mass ratio of the system and the sail position vectors are defined as $\mathbf{r}_1 = (x+\mu, y, z)$ and $\mathbf{r}_2 = (x-(1-\mu), y, z)$.

Equilibrium solutions are now required in the rotating frame of reference so that the first two terms of Eq. (7.16) vanish. The five classical Lagrange points are then obtained as the solutions to $\nabla U = 0$ with $\hat{\mathbf{r}}_1.\mathbf{n} = 0$ and so $\mathbf{a} = 0$. However, for $\hat{\mathbf{r}}_1.\mathbf{n} \succ 0$ there is an additional acceleration $\mathbf{a}$ which is a function of the lightness number β and the sail attitude $\mathbf{n}$ so that new artificial equilibrium solutions may be generated. Since the vector $\mathbf{a}$ is oriented in direction $\mathbf{n}$, taking the vector product of $\mathbf{n}$ with Eq. (7.16) it follows that

$$\nabla U \times \mathbf{n} = 0 \Rightarrow \mathbf{n} = \lambda \nabla U, \tag{7.18}$$

where λ is an arbitrary scalar multiplier. Using the normalization condition $|\mathbf{n}| = 1$, λ is identified as $|\nabla U|^{-1}$ so that the required sail attitude is defined by

$$\mathbf{n} = \frac{\nabla U}{|\nabla U|} \tag{7.19}$$

which can be used to obtain the sail pitch angle α, since $\cos\alpha = \hat{\mathbf{r}}.\mathbf{n}$. The required sail lightness number may also be obtained by taking a scalar product of Eq. (7.16) with $\mathbf{n}$. Again requiring an equilibrium solution it is found that

$$\beta = \frac{r_1^2}{(1-\mu)} \frac{\nabla U.\mathbf{n}}{(\hat{\mathbf{r}}_1.\mathbf{n})^2} \tag{7.20}$$

Since the sail lightness number and attitude can be selected, the set of five classical Lagrange points will be replaced by an infinite set of artificially generated equilibrium solutions. These solutions form enclosed, nested surfaces, parameterized by the sail lightness number β.

The regions in which these new solutions may exist are defined by the constraint $\hat{\mathbf{r}}_1.\nabla U \geq 0$ with a boundary surface defined by an equality. This constraint may be understood physically since the solar radiation pressure acceleration vector $\mathbf{a}$, and so the sail attitude vector $\mathbf{n}$, can never be directed sunward. The boundary surface has two topologically disconnected surfaces S_1 and S_2 which define the region of existence of equilibrium solutions near m_2, as shown in Figure 7.6. The classical equilibrium solutions lie on either S_1 or S_2 since they are the solutions to $\nabla U = 0$. Surfaces of constant sail lightness number generated from Eq. (7.20) for the Earth–Sun system are also shown in Figure 7.6. In general, the surfaces of constant sail lightness number approach these

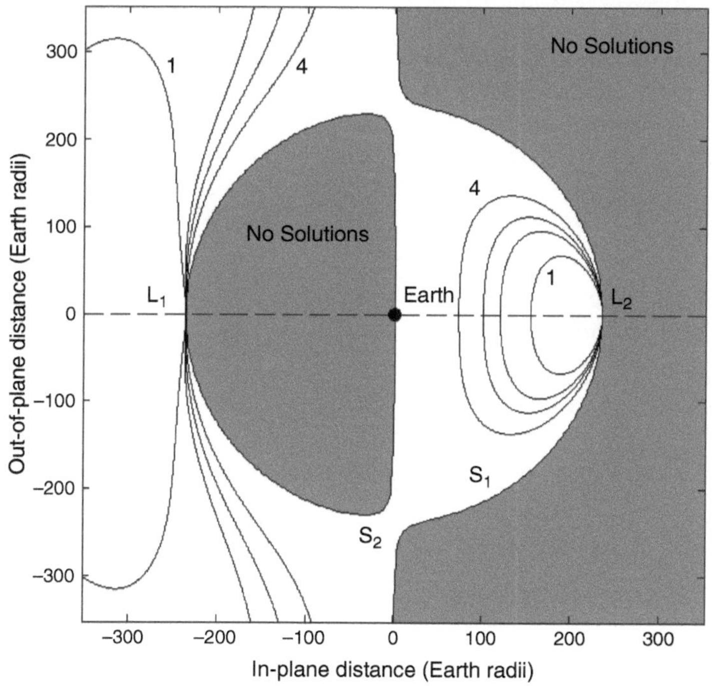

Fig. 7.6. Contours of sail loading in the x–z plan with reflectivity $\eta = 1$. Contours (gm^{-2}): [1] 30 [2] 15 [3] 10 [4] 5.

boundaries asymptotically with $\beta \to \infty$ when $\hat{\mathbf{r}}_1.\nabla U \to 0$, as is clear from Eq. (7.20). It can be seen that as the sail lightness number increases larger volumes of space are accessible for artificial equilibrium points. In particular, a solar sail of a given lightness number can be in equilibrium sunward or above/below the classical L_1 and L_2 points. In addition, in the limit $\mu \to 0$, displaced two-body orbits are obtained whose orbit period can be selected by an appropriate choice of sail lightness number and sail pitch angle [17].

7.4.3 *Artificial three-body equilibria: realistic solar sail*

A realistic solar sail force model which includes absorption will now be considered. To allow a closed-form solution, the solar sail will be assumed to have perfect specular reflectivity and no thermal re-emission but will still have an overall reflectivity η less than unity. Then, the radiation pressure acceleration will act in direction $\mathbf{m}$ and may be written as the sum of components normal $\mathbf{n}$ and transverse $\mathbf{t}$ to the sail surface such that

$$a\mathbf{m} = \frac{1}{2}\beta\frac{1-\mu}{r_1^2}(1+\eta)(\hat{\mathbf{r}}_1.\mathbf{n})^2\,\mathbf{n} + \frac{1}{2}\beta\frac{1-\mu}{r_1^2}(1-\eta)(\hat{\mathbf{r}}_1.\mathbf{n})(\hat{\mathbf{r}}_1.\mathbf{t})\,\mathbf{t}. \tag{7.21}$$

It can be seen that the main effect of the non-perfect reflectivity of the sail is to reduce the acceleration magnitude and to introduce an off-set in the direction of the radiation pressure acceleration. The acceleration $\mathbf{a}$ now acts in direction $\mathbf{m}$ rather than normal to the sail surface in direction $\mathbf{n}$. This off-set is defined by the centre-line angle ϕ, with the actual radiation pressure force direction defined by a cone angle θ, as shown in Figure 7.5.

The analysis presented in the previous Section will be repeated using the sail force model defined by Eq. (7.21) so that the equation of motion may now be written as

$$\frac{d^2\mathbf{r}}{dt^2} + 2\boldsymbol{\omega}\times\frac{d\mathbf{r}}{dt} + \nabla U = a\mathbf{m}. \tag{7.22}$$

For an equilibrium solution the first two terms of Eq. (7.22) will again vanish so that the sail attitude must be chosen as

$$\mathbf{m} = \frac{\nabla U}{|\nabla U|}. \tag{7.23}$$

The unit vector $\mathbf{m}$ can now be defined by the cone angle θ between the radial direction $\hat{\mathbf{r}}_1$ and $\mathbf{m}$ as

$$\tan\theta = \frac{|\hat{\mathbf{r}}_1\times\nabla U|}{\hat{\mathbf{r}}_1.\nabla U}. \tag{7.24}$$

In addition, using Eq. (7.21) the centre-line angle can be obtained from the ratio of the transverse and normal accelerations as

$$\tan\phi = \frac{1-\eta}{1+\eta}\tan\alpha, \tag{7.25}$$

where the sail pitch angle $\alpha = \theta + \phi$. Noting that $\mathbf{n}.\mathbf{t} = 0$ and taking a scalar product of Eq. (7.22) with the unit vector $\mathbf{n}$ gives the required sail lightness number as

$$\beta = \frac{2r_1^2}{1-\mu}\frac{\nabla U.\mathbf{n}}{(1+\eta)(\hat{\mathbf{r}}_1.\mathbf{n})^2} \tag{7.26}$$

The center-line angle may be obtained explicitly by again noting that $\alpha = \theta + \phi$. Then, after some reduction, Eq. (7.25) yields the center-line angle directly from the cone angle as

$$\tan\phi = \frac{\eta}{(1+\eta)\tan\theta}\left[1 - \left[1 - \frac{1-\eta^2}{\eta^2}\tan^2\theta\right]^{1/2}\right] \tag{7.27}$$

Lastly, using Eq. (7.26) it is found that the required sail lightness number may be obtained in terms of the lightness number for an ideal solar sail $\tilde{\beta}$ as

$$\beta = \frac{2}{(1+\eta)}\frac{\sqrt{1+\tan^2\phi}}{(1-\tan\theta\tan\phi)^2}\tilde{\beta} \tag{7.28}$$

where $\tilde{\beta}$ is defined by Eq. (7.20). Therefore, using Eqs. (7.24), (7.27) and (7.20) the sail orientation and sail lightness number required for an artificial equilibrium solution can be obtained.

The effect of a non-ideal solar sail is shown in Figure 7.7 for a reflectivity of 0.9, typical of a flat aluminized sail film. First, it can be seen that the volume of space available for equilibrium solutions about L_2 is significantly reduced. This is due to the center-line angle which limits the direction in which the radiation pressure force vector can be oriented. For solutions near L_1 the main effect of the non-ideal sail is to displace the equilibrium solutions towards the Earth. This is due to the reduction in the magnitude of the radiation pressure force, rather than the center-line angle. In general then, equilibrium solutions sunward of L_1 are not greatly affected by a realistic sail while equilibrium solutions about L_2 are severely restricted.

7.5 Mission applications

7.5.1 Geostorm mission

Currently, warnings of geomagnetic storms are made using terrestrial data and real-time solar wind data obtained from the Advanced Composition Explorer (ACE) spacecraft, stationed on a halo orbit [28] about the L_1 Lagrange point some 1.5 million km (0.01 AU) sunward of the Earth, as shown in Figure 7.8. Since the spacecraft is located sunward of the Earth, coronal mass ejections (CME) sensed by the suite of instruments on-board the ACE spacecraft can be used to provide early warning of impending geomagnetic storms. Typically, a prediction of 30–60 minutes can be made from the L_1 point, enhancing the quality of forecasts and alerts to operational user groups. These groups include civil and military satellite operators, electricity utility companies and airlines.

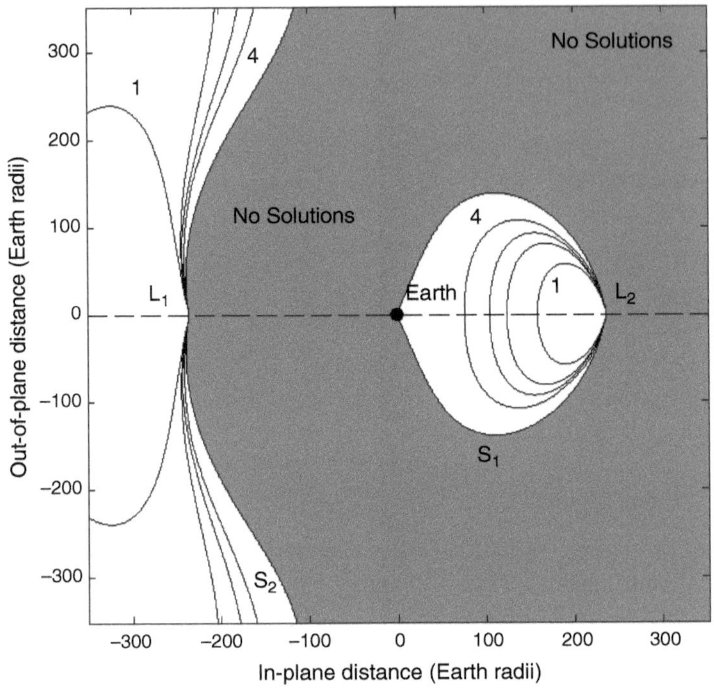

Fig. 7.7. Contours of sail loading in the x–z plan with reflectivity $\eta = 0.9$. Contours ($g\,m^{-2}$): [1] 30 [2] 15 [3] 10 [4] 5.

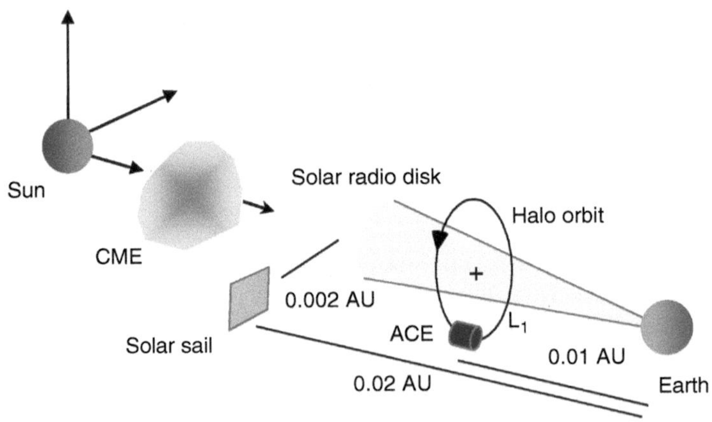

Fig. 7.8. Geostorm mission concept.

To enhance this warning time would require a spacecraft to be stationed at an artificial Lagrange point, sunward of the classical L_1 point, as discussed in Section 7.4. While this would require an unrealistic ΔV budget for a conventional spacecraft (of order 9 kms^{-1} per year of operation), a relatively small solar sail can be used to station a spacecraft approximately 3 million km (0.02 AU) from the Earth, again shown in Figure 7.8. This new artificial Lagrange point will double the warning time of impending geomagnetic storms [5]. The artificial Lagrange point must also be displaced away from the Sun–Earth line so that from the Earth, the spacecraft is viewed away from the solar radio disk to avoid interference with telemetry down-link. The volume of space accessible near L_1 in the ecliptic plane is shown in Figure 7.9, along with the Geostorm mission sub-L_1 design point. The Geostorm mission makes excellent use of solar sailing by only requiring a modest solar sail characteristic acceleration, but delivering an extremely high effective specific impulse for a multi-year mission duration, as discussed in Section 7.2.3. The solar sail can be transferred to the artificial Lagrange point by chemical kick-stages (and deployed on-station), or the solar sail can perform the transfer.

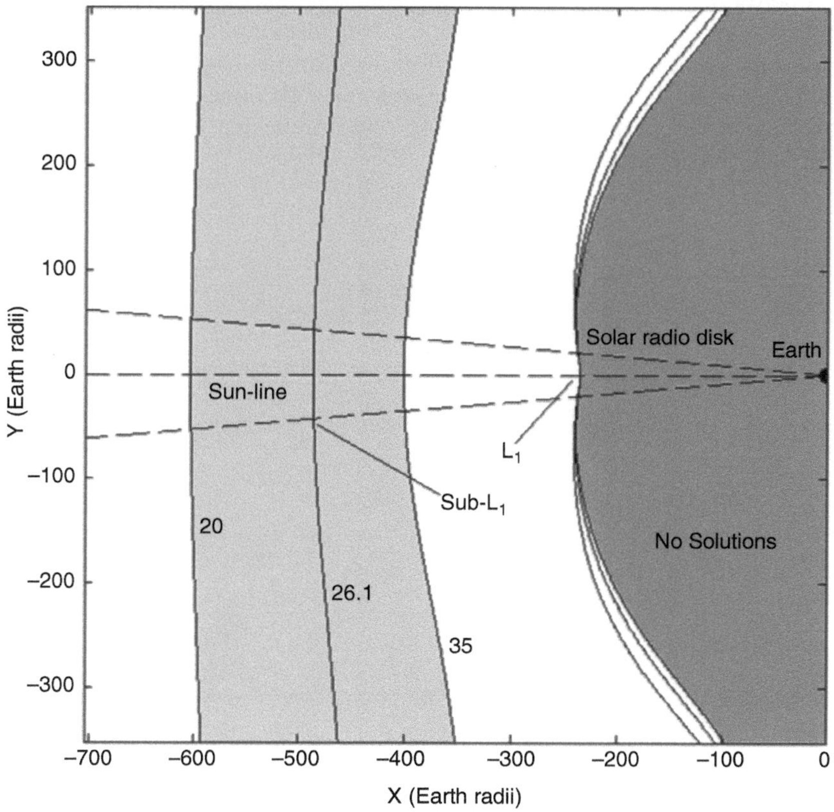

Fig. 7.9. Volume of space accessible in ecliptic plane for sail loadings 20–35 gm^{-2}.

7.5.2 Polar observer mission

It has been seen in Section 7.4 that solar sails may be used to generate artificial equilibrium solutions in the Sun–Earth three-body system. While in-plane equilibria have applications for missions such as Geostorm, out-of-plane equilibria may be utilised for continual, low-resolution imaging of the high latitude regions of the Earth. In fact, if the artificial Lagrange point is located high enough above or below the ecliptic plane, the solar sail may be stationed directly over the north pole, or indeed the south pole, during the summer solstice [24, 6]. The solar sail can be stationed directly over the north pole at the summer solstice, as shown in Figure 7.10, but will not remain over the pole during the entire year due to the tilt of the polar axis. From this unique vantage point a constant daylight view of the north pole is available at the summer solstice, however six months later at the winter solstice the polar regions are in permanent darkness. The volume of space accessible above L_1 is shown in Figure 7.11, along with the optimum Polar Observer mission design point. It is found that the required solar sail performance can be minimized by an appropriate selection of polar altitude. It can be shown that an equilibrium location some 3.8 million km (596 Earth radii) above the north pole will minimize demands on the solar sail performance. Closer equilibrium locations are possible using larger, or higher performance solar sails, or indeed selecting a less demanding viewing geometry.

Although the distance of the solar sail from the Earth is large for imaging purposes, there are potential applications for real-time, low-resolution images for continuous views of large scale polar weather systems along with Arctic ice and cloud coverage for

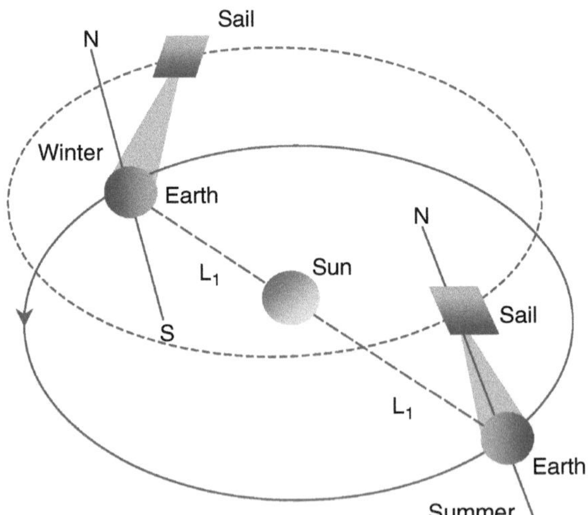

Fig. 7.10. Polar Observer mission concept.

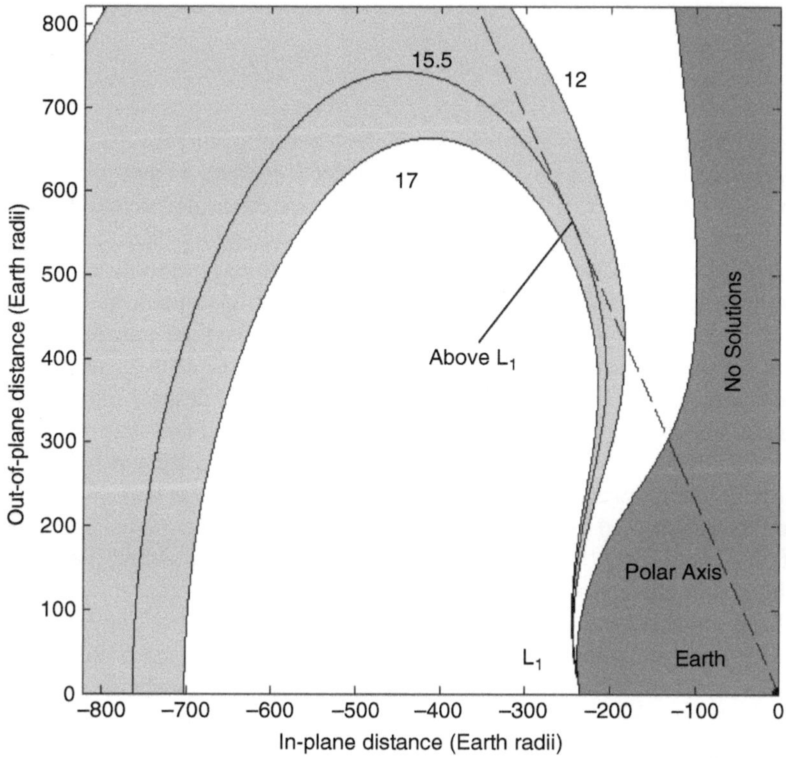

Fig. 7.11. Volume of space accessible out of ecliptic plane for sail loadings 12–17 gm^{-2}.

global climate studies. Although such images can be acquired by assembling a mosaic of instrument swaths from a conventional polar orbiting satellite, many high latitude passes are required to form a complete image. High resolution is then possible, but the completed image is not acquired in real-time and so dynamic phenomena cannot be captured.

For a 30 cm aperture instrument stationed 3.8 million km from the Earth and operating at optical wavelengths, a minimum ground resolution of order 10 km is possible, which is suitable for synoptic imaging. In practice though, the actual resolution obtained will be degraded due to factors such as the pointing stability of the camera. Higher resolution is possible if an equilibrium location closer to the pole is selected, at the expense of increased demands on the solar sail performance. Other applications of these orbits include line-of-sight, low-bandwidth communications to high-latitude users, such as Arctic or Antarctic stations. Applications for continuous data links to Mars polar landers and surface rovers have also been explored for a solar sail stationed high above the poles of Mars. Again, the Polar Observer mission makes excellent use of solar sailing by delivering an extremely high effective specific impulse for a multi-year mission duration.

7.6 Tethers in space

7.6.1 Introduction

The development of tether technology has had a strong international effort, with its origins over a century ago, and so only a limited selection of the many interesting concepts that have been proposed can be given here. The first serious reference to tether-like systems is usually attributed to Tsiolkovsky for proposals going back to 1895 [29] for a *space tower* linking a *celestial castle*, located at geostationary altitude and connected to a point on the surface of the Earth. The idea was based on an equatorial attachment point for the lower end of the tower which would be used for payload transfer into orbit and back to the Earth's surface. Now, as then, the construction of a tower tens of thousands of kilometres in length is beset with practical difficulties, despite major advances in modern materials science research. However the idea persisted and was taken up next by Artsutanov [30], who proposed the lowering of a cable from a geosynchronous satellite, with a suitable counterweight to be extended from the satellite into space to maintain tension in the system, and electric payload propulsion along the cable. The key problem of identifying commercially available material in realistic, and affordable quantities for the very high strength lightweight cable continued to be unsolved.

Roughly in parallel with this, Isaaks et al. [31] established the *Sky-hook* scenario in 1966, intended for the launch of payloads into space using the space elevator principle. The proposal was for an elevator to be deployed from the geostationary point in the form of two cables directed both towards the Earth and away from it. The cable design involved a taper in order to limit mass and to enhance strength as far as possible. The idea was based on harnessing the acceleration along the cable in order to generate payload lift from Earth, and so provide a purely mechanical launch capability for satellites and payloads into orbit. Identification of an appropriate material for the tapered cable still posed many challenges.

Applications of cables for payload movement received additional impetus in 1975 in the form of Grossi's original patented idea for a shuttle-borne orbiting *tether* [32] exploiting momentum exchange. Further work by Colombo, Grossi, and colleagues at the Smithsonian Astrophysical Laboratory has since been responsible for many major contributions to tether analysis and flights, several of which are summarized in the authoritative *Tethers in Space Handbook* [33]. Moravec [34] conceptualized the intriguing *Lunavator* tether proposal for surface payload collection and deposition on the Moon and other airless planets. This was based on a more generalized *rotovator* idea which enabled Moravec to show that limitations on current and future material strength, that would otherwise lead to an impractically high tether/payload mass for Earth surface contacts, would not necessarily impose the same restrictions for touch-downs on the Moon. The Lunavator concept has since been built into many tether mission concepts involving the Moon.

Although primary access to space using mechanical tethers is as practically challenging and potentially infeasible as it ever was, interest in tether systems has continued by moving towards rather less demanding, medium-term tether applications based on multi-strand polymeric tethers with small cable diameters and significant built-in redundancy.

The patented *Hoytether*™ is a well known example of such a design [35]. Typical applications are momentum exchange tethers for propellant-less de-orbiting of satellites, and sample return or waste disposal using re-entry vehicles deployed from space platforms. In addition to momentum exchange the electrodynamic effects of a gravity gradient stabilised conductive tether system can also be applied, either for the generation of electrical power, or as means for orbit raising and lowering. Carroll proposed a simple deployment-only tether known as the Small Expendable Deployer System (SEDS) in 1983 and then published a generic and very useful guidebook for the analysis of the wide range of tether applications studied up to that time [36].

In the mid to late 1990s Cartmell [37] showed independently that dumb-bell tether models could benefit from additional energy injected by a centrally located drive motor. This proposal enhanced the potential performance of spinning momentum exchange tethers as long as a suitable counter-inertia could be contrived for the motor to work against. Symmetry can be exploited in the form of two identical payloads and for such a system orbiting about Earth it has been shown that the outer payload can be boosted, potentially for Lunar Transfer, whilst the inner payload is de-boosted for return to the Earth's surface [38–41]. Staged Motorised Momentum Exchange Tethers (MMETs) are theoretically capable of generating ΔVs of up to 2 km/s, or more using conventional materials, although considerable research remains to be done on practically feasible orbital mechanics and mission logistics [42].

Tether flight experiments commenced in the mid 1960s with the manned Gemini 11/Agena mission in which the Gemini vehicle was linked to the Agena target vehicle by means of a 30 m tether. The first attempts to use a long deployed tether in space were the Tethered Satellite System (TSS) missions in which gravity gradient stabilized conductive tether systems, emanating from the Shuttle and deploying a satellite, were to be used for investigations in space physics and plasma-electrodynamics [33]. TSS-1 was the first of these, in July 1992, and was intended to explore the use of a retrievable tether. The tether was a 20 km Kevlar/Nomex conductive tether containing ten strands of 34 AWG copper wire (34 AWG is slightly less than 0.25 mm) but due to a protruding bolt the tether only actually deployed to about 256 m. However, it still verified some fundamental dynamics issues concerned with short deployment and gravity gradient stabilization, with implications for longer deployments. This led on to TSS-1R in February 1996, which successfully deployed to 19.7 km, just slightly short of the 20.7 km that had been planned. Plasma phenomena were observed with the conductive tether used and showed that currents significantly in excess of numerical predictions could be collected [33].

The Small Expendable Deployer missions, SEDS-1 and SEDS-2, were flown in March 1993 and 1994 showing that a small test payload of 25 kg could be de-orbited from LEO, and also that a closed loop controller could be used to deploy a tethered payload along a limited angle with respect to the local vertical. The SEDS-2 mission deployed to 19.7 km, and utilized a friction multiplier brake during deployment, an interesting concept which is currently being incorporated in the form of a *barberpole* design within the forthcoming YES2 mission [43] and which was theoretically and experimentally investigated by Lennert and Cartmell [44]. This work, along with other theoretical and experimental studies [43], showed that the friction characteristics between practical tethers wrapped in a spiral around a metallic cylindrical surface can generate useful braking forces involving significant speed-dependencies. There are some phenomenological analogies with literature observations in automotive disk-pad brake systems [45].

Dynamic analysis was one of the goals of the OEDIPUS-C flight in November 1995 in which two fore and aft payloads were used to implement and demonstrate spin stabilization by means of the so-called Tether Dynamics Experiment (TDE). The dynamics of spinning tethered two-body *dumb-bell* systems have received considerable literature attention and this experiment provided valuable in-service tether force data and payload nutation responses within the time domain. Another notable mission was the Tether Physics and Survivability Experiment (TiPS) in 1996 based on two end-bodies known as Ralph and Norton (after Ralph Kramden and Ed Norton, who starred in the 1950s comedy show *The Honeymooners*). Ralph and Norton were connected together by a 4 km insulating tether. Norton, at 10.8 kg, was mounted closest to the host vehicle and contained no electronics whereas Ralph, at 37.7 kg and at the other end, contained the electronics, instrumentation, and the tether deployer, as shown in Figure 7.12. The TiPS system was ejected from the host spacecraft in June 1996 and the objectives were to investigate long-term orbit and attitude dynamics and survivability. This successful flight provided data which suggested that reasonable long-term survivability could be expected and that predicted stabilizations of small angle libration are practically achievable [33]. Subsequent tether missions have been subject to various delays and cancellations (ATEx, ProSEDS, TSE, STEP-AirSEDS, and ASTOR, are notable examples), but at the time of writing the Young Engineers' Satellite 2 (YES2) sample return mission from the ISS is still under active technology development [43]. A major tether project which is also currently under intensive development is NASA's *Momentum Exchange Electrodynamic Reboost* (MXER) concept in which momentum exchange is to be used to transfer payloads from LEO to geosynchronous transfer orbit and beyond, after which electricity from solar panels fitted to the system would be used to drive current through the tether in order to re-boost the tether by means of an interaction against the geomagnetic field, thereby restoring the energy that was transferred to the payload [46].

7.6.2 Hanging, swinging, and spinning tethers

7.6.2.1 Hanging tethers

One of the fundamentals of space tether applications is that two tethered masses orbiting a source of gravity in space must possess the same orbital angular velocity as the overall centre of mass (CoM) [38]. Figure 7.13 shows the case of a hanging dumb-bell tether in which the upper payload (UP) is released from a hanging tether and then onto an elliptical orbit. This is because the upper payload carries more velocity than required for that orbit, but not enough to escape the influence of the Earth. The payload's release point is then the perigee of that elliptical orbit. On release of the upper payload the lower payload (LP) and the tether do not have enough velocity to stay on the original orbit so they also go into an elliptical orbit, but with the release point at the apogee. Half an orbit later the UP reaches its apogee, and so it is further away from the Earth than it was when it was released. Similarly the LP and the tether reach their perigee and are therefore closer to the Earth than they were when the UP was released. This means that the upper and lower payloads of a hanging tether are, respectively, raised and lowered.

Fig. 7.12. Artist's rendition of TiPS tether in orbit configuration (Naval Research Laboratory).

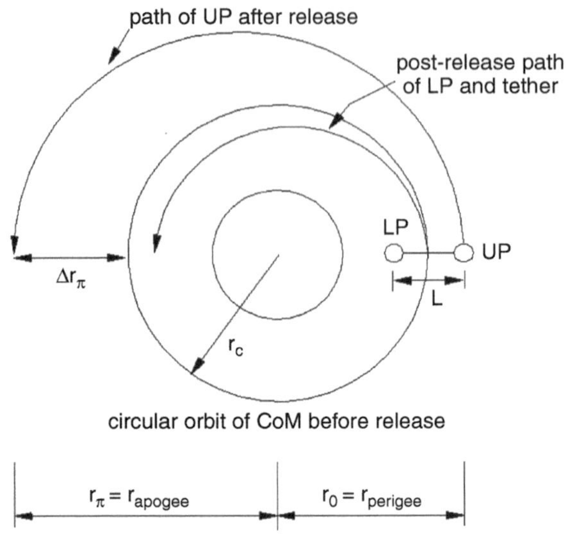

Fig. 7.13. Upper and lower payload separations and definition of Δr_{π} [38, 39].

7.6.2.2 Adding swinging or spin motion

A prograde swing or spin will add velocity to the upper payload and will subtract velocity from the lower payload. Conversely a retrograde swing or spin will subtract velocity from the UP and add velocity to the LP. So, retrograde swing or spin could be used to maintain the original orbit of the UP on release, for example. For maximum apogee altitude gain of the UP, and perigee altitude loss of the LP, the most desirable tether motion has to be either a prograde swing or spin. It appears that the optimum payload release point for a swinging or spinning tether is when it is aligned along the local gravity vector and when the motion is coplanar with the orbital plane [38]. The radial separation, Δr_{π}, between a payload half an orbit after release, and the tether CoM's circular orbital radius at release, is usually defined as being greater than the sub-span length l for orbit raising and less than l for orbit lowering (where the end-to-end length equals twice the sub-span, i.e., $L = 2l$).

7.6.2.3 Literature applications and proposals

The first well-known proposal for payload orbit raising or lowering using momentum exchange tethers was made by Colombo et al. [47]. Soon after that a system based on a hanging tether in a circular orbit was proposed by Bekey and Penzo [48], for payload transfer from LEO to Geostationary orbit. This also raised the problem of unwanted lowering (effectively de-orbiting) of the mass at the other end (in this case the shuttle) and also potential problems with excessive tether tension. Kelly [49] suggested that the shuttle could be tethered to its external fuel tank at separation in order to raise the shuttle's orbit and to de-orbit the tank. Lorenzini et al. [50] suggested propelling a payload from LEO to GEO using a two-stage tether system, with transfer times that were found to be potentially comparable to those for a conventional chemically propelled upper stage. The tethers were shown to be better in terms of mass reduction.

7.6.2.4 Literature analyses and results

The commonly quoted literature result for an UP released from a hanging tether is that it rises approximately seven times its sub-span length half an orbit later, therefore,

$$\Delta r_\pi \approx 7l. \tag{7.29}$$

In the case of swinging and spinning tethers, respectively, it has been shown that,

$$\Delta r_\pi < 14l \tag{7.30}$$

$$\Delta r_\pi > 14l \tag{7.31}$$

[33, 48, 50–52]. Bekey [53], also claimed that Eq. (7.31) could be extended as far as,

$$\Delta r_\pi > 25l. \tag{7.32}$$

It was shown by Cosmo and Lorenzini [33] that the swing angle of a swinging tether could be incorporated into such expressions and they went on to show that a swinging tether could generate a Δr_π of,

$$\Delta r_\pi \approx (7 + \sqrt{48}\sin\theta_{\max})l, \tag{7.33}$$

where $\theta_{\max}$ was defined as the maximum swing angle attainable and considered to be positive for prograde motion and negative for retrograde motion. This analysis assumed that the tether and LP remained on their original circular orbit. According to Cosmo and Lorenzini [33] the equations above only hold for $\Delta r_\pi \ll r_c$. Finally, in this short summary, we note another expression, this time due to Kumar et al. [54] which holds for both swinging and spinning systems on the assumption that $l \ll r_c$, where here $\dot{\theta}$ and $\dot{\Psi}$ are the angular orbital and pitch velocities, both in rad/s,

$$\Delta r_\pi \approx \left(7 + 4\frac{\dot{\Psi}}{\dot{\theta}}\right)l. \tag{7.34}$$

One can readily see from this that although Eq. (7.33) is independent of both the orbital radius of the facility (central system) and the payload mass these quantities will clearly affect the momentum of the payload in practice. Similarly, although Eq. (7.34) takes the angular pitch and orbital velocities into account it does not incorporate orbital radius, which is also obviously relevant. In order to address these issues the problem was reconsidered by Ziegler and Cartmell [summarised in Ref. [38] from work originally carried out by Ziegler and subsequently reported in full in Ref. [39]], with the overall geometry given in Figure 7.14(a). The general equation for the radius r on an elliptical orbit is given by,

$$r = \frac{p}{1 + e\cos\Theta} \tag{7.35}$$

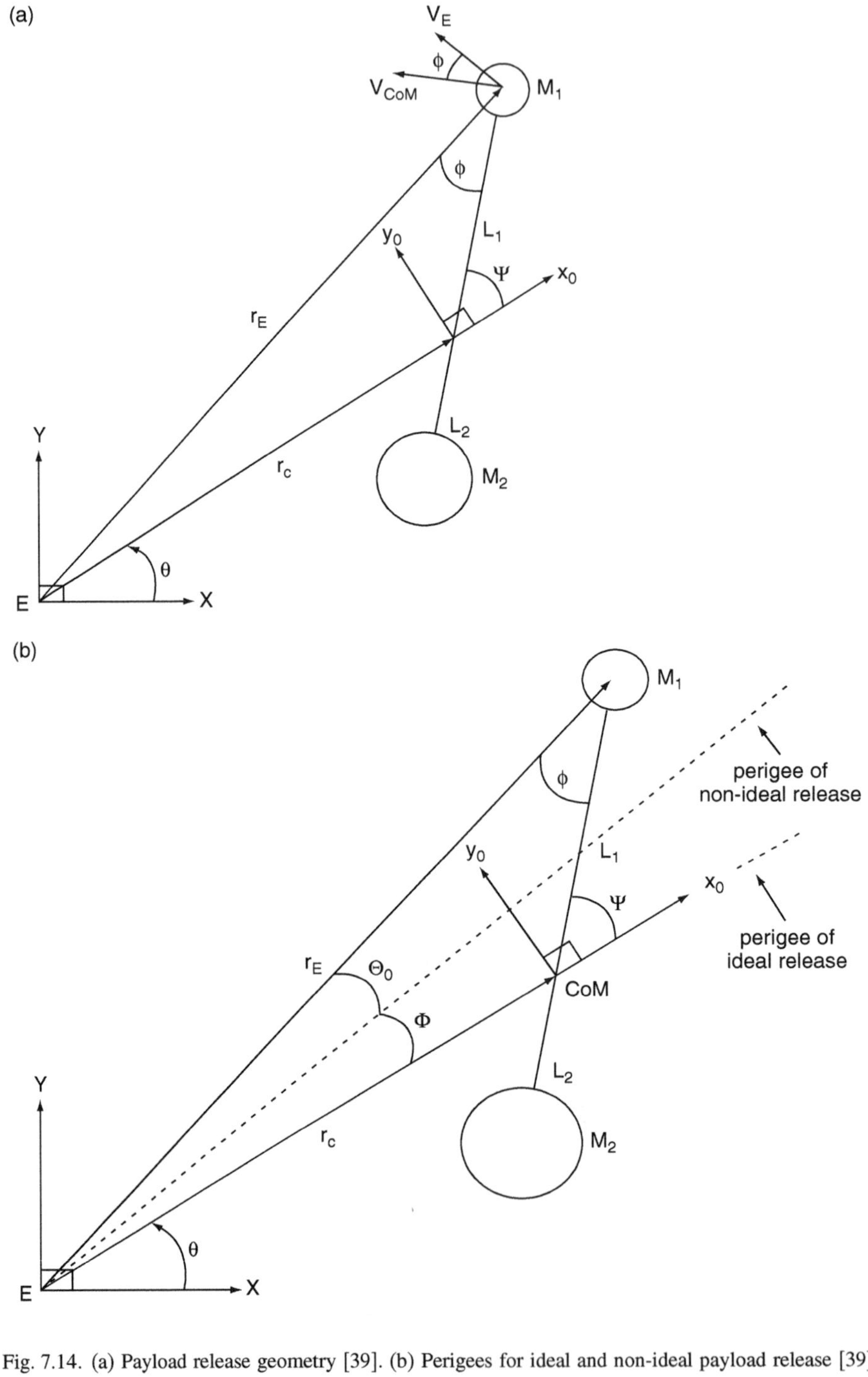

Fig. 7.14. (a) Payload release geometry [39]. (b) Perigees for ideal and non-ideal payload release [39].

[55] from which the semilatus rectum, p, is given by the following,

$$p = a(1-e^2) = \frac{b^2}{a} = r_p(1+e) = r_a(1-e), \tag{7.36}$$

where a is the semi-major axis, b is the semi-minor axis, e is the orbital eccentricity, Θ is the true anomaly, r_a is the apogee radius, and r_p is the perigee radius (or *apoapsis* and *periapsis* in a non-Earth centred system). The *total* release velocity, V_{Tot}, for the payload can be obtained in terms of the distance from the Earth's centre, E, to the payload, r_E, the semi-major axis of the orbit, a, and μ (where $\mu = GM_{\text{Earth}}$), from the vis-visa equation,

$$V_{\text{Tot}} = \sqrt{\mu\left(\frac{2}{r_E} - \frac{1}{a}\right)}. \tag{7.37}$$

This can be rearranged to obtain a for later use,

$$a = \frac{\mu r_E}{2\mu - V_{\text{Tot}}^2 r_E}. \tag{7.38}$$

The distance from the centre of the Earth, E, to the payload, M_1 is directly obtained from the geometry of Figure 7.14(a),

$$r_E = \sqrt{r_c^2 + L_1^2 + 2r_c L_1 \cos\Psi}, \tag{7.39}$$

noting that r_c is the circular orbit radius from E to the CoM and that L_1 is not necessarily equal to L_2. The total payload velocity comprises the following components,

$$V_{\text{Tot}} = \sqrt{V_R^2 + V_N^2} \tag{7.40}$$

where V_R and V_N are the radial and normal components, respectively. The radial velocity (pointing inwards to E) is given by,

$$V_R = V_{\text{CoM}} \sin\phi \tag{7.41}$$

The normal velocity component is the sum of the payload's orbital velocity, V_E, and a component of the tangential payload velocity relative to the CoM, V_{CoM}, and this acts in the direction of V_E. Therefore we obtain,

$$V_N = V_E + V_{\text{CoM}} \cos\phi \tag{7.42}$$

and the velocities V_E and V_{CoM} are given by,

$$V_E = r_E\dot{\theta} \quad \text{and} \tag{7.43}$$

$$V_{\text{CoM}} = L_1\dot{\Psi}. \tag{7.44}$$

Since $\dot{\theta} = \sqrt{\mu/r_c^3}$ [55] this can be substituted into Eq. (7.43). The angle, ϕ, between V_{CoM} and V_E is obtained from the sine rule, as, $\frac{r_E}{\sin(\pi-\Psi)} = \frac{r_c}{\sin\phi}$, therefore,

$$\phi = \sin^{-1}\left(\frac{|r_c|}{|r_E|}\sin\Psi\right). \tag{7.45}$$

Now, substituting (7.43–7.45) into (7.41), the equation for V_R, and then into (7.42), the equation for V_N, those two velocities emerge as follows, after dropping the moduli for brevity,

$$V_R = \frac{L_1 r_c}{r_E}\dot{\Psi}\sin\Psi \tag{7.46}$$

$$V_N = r_E\sqrt{\frac{\mu}{r_c^3}} + L_1\dot{\Psi}\cos\left[\sin^{-1}\left(\frac{r_c}{r_E}\sin\Psi\right)\right] \tag{7.47}$$

From Eq. (7.36) we can obtain e directly,

$$e = \sqrt{1-\frac{p}{a}} \tag{7.48}$$

Also, the semilatus rectum, p, can be defined from standard analysis in terms of specific angular momentum, H, [55] thus,

$$p = \frac{H^2}{\mu} = \frac{(V_N r_E)^2}{\mu} \tag{7.49}$$

noting that H is the magnitude of the specific angular momentum given by $H = r_E^2\dot{\theta}$ [55]. It has already been stated that the optimum release for a payload is when the tether is aligned along the local gravity vector, and therefore for the UP at the perigee (periapsis) of the released payload's elliptical orbit. This is along the ECoMx_0 line in Figure 7.14(b). The dotted line shows an alternative, non-ideal release orientation when the payload, M_1 lies on that line. Clearly, $\Psi - \phi = \Theta_0 + \Phi$, so we rearrange to get,

$$\Phi = \Psi - \phi - \Theta_0. \tag{7.50}$$

The apogee (or, more generally, the apoapsis) of an orbit-raised payload represents the location between an incoming payload and possibly the catching end of another tether. In order to obtain the orbital radius of a non-optimally released payload at the location of the required ideal apogee (apoapsis) the orbit's true anomaly defining this position will be,

$$\Theta_{\Delta\pi,apo} = \pi - \Phi. \tag{7.51}$$

7.6.2.5 Hanging tether

The optimum release position for a hanging tether is when the system is aligned along the local gravity gradient and because it is not librating or spinning, $\Psi = \dot{\Psi} = 0$. Calculating r_π from Eq. (7.35) allows us to obtain,

$$\Delta r_\pi = r_\pi - r_c, \tag{7.52}$$

where

$$r_\pi = r_{apo}. \tag{7.53}$$

After some algebra (inserting Eq. (7.36), Eqs. (7.38–7.40), and Eqs. (7.46–7.49) into Eq. (7.35)) the altitude gain for the upper payload for the hanging tether is given by,

$$\Delta r_{\pi} = \frac{(r_c + L)^4}{2r_c^3 - (r_c + L)^3} - r_c. \tag{7.54}$$

The altitude loss for the lower payload for the hanging tether is,

$$\Delta r_{\pi} = \frac{(r_c - L)^4}{2r_c^3 - (r_c - L)^3} - r_c. \tag{7.55}$$

For the case when the tether is swinging or spinning (both prograde), i.e., when $\dot{\Psi} \neq 0$, and when it is instantaneously coincident with the local gravity vector, then the altitude gain and loss, respectively, are given by,

$$\Delta r_{\pi} = \frac{(r_c + L)^2 [(r_c + L)\dot{\theta} + L\dot{\Psi}]^2}{2\mu - (r_c + L)[(r_c + L)\dot{\theta} + L\dot{\Psi}]^2} - r_c \tag{7.56}$$

$$\Delta r_{\pi} = \frac{(r_c - L)^2 [(r_c - L)\dot{\theta} - L\dot{\Psi}]^2}{2\mu - (r_c - L)[(r_c - L)\dot{\theta} - L\dot{\Psi}]^2} - r_c \tag{7.57}$$

We continue to assume that $L << r_c$, so if $\delta = \frac{L}{r_c}$ then $\delta << 1$, and applying a binomial expansion to Eqs. (7.54–7.57) can lead to simplified results identical to the literature results quoted in Eqs. (7.29) and (7.34); refer to Ziegler [39] for full details and to Ziegler and Cartmell for a summary [38]. From this it becomes clear that the best payload raising performance for a hanging tether is when it is closest to the Earth and as long as possible. Conversely the best payload de-boosting performance for a hanging tether is when it is as far as possible from the Earth and the tether is very long. The best performance obtainable from a prograde librating tether used for payload raising is for the largest possible libration angle and tether length, and for the system located as close to the Earth as possible. De-boost is optimized for the same system but located as far as possible from the Earth. In the case of a spinning (possibly motorized) tether then the situation is more complex, with various possibilities arising. The most important general conclusion is that an UP could be propelled from circular LEO to GEO using a long motorized system [39].

7.7 Tethers in orbit

7.7.1 Strength & materials

7.7.1.1 Terrestrially located hanging tether

We start off by considering a simple hanging tether located on Earth. Full weight is experienced at the support and zero weight at the free end below, so, the tether tension is given by $T = \left(\frac{m}{l}\right) zg$, where $0 \leq z \leq l$, and m/l equals mass per unit length, which is constant for a chosen material. The tension varies linearly along the length of the tether, and if we write $\frac{m}{l}$ as $\hat{\rho}$, then,

$$T = \hat{\rho} l g. \tag{7.58}$$

This implies that there must be a critical value of length beyond which the tether will break due to its own weight. This is known as the *break-length*, l_*. Using this allows us to re-write Eq. (7.58) as,

$$T_* = \hat{\rho} l_* g, \tag{7.59}$$

where T_* is the *break tension.* These quantities were formally introduced by Beletsky and Levin in their seminal text [56]. So, re-arranging Eq. (7.59) leads to an expression for break-length,

$$l_* = \frac{T_*}{\hat{\rho} g} \tag{7.60}$$

and if we also introduce the *specific strength*, σ_*,

$$\sigma_* = \frac{T_*}{A}, \tag{7.61}$$

where A is the cross-sectional area of the tether, then we get the following,

$$l_* = \frac{\sigma_*}{\left(\frac{\hat{\rho}}{A}\right) g}, \tag{7.62}$$

$$\text{where } \rho = \frac{\hat{\rho}}{A}. \tag{7.63}$$

$$\text{So, } l_* = \frac{\sigma_*}{\rho g} \tag{7.64}$$

The specific strength, σ_*, has units of $\mathrm{N/m^2}$ or Pa, and ρ is the material density in $\mathrm{kg/m^3}$. Therefore, l_*, the break length, is a function of specific strength and density, and is effectively a *strength-to-density* ratio. We have assumed that stress is distributed uniformly over the tether cross-section but this depends on the homogeneity of the material and the design although it is a generally accepted assumption. Note also that the mechanics of the preceding analysis are purely terrestrial, therefore l_* is the break length *on Earth.*

7.7.1.2 Hanging tether in space

The next stage in the development is to consider a massive satellite in a circular orbit deploying a tether whose other end is free and orientated 'downwards' towards the Earth. For an orbital angular velocity of Ω, Beletsky and Levin [56] give the tension at the attachment point as,

$$T = \frac{3}{2}\hat{\rho}\Omega^2 l^2 = \frac{3}{2}\frac{m}{l} l^2 \Omega^2 = \frac{3}{2} m l \Omega^2 \tag{7.65}$$

with $l = 2l_{\text{subspan}}$, where l_{subspan} is half the total length of the tether, so this could also be expressed as, $T = 3ml_{\text{subspan}}\Omega^2$, to which we return at the end of Section 7.1.2. Adopting the same approach as before leads to an *orbital break tension* given by,

$$T_* = \frac{3}{2}\hat{\rho}\Omega^2 L_*^2 \tag{7.66}$$

reverting to Beletsky and Levin's notation [56].

Rearranging this gives the *orbital break length*,

$$L_* = \frac{1}{\Omega}\sqrt{\frac{2T_*}{3\hat{\rho}}}. \tag{7.67}$$

So if we re-use Eq. (7.61) we get,

$$L_* = \frac{1}{\Omega}\sqrt{\frac{2\sigma_* A}{3\hat{\rho}}} = \frac{1}{\Omega}\sqrt{\frac{2\sigma_*}{3\left(\frac{\hat{\rho}}{A}\right)}} = \frac{1}{\Omega}\sqrt{\frac{2\sigma_*}{3\rho}}. \tag{7.68}$$

Clearly, $L_* > l_*$, and in geostationary orbit it is even bigger and defined as $L_{*(\text{geo})}$, where $\Omega = \Omega_{\text{geo}}$.

Geostationary orbit is of interest for the *Space Elevator* application (a tether from the surface of the Earth to geostationary orbit). Table 7.1 is adapted from data given by Beletsky and Levin [56] and shows increasing break-lengths in descending order for four selected materials.

Note that a tether longer than L_*, for some chosen material, cannot even carry its own weight, let alone any cargo or payload. A relatively new material which can now be produced in quantity is *Spectra 2000*, which has a lower density than any of the above (970 kg/m^3) and a specific strength of about 3.5 GPa. For geostationary orbit, where the orbital rate, Ω_{geo}, is 0.0000727 rad/s, the orbital break length of *Spectra 2000* is 21461 km. This is approximately half-way back from geostationary orbit to Earth, which compares to only a third of the way back for *Kevlar*. In fact there are no mass produced materials yet capable of supporting the Space Elevator. Carbon nanotubes are of considerable speculative interest, but economical mass production technologies have yet to be developed.

Table 7.1
Potential tether material properties.

Material	ρ [kg/m^3]	σ_* [GPa]	E [GPa]	l_* [km]	L_* [km]	$L_{*(\text{geo})}$ [km]
Tungsten	19300	4.0	410	21	320	5100
HP steel	7900	4.0	210	52	502	8000
Kevlar	1450	2.8	130	197	981	15700
Graphite	2200	20	690	928	2129	34100

7.7.1.3 Mass considerations

The *breaking mass* M_* can be defined as an upper limit for an application, therefore $M_T \leq M_*$. The mass of the tether can be defined by,

$$M_T = \rho k_a A l \tag{7.69}$$

for which Beletsky and Levin [56] define the constant k_a as,

$$k_a = \frac{d_T^2}{d_f^2 n} \tag{7.70}$$

noting that d_T and d_f are the diameters of the overall tether and the individual fibers, respectively, and n is the number of fibers that make up the multi-line tether. Equation (7.69) can be slightly revised to define the breaking-mass, $M_* = \rho k_a A L_*$, which rearranges to $\frac{M_*}{A} = \rho L_*$ if $k_a = 1$ (for the conceptually simple single line tether case). The previous table allows us to calculate this ratio of break-mass to tether cross-sectional area based on the values given for the density, ρ, and break-length, L_* for selected materials. Assuming a 1000 kg break-mass for an installation using *Kevlar*, we get a commensurate cross-sectional area of $0.7\,\text{mm}^2$ for the tether ($d_T = 0.944\,\text{mm}$). Preserving this area, and looking at high performance steel and graphite, their associated break-masses for the same cross-section can be evaluated, $M_{*(\text{highperfsteel})} = 2776\,\text{kg}$ and $M_{*(\text{graphite})} = 3278\,\text{kg}$. Clearly Spectra 2000 scores highly in comparison, with a break-length between *Kevlar* and graphite, but a density considerably lower than *Kevlar*.

7.7.2 Gravity gradient stabilisation for hanging tethers

An object of mass m, and radial position r from the centre of the Earth, and orbiting the Earth, is subjected to a gravitational force,

$$\mathbf{F}_G = -\frac{\mu m}{r^2}\mathbf{u}_r, \text{ with scalar form, } F_G = -\frac{\mu m}{r^2}, \tag{7.71}$$

where the unit vector -$\mathbf{u}_r$ refers to the fact that the gravity force is directed radially inwards. The inward acting centripetal force on the body is,

$$\mathbf{F}_C = -mr\Omega^2\mathbf{u}_r, \text{ with scalar form, } F_C = -mr\Omega^2. \tag{7.72}$$

A *hanging* tethered dumb-bell in a circular orbit travels along the orbital path with all points on the tether moving with the same velocity. The Centre of Mass (COM) of the tether will experience exactly enough gravitational pull to provide the centripetal force necessary for that radial position, therefore the gravitational force from Eq. (7.71) produces the centripetal acceleration. For a dumb-bell the outer end-mass is at a higher altitude and will therefore not feel sufficient gravity to provide the centripetal force consistent with the orbital velocity (which is the same as that of the COM due to the tether). Therefore $F_G < F_C$ and the outer end-mass will try to move outwards, generating a tension in the outer sub-span of the tether. The inner end-mass, however, feels too much gravity for the centripetal force required at that point, so $F_G > F_C$, and the inner end-mass tends to try to move further inwards. This generates a tension in the inner sub-span. Therefore, at the COM ($r = r_0$) we can equate the scalar forms of Eqs. (7.71) and (7.72) to get,

$$-\frac{\mu m}{r_o^2} = -mr_0\Omega^2, \text{ and rearrangement gives, } \mu = \Omega^2 r_0^3. \tag{7.73}$$

Taking two cases either side of the COM for the dumb-bell, starting first with the outer case, for which, in general, $|F_G| < |F_C|$ and $r = r_0 + z$.

$$\therefore -\frac{\mu m}{r^2} = -mr\Omega^2 + F_{Zout}, \tag{7.74}$$

where m is the mass at r. Substituting for r and μ (from Eq. (7.73)) leads to,

$$F_{Zout} = m\Omega^2\left((r_0 + z_{out}) - \frac{r_0^3}{(r_0 + z_{out})^2}\right) \tag{7.75}$$

and $z = z_{out}$, which is a point along the tether on the outer side of the COM. This resultant force is F_{Zout}. The quantities in the brackets in Eq. (7.75) can be expanded, and as $r_0 + z_{out} = r$ then the expansion becomes, $z_{out} + \frac{r_0 z_{out}(r_0+r)}{r^2}$. If $r \to r_0$, implying a relatively short tether, then the bracketed term becomes, $z_{out} + \frac{r z_{out}(2r)}{r^2}$. This is equal to $3z_{out}$. The net tensile force F_{Zout} is outwards acting and so Eq. (7.75) reduces to,

$$F_{Zout} = 3mz_{out}\Omega^2, \text{ in the limit } r \to r_0 \text{ (i.e., for short tethers).} \tag{7.76}$$

For the inner case we have, $|F_G| > |F_C|$ and $r = r_0 - z$, but as we are using the COM as the datum for z_{out} and z_{in} we can put $-z = z_{in}$, since z_{in} is measured inwards, so $r = r_0 + z_{in}$.

$$\therefore -\frac{\mu m}{r^2} = -mr\Omega^2 - F_{Zin}. \tag{7.77}$$

Substituting for r and μ (from Eq. (7.73)) leads to,

$$F_{Zin} = m\Omega^2\left(\frac{r_0^3}{(r_0 + z_{in})^2} - (r_0 + z_{in})\right). \tag{7.78}$$

The quantities in the brackets can be expanded by using $r_0 + z_{in} = r$, and then letting $r \to r_0$, so that the expansion simplifies to $-3z_{in}$. This means that the net tensile force given in Eq. (7.78) can be reduced to,

$$F_{Zin} = -3mz_{in}\Omega^2, \text{ in the limit } r \to r_0 \text{ (i.e., for short tethers)} \tag{7.79}$$

Note that these are approximations which assume 'short' tethers and are commonly used in the tether literature [51, 56]. These *gravity gradient forces*, $F_{Zout} = 3mz_{out}\Omega^2$ and $F_{Zin} = -3mz_{in}\Omega^2$ define the tensile forces in each sub-span from $0 \le |z_{out}| \le l$ and $0 \le |z_{in}| \le l$. Therefore, for a symmetrical system, at any point equidistant from the COM on either side, $|F_{Zout}| = |F_{Zin}|$, and so the tensions are the same in each sub-span. Note here that total (payload-to-payload) tether length $L = 2l$, which is directly analogous to $l = 2l_{\text{subspan}}$ when referring to Beletsky and Levin's analysis [56] (see Section 7.7.1).

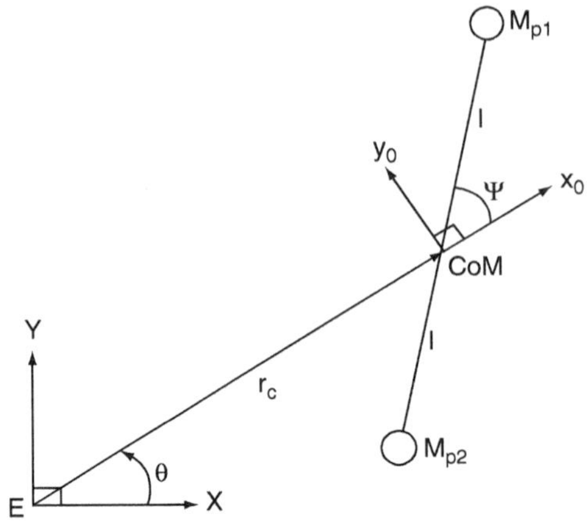

Fig. 7.15. Planar tether co-ordinate system.

7.7.3 *Fundamental dynamical models for dumb-bell tethers*

7.7.3.1 Planar tether on a circular orbit

Figure 7.15 shows a planar dumb-bell tether in the orbital plane. EXY is an Earth centered frame which is assumed to be inertial for this analysis, and $CoMx_0y_0$ is a rotating frame whose origin is at the geometrical center, assumed coincident with the CoM. The tether sub-span lengths are equal and defined by l. The payload masses are equal such that $M_{P1} = M_{P2} = M_P$. The distance from the centre of the Earth E to the tether CoM is defined by the position vector, of magnitude r_c and true anomaly θ. The planar spin angle, or *pitch-angle*, is Ψ. This system is sufficient to construct a basic equation of motion for a passive planar dumb-bell tether, but by means of a relatively small extension to the modelling it is also possible to introduce an active motor drive to the tether by means of a conceptualized motor, gearbox, and counter inertia. On that basis the mass of a central facility containing the motor drive, gearboxes, power supplies, control systems, etc., needs to be defined and M_M is introduced for this. However, the counter-inertia does not need to be considered explicitly here, as long as it is appreciated that it would be necessary for any motor driven tether, and that it could take the form shown in the *outriggers* of Figure 7.16. Irrespective of whether or not the passive or motorized tether is pursued, the equation of motion for spin about the CoM can readily be obtained by deriving the system energies and then applying Lagrange's Equation.

The Cartesian positions of the principal components with respect to E are given as follows, assuming that the tether mass is located at $\frac{l}{2}$ on each side of the CoM,

$$x_{P1} = r_c \cos\theta + l\cos(\Psi+\theta), \quad y_{P1} = r_c \sin\theta + l\sin(\Psi+\theta),$$

$$x_{P2} = r_c \cos\theta - l\cos(\Psi+\theta), \quad y_{P2} = r_c \sin\theta - l\sin(\Psi+\theta),$$

$$x_M = r_c \cos\theta, \; y_M = r_c \sin\theta, \quad x_{T1} = r_c \cos\theta + \frac{l}{2}\cos(\Psi+\theta), \tag{7.80}$$

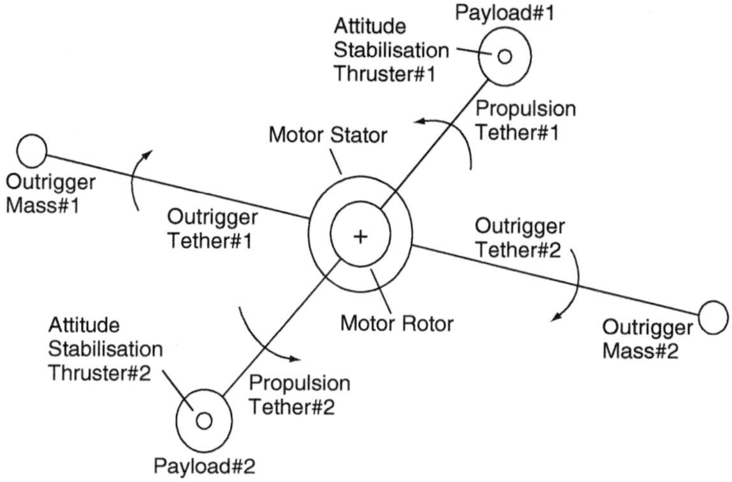

Fig. 7.16. Plan schematic of contra-rotating motorised tether components.

$$y_{T1} = r_c \sin\theta + \frac{l}{2}\sin(\Psi+\theta), \quad x_{T2} = r_c \cos\theta - \frac{l}{2}\cos(\Psi+\theta),$$

$$y_{T2} = r_c \sin\theta - \frac{l}{2}\sin(\Psi+\theta),$$

where subscripts $T1$ and $T2$ refer to positions halfway along the tether from the COM on each side and $P1$ and $P2$ denote payloads 1 and 2, respectively.

The system kinetic energy is easily constructed,

$$\begin{aligned} T =& \frac{1}{2}M_P(\dot{x}_{P1}^2+\dot{y}_{P1}^2)+\frac{1}{2}M_P(\dot{x}_{P2}^2+\dot{y}_{P2}^2)+\frac{1}{2}M_M(\dot{x}_M^2+\dot{y}_M^2)+\frac{1}{2}\rho Al(\dot{x}_{T1}^2+\dot{y}_{T1}^2) \\ &+\frac{1}{2}\rho Al(\dot{x}_{T2}^2+\dot{y}_{T2}^2)+\frac{1}{2}(2I_P+2I_T+I_M)(\dot{\theta}+\dot{\Psi})^2, \end{aligned} \tag{7.81}$$

where the mass moments of inertia about the local z-axes (component z-axes normal to the orbital plane of the system) are given by,

$$I_P = \frac{1}{2}M_P r_P^2, \tag{7.82}$$

$$I_T = \frac{1}{12}\rho Al(3r_T^2+l^2), \tag{7.83}$$

$$I_M = \frac{1}{2}M_M r_M^2 \tag{7.84}$$

noting that I_T is the standard result for a solid circular section tether with axis normal to the orbital plane at $\frac{l}{2}$. Differentiating Eqs. (7.80) with respect to time and then substituting these velocities, and also Eqs. (7.82)–(7.84), as appropriate, into Eq. (7.81) leads to,

$$T = \left(M_P + \rho A l + \frac{M_M}{2} \right) (\dot{r}_c^2 + r_c^2 \dot{\theta}^2) + \left(M_P \left[l^2 + \frac{r_P^2}{2} \right] + \frac{M_M r_M^2}{4} + \frac{\rho A l}{12} [4L^2 + 3r_T^2] \right) (\dot{\theta} + \dot{\Psi})^2 \tag{7.85}$$

Note that for a circular orbit the quantities r_c and $\dot{\theta}$ represent a constant orbital radius and angular velocity, so these are not generalized coordinates in the Lagrangian sense and therefore $r_c = C_1$, $\dot{r}_c = 0$, $\dot{\theta} = C_2$, where C_1 and C_2 are constants, with C_2 disappearing from the analysis from hereon. C_1 is retained as r_c. The potential energy is based on the local position vectors from E to M_{P1} and M_{P2}, the position vector, r_c, between E and the CoM, and an appropriate integration along the sub-span lengths of the tether, thus,

$$U_{P1} = -\frac{\mu M_P}{\sqrt{r_c^2 + l^2 + 2r_c l \cos\Psi}}, \tag{7.86}$$

$$U_{P2} = -\frac{\mu M_P}{\sqrt{r_c^2 + l^2 - 2r_c l \cos\Psi}} \tag{7.87}$$

$$U_M = -\frac{\mu M_M}{r_c} \tag{7.88}$$

$$U_{T1} = -\mu \rho A \int_0^l (r_c^2 + z^2 + 2r_c z \cos\Psi)^{1/2} \mathrm{d}z \tag{7.89}$$

$$U_{T2} = -\mu \rho A \int_0^l (r_c^2 + z^2 - 2r_c z \cos\Psi)^{1/2} \mathrm{d}z, \tag{7.90}$$

where z is distance along each sub-span from the CoM outwards. Performing the integrations in Eqs. (7.89) and (7.90) leads to analytical forms for the potential energy contributions of the tether sub-spans, as follows,

$$U_{T1} = \mu \rho A \ln \frac{r_c(1 + \cos\Psi)}{l + r_c \cos\Psi + \sqrt{r_c^2 + l^2 + 2r_c l \cos\Psi}} \tag{7.91}$$

$$U_{T2} = \mu \rho A \ln \frac{r_c(1 - \cos\Psi)}{l - r_c \cos\Psi + \sqrt{r_c^2 + l^2 - 2r_c l \cos\Psi}} \tag{7.92}$$

Differentiating the energy functions for use with Lagrange's equation generates the equation of motion. Lagrange's equation is applied in the form,

$$\frac{\mathrm{d}}{\mathrm{d}t}\left(\frac{\partial T}{\partial \dot{q}}\right)-\frac{\partial T}{\partial q}+\frac{\partial U}{\partial q}=Q_q, \tag{7.93}$$

where Q_q is the generalized force associated with coordinate q, which in this case is $q\equiv\Psi$. Note that at this level of modelling if we require to incorporate the motor drive then the generalized force term is given simply by,

$$Q_\Psi=\tau \tag{7.94}$$

where τ is the applied torque in Nm. Therefore, applying Eq. (7.93) leads to the equation of motion for the motor driven system,

$$\begin{aligned}&\left(\frac{M_m r_m^2}{2}+M_p(2l^2+r_p^2)+\frac{\rho A l[4l^2+3r_T^2]}{6}\right)\ddot{\Psi}\\&+\frac{\mu M_p r_c l\sin\Psi}{(r_c^2+l^2-2r_c l\cos\Psi)^{3/2}}-\frac{\mu M_p r_c l\sin\Psi}{(r_c^2+l^2+2r_c l\cos\Psi)^{3/2}}\\&-\mu\rho A\frac{r_c^2+l^2-r_c l(1-\cos\Psi)+(l-r_c)\sqrt{r_c^2+l^2+2r_c l\cos\Psi}}{r_c^2+l^2+2r_c l\cos\Psi+(l+r_c\cos\Psi)\sqrt{r_c^2+l^2+2r_c l\cos\Psi}}\tan\left(\frac{\Psi}{2}\right)\\&+\mu\rho A\frac{r_c^2+l^2-r_c l(1+\cos\Psi)+(l-r_c)\sqrt{r_c^2+l^2-2r_c l\cos\Psi}}{r_c^2+l^2-2r_c l\cos\Psi+(l-r_c\cos\Psi)\sqrt{r_c^2+l^2-2r_c l\cos\Psi}}\cot\left(\frac{\Psi}{2}\right)\\&=\tau\end{aligned} \tag{7.95}$$

This is a non-linear ordinary differential equation, with the non-linearities coming in from the potential terms. The inertia term, although algebraically complicated, is in fact linear in this model. This equation can be solved by numerical integration and this can be readily demonstrated using the *TetherSim* animated simulation [57]. In *TetherSim* the tether cross-section is assumed to be tubular (and not solid as defined above), so there is an inner radius, r_{Ti}, and an outer radius, r_{To}, within the expression for I_T in Eq. (7.83), so $(r_{TO}^2+r_{Ti}^2)$ replaces r_T^2. *TetherSim* is based on a set of user-definable data with defaults given by the following, $\mu=3.9877848*10^{14}\,\mathrm{m^3s^{-2}}$, $M_p=1000\,\mathrm{kg}$, $M_M=5000\,\mathrm{kg}$, $r_M=r_p=0.5\,\mathrm{m}$, $l=50\,\mathrm{km}$, $r_c=6870\,\mathrm{km}$, $r_{To}=0.006\,\mathrm{m}$, $r_{Ti}=0.004\,\mathrm{m}$, $A=62.83\,\mathrm{mm}^2$, $\rho=970\,\mathrm{kg/m^3}$ (Spectra 2000), $\sigma=3.25\,\mathrm{GPa}$ (Spectra 2000), and a safety factor of 2 for the allowable tether strength. The initial conditions that are used are taken from Ziegler and Cartmell [38],

$$\Psi(0)=-0.9\ \text{rad},\qquad \dot{\Psi}(0)=0\ \text{rad/s}. \tag{7.96}$$

It should be noted that the *TetherSim* graphics may occasionally run slowly, dependent on the connection to the server. The interested user is recommended to start with the default data, for which the motorized tether will librate, and then to change one parameter at a time in order to explore other aspects of the dynamics. Clearly the easiest way of getting monotonic spin-up for a chosen set of design data is to merely increase the motor torque, but this may not always be the most desirable approach and in such situations a multi-variable optimization of the design may well be preferable.

7.7.3.2 Non-planar dumb-bell model

Having discussed the planar model on a circular orbit in Section 7.7.3.1, a non-planar model can be proposed next, once again using the *geocentric co-ordinate system*, as shown in Figure 7.17, and the governing equations of motion for a dumb-bell tether capable of operating out of the orbital plane can be derived. This is summarized from original work by Ziegler, first published within work by Cartmell et al. [58] and then in full by the originator [39]. In this section, R is used to represent the radius vector to the centre of mass of the orbiting tether system, and r_P, can be used to define the radius vector to the perigee of the orbit. The representation of Figure 7.17 is sufficient to define completely the system based on the Earth centre location, E, representing the origin of the X, Y, Z system, and the origin of the relative rotating x_0, y_0, z_0 co-ordinate system located at the centre of mass of the tether system, as in section 7.7.3.1. The X, Y plane and the x_0, y_0 plane lie on the orbital plane and the Z and z_0 axes are perpendicular to this. The X axis is directed to the orbit perigee and x_0 rotates in axial alignment with R. The in-plane angle, previously defined as the planar spin, or pitch angle, ψ, is now re-defined as the angle from the x_0 axis *to the projection of the tether onto the orbital*

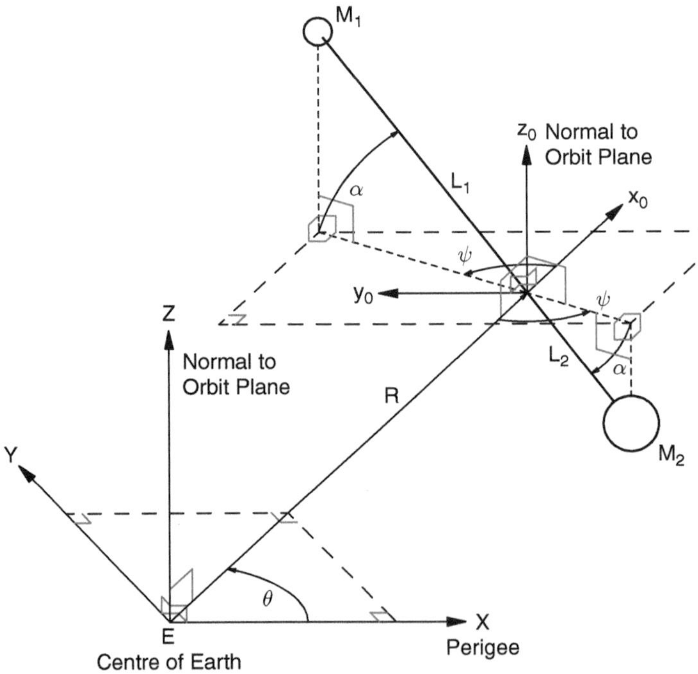

Fig. 7.17. Non-planar tether co-ordinate system [39, 58].

plane. In addition to this an out-of-plane angle, α, goes from the tether to its projection onto the orbital plane, and is always normal to the orbital plane.

The equations of motion are once again obtained by means of the Lagrangian approach. The system consists of two end masses, M_1 and M_2, connected by a tether with sub-span lengths denoted by L_1 and L_2. The Earth's gravitational field is assumed to be spherical, and there are negligible environmental perturbations, rigid tethers with constant cross-sectional area and negligible mass compared to that of the payloads (for simplicity). As in Section 7.7.3.1 the Cartesian positions are obtained, this time in the following forms,

$$\begin{aligned} x_1 &= R\cos\theta + L_1\cos\alpha\cos(\psi+\theta) \\ y_1 &= R\sin\theta + L_1\cos\alpha\sin(\psi+\theta) \\ z_1 &= L_1\sin\alpha \\ x_2 &= R\cos\theta - L_2\cos\alpha\cos(\psi+\theta) \\ y_2 &= R\sin\theta - L_2\cos\alpha\sin(\psi+\theta) \\ z_2 &= -L_2\sin\alpha. \end{aligned} \tag{7.97}$$

In this case R, θ, α, ψ are regarded as the generalized co-ordinates, and the positions of the payloads with respect to the centre of the Earth are given by,

$$R_1 = \sqrt{x_1^2+y_1^2+z_1^2} = \sqrt{L_1^2+R^2+2L_1R\cos\alpha\cos\psi} \tag{7.98}$$

$$R_2 = \sqrt{x_2^2+y_2^2+z_2^2} = \sqrt{L_2^2+R^2-2L_2R\cos\alpha\cos\psi}. \tag{7.99}$$

The kinetic energy of the system in terms of the payloads and neglecting the tether can be expressed by translational point-mass payload components,

$$T = \frac{1}{2}M_1\left(\dot{x}_1^2+\dot{y}_1^2+\dot{z}_1^2\right) + \frac{1}{2}M_2\left(\dot{x}_2^2+\dot{y}_2^2+\dot{z}_2^2\right), \tag{7.100}$$

where the dot once again denotes differentiation with respect to time. Differentiating equations (7.97) with respect to time and substitution, as appropriate, into Eq. (7.100) gives the following, assuming moment equilibrium $M_1L_1 = M_2L_2$,

$$T = \frac{1}{2}(M_1+M_2)(\dot{R}^2+R^2\dot{\theta}^2) + \frac{1}{2}(M_1L_1^2+M_2L_2^2)\times[\,\dot{\alpha}^2+\cos^2\alpha\;(\dot{\psi}+\dot{\theta})^2] \tag{7.101}$$

Neglecting the tether mass, the potential energy of the payloads is given by,

$$U = -\frac{\mu M_1}{R_1} - \frac{\mu M_2}{R_2}. \tag{7.102}$$

Substituting (7.98) and (7.99) into Eq. (7.102) gives the system's potential energy,

$$U = -\frac{\mu M_1}{\sqrt{L_1^2+R^2+2L_1R\cos\alpha\cos\psi}} - \frac{\mu M_2}{\sqrt{L_2^2+R^2-2L_2R\cos\alpha\cos\psi}}. \tag{7.103}$$

Given that the length of a typical tether is likely to be two to three orders of magnitude less than the orbital radius then its potential energy can be expanded with $\delta_1 = \frac{L_1}{R}$ and $\delta_2 = \frac{L_2}{R}$, where $\delta_1 \ll 1$ and $\delta_2 \ll 1$. On the assumption that $M_1 L_1 = M_2 L_2$, expansion up to $O(\delta_1^3)$ and $O(\delta_2^3)$ leads to,

$$U = -\frac{\mu (M_1 + M_2)}{R} + \frac{\mu \left(M_1 L_1^2 + M_2 L_2^2\right)\left(1 - 3\cos^2\alpha\cos^2\psi\right)}{2R^3} - \frac{\mu\left(M_1 L_1^3 - M_2 L_2^3\right)\left(3\cos\alpha\cos\psi - 5\cos^3\alpha\cos^3\psi\right)}{2R^4} \tag{7.104}$$

The kinetic and potential energies in Eqs. (7.101) and (7.104) are substituted into Lagrange's equation in the required multi-degree of freedom form,

$$\frac{\mathrm{d}}{\mathrm{d}t}\left(\frac{\partial T}{\partial \dot{q}_i}\right) - \frac{\partial T}{\partial q_i} + \frac{\partial U}{\partial q_i} = Q_{q_i} \quad i = 1, 2, \ldots, N, \tag{7.105}$$

where, in this case, the generalized forces Q_{q_i} are zero because no external forces or torques are assumed here (such as were provided by the motor drive in Section 7.7.3.1). Applying Eq. (7.105) leads to four equations of motion governing motion through the in and out-of-plane angles defined for the tether, the true anomaly, and the radial distance from the focus of the elliptical orbit to the center of mass of the tether. Alternatively, standard orbital mechanics relationships [55] can be used to transform the equations from the time domain to the true anomaly domain, thereby simplifying them algebraically, and also reducing the number of equations down from four to two by assuming a circular orbit, for which $e = 0$, resulting in,

$$\psi'' - 2\alpha' \tan\alpha\,(\psi' + 1) + \frac{3}{2}\sin(2\psi) + \frac{3(L_1 + L_2)\sin\psi\sec\alpha}{2r_C} \times \left(1 - 5\cos^2\psi\cos^2\alpha\right) = 0, \tag{7.106}$$

$$\alpha'' + \frac{1}{2}\sin(2\alpha)\left[(\psi' + 1)^2 + 3\cos^2\psi\right] + \frac{3(L_1 - L_2)}{2r_C}\cos\psi\sin\alpha \times \left(1 - 5\cos^2\psi\cos^2\alpha\right) = 0. \tag{7.107}$$

It is important to note that the prime represents differentiation with respect to the true anomaly and the dot denotes differentiation with respect to time, and that this is opposite to the convention used in Cartmell et al. [58]. Equations (7.106) and (7.107) could be solved numerically, however Ziegler offers approximate analytical solutions (within the review by Cartmell et al. [58] and in full within Ref. [39]) by means of the multiple scales perturbation scheme applied up to third-order perturbation accuracy and also by expanding the trigonometrical functions by means of the Taylor series and retaining the first two terms. The multiple scales solution is uniformly valid close to $\psi = \alpha = 0$. In the case of numerical integration solutions the simplifying expansion of the potential energy expression is not required and so Eq. (7.103) can be used instead to lead to alternative forms for the two governing equations, as follows,

$$\psi'' - 2\tan\alpha\,\alpha'\,(\psi'+1) + \frac{r_C^4}{L_1+L_2}\sin\psi\sec\alpha\left[\left(r_C^2+L_2^2-2r_CL_2\cos\psi\cos\alpha\right)^{-3/2} - \left(r_C^2+L_1^2+2r_CL_1\cos\psi\cos\alpha\right)^{-3/2}\right] = 0 \tag{7.108}$$

$$\alpha'' + \frac{1}{2}\sin(2\alpha)\,(\psi'+1)^2 + \frac{r_C^4}{L_1+L_2}\cos\psi\sin\alpha\left[\left(r_C^2+L_2^2-2r_CL_2\cos\psi\cos\alpha\right)^{-3/2} - \left(r_C^2+L_1^2+2r_CL_1\cos\psi\cos\alpha\right)^{-3/2}\right] = 0. \tag{7.109}$$

Equations (7.108) and (7.109) can be numerically integrated and their solutions compared with the approximate analytical solutions to Eqs. (7.106) and (7.107) (which are not given here but discussed in full by Ziegler [39] and partly reproduced by Cartmell et al. [58]). The approximate solution to third-order perturbation must be differentiated with respect to the true anomaly to evaluate the constants of integration at $\theta = 0$ for zero initial velocity conditions for each angle [39, 58]. The value chosen for the low Earth orbit radius, r_C, is taken to be 7000 km, and very short tether lengths are used, $L_1 = L_2 = 500$ m, thereby ensuring that the potential energy expansion conditions are properly satisfied. Figures 7.18 and 7.19 illustrate the response obtained between the fourth and fifth orbit in an attempt to magnify the discrepancies between the solutions. It can be seen that these are relatively marginal and that for the data used the third-order multiple scales solutions approximate the numerical integrations results quite well. Further work on non-planar models using non-equatorial elliptical orbits has shown the potential for extremely complex three-dimensional motions [40] in which there is scope for very considerable inter-coordinate coupling and highly non-linear dynamical behavior. The circular restricted three-body

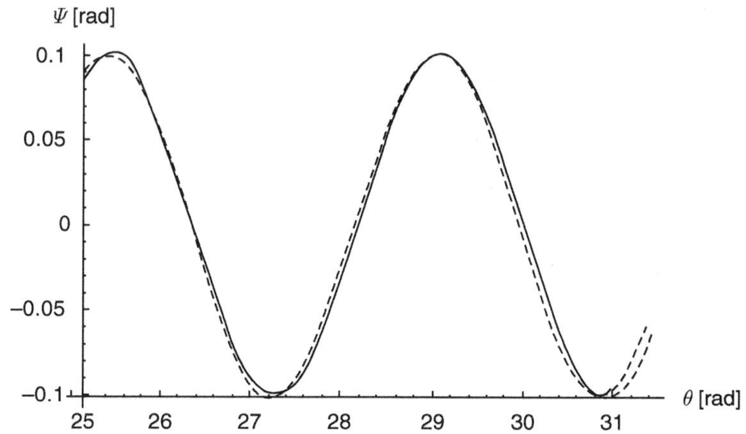

Fig. 7.18. Response of angle ψ as a function of true anomaly θ, solid line—numerical solution, dashed line—second-order multiple scales solution, chain dashed line—third-order multiple scales solution [39, 58].

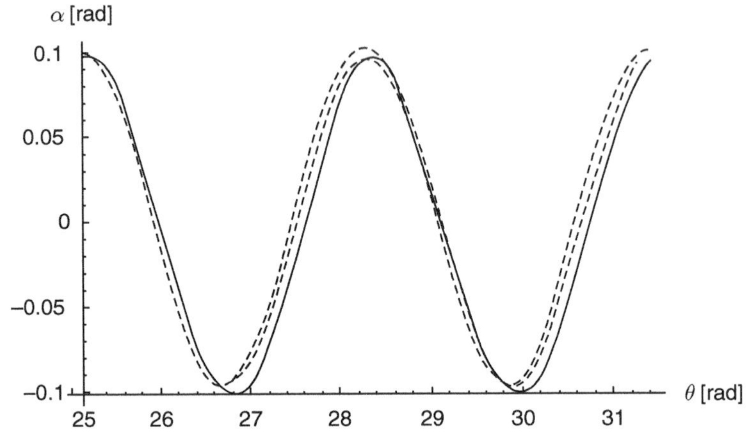

Fig. 7.19. Response of angle α as a function of true anomaly θ, solid line—numerical solution, dashed line—second-order multiple scales solution, chain dashed line—third-order multiple scales solution [39, 58].

problem can be used (the third body is the Sun), in order to permit a weak stability boundary transfer via Lagrange point L_1 and this in conjunction with a non-equatorial orbit for a pair of staged passive tethers is shown to be potentially capable of Earth–Moon payload transfer for realistic tethers and significant payload masses (typically 500 kg each). The work reported in [40] is based on pragmatically sized tethers working in a staged configuration at the Earth end (see next section for a discussion on staging) and a long Moravec Lunavator tether depositing and retrieving payloads to and from the lunar surface.

7.7.4 Payload exchange concepts

Hoyt and Forward [59] first suggested that doubled-up, staged, tethers could be used from SEO to LTO for transfers of payloads between the Earth and the Moon by using mass balance to ensure conservation of momentum throughout. The method links the *Sub-Earth-orbit to LEO* method of Carroll [60] to the *EEO to LTO* proposal of Stern [61]. Cartmell & Ziegler [42] applied this to the motorized momentum exchange concept and then Cartmell et al. [40] presented some preliminary calculations for a lunar transfer through L_1 on the basis of non-equatorial tether orbits around Earth in order to obviate the need for a plane change at the Moon end. On reaching the Moon payloads are captured by a Moravec Lunavator [34], which brings the payload round from LLO and down to the Lunar surface. The reverse is also possible in principle. Clearly the appropriate orbital elements must be calculated to make payload handovers occur when there is instantaneous zero relative velocity at the points where the payload is to be transferred during its journey. Initially, the payload is boosted up to SEO by means of chemical propulsion to rendezvous with the LEO tether in a slightly elliptical orbit, after which tether propulsion takes over, first by means of the staged tether rendezvous and boost through to LTO, passage through

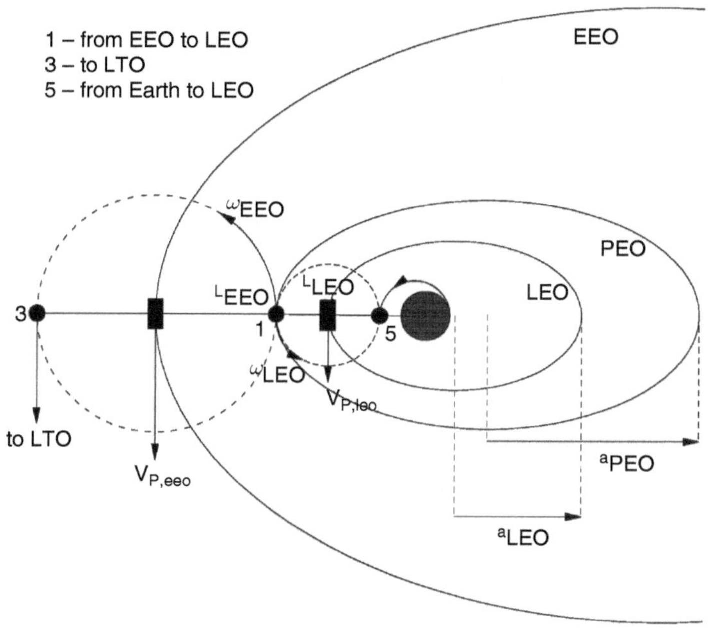

Fig. 7.20. Staged tethers shown at alignment [42].

L_1, with sufficient energy to continue to the Moon and capture by a Moravec Lunavator. Effects of gravitational perturbations on the tethers due to other bodies, and the Earth's oblateness, have both been neglected in analyses to date. Figure 7.20 shows the instant when the two staged tethers are in alignment. The inner tether is on LEO and the outer tether is on EEO. There are two emergency payload capture orbits, PEO, and PEO_1 (not shown). In principle this approach could be used with passive momentum exchange tethers, but the control and logistics required, and the maximizing of available energy levels for optimal performance, suggest that motorized tethers would be the best solution. The concept is based on the continual and synchronized arrival and departure of payloads of identical mass flowing in and out of the system, thereby ensuring that each tether is alternately either fully laden with two payloads, or completely unladen on both sides. It also assumes that suitable technologies will be in place for payload handovers, major orbital maintenance in the event of missed handovers (and therefore asymmetrically laden tethers and unattached payloads moving along their rescue orbits), solar electric power supplies, control and logistics, and of course an appropriate manufacturing or habitation infrastructure which requires to exploit two-way mass movement between the planets. Figure 7.20 shows the instant at which the payload labelled 5 has just been delivered from SEO to the right-hand end of the LEO tether (looking down onto the system) under conditions of zero relative velocity at that point. At this precise moment the EEO tether passes payload 1 inwards to the left-hand side of the LEO tether, so that the LEO tether is simultaneously loaded at each end. At that same moment the EEO tether releases payload 3 into LTO. This means that the EEO tether is unladen at both ends at exactly the same time that the LEO tether is fully laden at both ends. The step changes in mass moment of inertia of the two tethers will have a potential de-orbiting effect and work is in hand to quantify this at the time of writing. The orbit periods are harmonic, and this and the facility spin rate determines when the necessary handover conditions can occur. If the handover fails another attempt may be possible the next time they come round. This potential problem arises for transfers from SEO to LEO, and vice versa, and also from LEO to EEO, and vice versa. In the next stage payload 1 comes to rendezvous with the Earth transfer vehicle on a return SEO. Meanwhile payload 5 is handed over from the LEO tether to the EEO tether to collect payload 6 (not shown in Figure 7.20). Payload 6 could have been in PEO_1 instead, perhaps due to a missed capture first time round. Assuming handovers occur as planned the LEO tether becomes empty as the EEO tether becomes full. The SEO booster brings payload 7 (also not shown in Figure 7.20) up to the LEO tether and the EEO tether hands over payload 6. The EEO tether also releases payload 5 into LTO. Thus, the EEO tether empties and the LEO tether fills. This all charts the progress of payload 5 from SEO to LTO, and the logical process can continue indefinitely as long as payloads continue to arrive at the right time and place. The above shows that flows of payload can be arranged into and out of the staged tether system so that neither of the two tethers is ever asymmetrically laden. A full analysis of this is given by Cartmell and Ziegler [42] and Cartmell et al. [40] in which it is shown that this concept, when used with pragmatic data can readily generate Earth escape velocities for reasonable tether geometries and mass characteristics.

7.8 Conclusions

Solar sailing and space tethers both offer exciting new possibilities for the future by either extracting momentum from the environment, or balancing momentum through payload exchanges. These novel propulsion technologies offer many interesting new avenues of research in orbital dynamics, some of which have been discussed here. For solar sailing, the continuous thrust generated by the sail allows exotic new families of highly non-Keplerian orbits, while staged tether systems allow intriguing means of transferring payloads at essentially no cost. While some of these possibilities have been discussed here, it is clear that there are many unexplored opportunities which await investigation in the future. In particular, future technology developments may allow solar sails to be fabricated with a sail lightness number of order one. This would provide a continuous thrust of the same order as the local solar gravitational acceleration, surely enabling interesting new families of orbit. Similarly, developments in carbon nanotubes may provide tethers with remarkable mechanical properties, allowing extremely large loads to be transferred to high energy orbits.

Lastly, it is surely through the novel features of the orbital mechanics of these propellantless propulsion systems that further compelling practical applications will arise, which will in turn unlock the resources to meet the engineering challenges they present. While the underlying orbital mechanics may be interesting and exciting, implementation is the key to the future.

References

1. Tsiolkovsky, K.E. (1921). *Extension of Man into Outer Space* [also, Tsiolkovsky, K.E. (1936), Symposium Jet Propulsion, No. 2, United Scientific and Technical Presses].
2. Tsander, K. (1924). *From a Scientific Heritage*, NASA Technical Translation TTF-541, 1967 [quoting a 1924 report by the author].
3. Friedman, L., et al. (1978). Solar sailing—the concept made realistic. AIAA-78-82, AIAA Aerospace Sciences Meeting, Huntsville.
4. McInnes, C.R., Macdonald, M., Angelopolous, V. and Alexander, D. (2001). GeoSail: exploring the geomagnetic tail using a small solar sail. *Journal of Spacecraft and Rockets*, **38**, pp. 622–629.
5. West, J.L. and Derbes, B. (2000). Solar sail vehicle system design for the geostorm warning mission. AIAA-2000-5326, AIAA Structures, Structural Dynamics and Materials Conference and Adaptive Structures Forum, Atlanta.
6. Macdonald, M., McInnes, C.R. and Hughes, G.W. (2004). A near-term roadmap for solar sailing. IAC 04-U.1.09, 55th International Astronautical Congress, Vancouver.
7. Sauer, C.G. (1999). Solar sail trajectories for solar polar and interstellar probe missions. AAS-99-336, AAS/AIAA Astrodynamics Specialist Conference, Gridwood, Alaska.
8. Dachwald, B. (2004a). Optimal solar sail trajectories for missions to the outer solar system. AIAA-2004-5406, AIAA/AAS Astrodynamics Specialist Conference, Providence.
9. McInnes, C.R., Hughes, G. and Macdonald, M. (2003), Low cost mercury orbiter and mercury sample return missions using solar sail propulsion. *The Aeronautical Journal*, **107**, pp. 469–478.
10. Hughes, G.W. and McInnes, C.R. (2004). Small-body encounters using solar sail propulsion. *Journal of Spacecraft and Rockets*, **41**, pp. 140–150.
11. McInnes, C.R. (2004). Deflection of near-earth asteroids by kinetic energy impacts from retrograde orbits. *Planetary and Space Science*, **52**, pp. 587–590.
12. McInnes, C.R. (1999a). *Solar Sailing: Technology, Dynamics and Mission Applications*, Springer-Verlag, London.
13. Murphy, D.M. and Murphey, T.W. (2003). Scalable solar-sail subsystem design concept. *Journal of Spacecraft and Rockets*, **40**, pp. 539–547.
14. Salama, M., White, C. and Leland, S. (2003). Ground demonstration of a spinning solar sail deployment concept. *Journal of Spacecraft and Rockets*, **40**, pp. 9–14.
15. McNeal, R., Hedgepeth, J.M. and Schuerch, H.U. (1969). Heliogyro solar sailer summary report. NASA Contractor Report CR-1329.
16. Coverstone-Carroll, V. L. and Prussing, J. E. (2003). Technique for Escape from Geosynchronous Transfer Orbit Using a Solar Sail. *Journal of Guidance, Control, and Dynamics*, **26**(4), pp. 628–634.
17. McInnes, C.R. (2002). Solar sailing: orbital mechanics and mission applications. COSPAR02-A-03497, 53rd International Astronautical Federation Congress, Houston.
18. Bacon, R.H. (1959). Logarithmic spiral—an ideal trajectory for an interplanetary vehicle with engines of low sustained thrust. *American Journal of Physics*, **27**, pp. 12–18.
19. Tsu, T.C. (1959). Interplanetary travel by solar sail. *American Rocket Society Journal*, **29**, pp. 422–427.
20. London, H.S. (1960). Some exact solutions of the equations of motion of a solar sail with a constant setting. *Journal of the American Rocket Society*, **30**, pp. 198–200.
21. Sauer, C.G. (1976). Optimum solar sail interplanetary trajectories. AIAA-76-792, AAS/AIAA Astrodynamics Specialist Conference, San Diego.
22. Dachwald, B. (2004b). Optimization of interplanetary solar sailcraft trajectories using evolutionary neurocontrol. *Journal of Guidance, Control, and Dynamics*, **27**(1), pp. 66–72.
23. Powers, B. and Coverstone-Carroll, V. (2001). Optimal solar sail orbit transfers to synchronous orbits. *Journal of the Astronautical Sciences*, **49**(2), pp. 269–281.
24. Forward, R.L. (1991). Statite: A Spacecraft that Does Not Orbit, *Journal of Spacecraft and Rockets*, **28**(5), pp. 606–611.
25. McInnes, C.R., McDonald, A.J.C., Simmons, J.F.L. and MacDonald, E.W. (1994). Solar sail parking in restricted three-body systems. *Journal of Guidance Dynamics and Control*, **17**, pp. 399–406.
26. McInnes, C.R. (1999b). Artificial lagrange points for a non-perfect solar sail. *Journal of Guidance, Control and Dynamics*, **22**, pp. 185–187.

27. McInnes, C.R. (2003). Solar sailing: mission applications and engineering challenges. Philosophical Transactions of the Royal Society A, **361**, pp. 2989–3008.
28. Farquhar, R.W., Muhonen, D.P. and Richardson, D.L. (1977). Mission design for a halo orbiter of the earth. *Journal of spacecraft and rockets*, **14**(3), pp. 170–177.
29. Tsiolkovsky, K.E. (1959). *Grezi o zemle i nene*, USSR Academy of Sciences Edition, p. 35. [also, Tsiolkovsky, K.E. (1961). *A Way to the Stars*, Izdatelstvo, AN SSSR], both in Russian.
30. Artsutanov, Yu. (1960). V kosmos na elektrovoze. *Komsomolskaya Pravda*.
31. Isaaks, J.D., *et al.* (1966). Satellite elongation into a true 'skyhook'. *Science*, **151**, pp. 682–683.
32. Colombo, G., Gaposchkin, E.M., Grossi, M.D. and Weiffenbach, G.C. (1975). The 'Skyhook': a shuttle-borne tool for low-orbital-altitude research. *Meccanica*, **10**, pp. 3–20.
33. Cosmo, M.L. and Lorenzini, E.C. (1997). Tethers in space handbook. 3rd edition, Smithsonian Astrophysical Observatory.
34. Moravec, H. (1977), A Non-synchronous Orbital Skyhook. *Journal of the Astronautical Sciences*, **25**, pp. 307–332.
35. Tethers Unlimited Inc. (2005). http://www.tethers.com/
36. Carroll, J.A. (1985). Guidebook for Analysis of Tether Applications. Final Report on Contract RH4-394049 for the Martin Marietta Corp.
37. Cartmell, M.P. (1998). Generating velocity increments by means of a spinning motorised tether. 34th AIAA/ASME/SAE/ASEE Joint Propulsion Conference and Exhibit, Cleveland, Ohio, USA, paper AIAA 98-3739.
38. Ziegler, S.W. and Cartmell, M.P. (2001). Using motorised tethers for payload orbital transfer. *AIAA Journal of Spacecraft and Rockets*, **38**(6), 904–913.
39. Ziegler, S.W. (2003). *The Rigid-Body Dynamics of Tethers in Space*, Ph.D. thesis, Department of Mechanical Engineering, University of Glasgow.
40. Cartmell, M.P., McInnes, C.R. and McKenzie, D.J. (2004). Proposals for an earth–moon mission design based on motorised momentum exchange tethers. XXXII Summer School on 'Advanced Problems in Mechanics', Russian Academy of Sciences, St. Petersburg, Russia.
41. McKenzie, D.J. and Cartmell, M.P. (2004). On the performance of a motorized tether using a ballistic launch method. 55th International Astronautical Congress 2004, Vancouver Convention and Exhibition Centre, Vancouver.
42. Cartmell, M.P. and Ziegler, S.W. (1999). The use of Symmetrically Laden Motorised Tethers for Continuous Two-way Interplanetary Payload Exchange. 35th AIAA/ASME/SAE/ASEE Joint Propulsion Conference and Exhibit, Bonaventure Hotel and Conference Center, Los Angeles, California, USA, paper AIAA 99-2840.
43. Delta Utec SRC, 2005, http://www.delta-utec.com/.
44. Lennert, S. and Cartmell, M.P. (2003). Analysis and design of a friction brake for momentum exchange propulsion tethers. Fifth IAA International Conference on Low-Cost Planetary Missions, ESA/ESTEC.
45. Ouyang, H., Mottershead, J.E., Cartmell, M.P. and Friswell, M.I. (1997). Parametric Resonances in an Annular Disc with a Rotating System of Distributed Mass and Elasticity; and Effects of Friction and Damping, *Proceedings of the Royal Society of London*, Series A, **453**, pp. 1–19.
46. NASA Science Mission Directorate (2005). http://www.inspacepropulsion.com/
47. Colombo, G., Martinez-Sanchez, M. and Arnold, D. (1982). The use of tethers for payload orbital transfer. Smithsonian Astrophysical Observatory, NAS8-33691, Cambridge, MA, USA.
48. Bekey, I. and Penzo, P.A. (1986). Tether propulsion. *Aerospace America*, **24**(7), pp. 40–43.
49. Kelly, W.D. (1984). Delivery and Disposal of a Space-Shuttle External Tank to Low Earth Orbit. *Journal of the Astronautical Sciences*, **32**(3), pp. 343–350.
50. Lorenzini, E.C., Cosmo, M.C., Kaiser, M., Bangham, M.E., Vonderwell, D.J. and Johnson, L. (2000). Mission Analysis of Spinning Systems for Transfers from Low Orbits to Geostationary. *Journal of Spacecraft and Rockets*, **37**(2), pp. 165–172.
51. Arnold, D.A. (1987). The behaviour of long tethers in space. *Journal of the Astronautical Sciences*, **35**(1), pp. 3–18.
52. Carroll, J.A. (1986). Tether Applications in Space Transportation. *Acta Astronautica*, **13**(4), pp. 165–174.
53. Bekey, I. (1983). Tethers open new space options. *Astronautics and Aeronautics*, **21**(4), pp. 32–40.
54. Kumar, K., Kumar, R. and Misra, A.K. (1992), Effects of deployment rates and librations on tethered payload raising. *Journal of Guidance, Control, and Dynamics*, **15**(5), pp. 1230–1235.

55. Chobotov, V.A. (1996). *Orbital Mechanics*, 2nd edition, AIAA Education Series, AIAA, Reston, VA, USA.
56. Beletsky, V.V. and Levin, E.M. (1993). *Dynamics of Space Tether Systems*, Advances in the Astronautical Sciences, **83**, American Astronautical Society, San Diego, CA, USA.
57. TetherSim (2002). Available by request from colin.mcinnes@strath.ac.uk.
58. Cartmell, M.P., Ziegler, S.W., Khanin, R. and Forehand, D.I.M. (2003). Mechanics of Systems with Weak Nonlinearities. Transactions of the ASME, Applied Mechanics Reviews, **56**(5), pp. 455–492.
59. Hoyt, R.P. and Forward, R.L. (1997). Tether transport from sub-earth-orbit to the lunar surface . . . and back!. International Space Development Conference, Orlando, Florida, USA.
60. Carroll, J.A. (1991). Preliminary design of a 1 km/s tether transport facility. Tether Applications Final Report on NASA Contract NASW-4461, with NASA/HQ.
61. Stern, M.O. (1988). Advanced propulsion for LEO-Moon transport. NASA CR-172084, California Space Institute progress report on NASA grant NAG 9-186, with NASA/JSC.

8 Cooperative Spacecraft Formation Flying: Model Predictive Control with Open- and Closed-Loop Robustness

Louis Breger, Gokhan Inalhan, Michael Tillerson, and Jonathan P. How
Aerospace Controls Laboratory, Massachusetts Institute of Technology

Contents

8.1 Introduction

Formation flying of multiple spacecraft is an enabling technology for many future space science missions including enhanced stellar optical interferometers and virtual platforms for Earth observations [1, 2]. Controlling a formation will require several considerations beyond those of a single spacecraft. Key among these is the increased emphasis on fuel savings for a fleet of vehicles because the spacecraft must typically be kept in an accurate formation for periods on the order of hours or days [3–5], and the performance of the formation should degrade gracefully as one or more of the spacecraft runs out of fuel [6]. This chapter presents a model predictive controller that is particularly well-suited to formation flying spacecraft because it explicitly minimizes fuel use, exploits the well-known orbital dynamics environment, and naturally incorporates constraints (e.g., thrust limits, error boxes). This controller is implemented using Linear Programming (LP) optimization, which can be solved very rapidly and has always-feasible formulations. The resulting algorithms can be solved in real-time to optimize fuel use and are sufficiently robust that they can be embedded within an autonomous control system.

Efficient execution of precise formation flying relies on both accurate descriptions of the fleet dynamics and accurate knowledge of the relative states. Navigational errors [7, 8] and inaccurate physical models (such as ignored non-linearities, thruster misalignments, and differential disturbances such as J_2 and drag) can be significant sources of error [9].

Even though an open-loop analysis can model and numerically predict the impact of navigation and modeling errors on the fuel usage, there still remains the challenge of designing a control systems to produce fuel efficient and robust command inputs when subject to these errors. This chapter focuses on techniques to robustify the formation flying control system to handle these types of errors.

Our treatment of the control system design begins with the development of dynamic models in Section 8.2. In this section, starting from the general nonlinear model, we present the linearized dynamics for relative motion around a circular and eccentric orbit. In addition, a survey of dynamics used in formation flight literature, extensions to cases with disturbances and the new directions such as Gauss Variational Equations (GVEs) are reviewed. Section 8.3 provides a survey of recent control techniques used in formation flying; a more extensive survey on the state-of-art in formation flying is provided in Refs. [10, 11]. The last part of Section 8.3 covers the basics of the model predictive controller (MPC) developed in our previous work [12], and provides the mathematical basis for the formation flying control algorithms for multiple spacecraft. This mathematical formulation provides the flexibility of enabling state constraints, disturbance and sensor noise models, and actuator limitations. In addition, the formulation is directly applicable to a variety of linearized discrete dynamics matrices, such as the ones obtained via approximation methods [13], direct integration [14], and closed-form solutions [15]. In this work, the controller proposed in Ref. [12] is used as a baseline design and is modified to address formation-wide cooperation and robust formation keeping aspects of fuel critical missions.

With these goals in mind, the fourth section presents a new coordination method, *virtual center* which enables the model predictive controllers on each satellite to cooperate to achieve formation wide fuel savings [16]. The virtual center represents the weighted average motion of the fleet, including an average of the predicted disturbances. Since the formation tracking through the virtual center results in each spacecraft using its control effort against the disturbances relative to the fleet average, the weights in the average can be updated to balance the fuel usage across the fleet. This approach is compared to the widely used leader–follower coordination mechanism, and the results demonstrate formation-wide fuel savings. However, even though fleet coordination provides a balanced mission execution, the accuracy of this execution relies on path planning that is robust to disturbances and sensor noise. Thus, it is crucial to devise robust planning methods within the cooperation structure to enable formation flight for extended periods of time under realistic disturbances and sensor noise.

The final two sections provide two complementary approaches to achieve this objective: open-loop control objective selection and closed-loop robust planning. Section 8.5 examines the effect of changing the terminal condition of the formation keeping optimization to include both the original robust error-box constraint and also a requirement that the spacecraft nominally enter a closed orbit inside the error box. This method, called the *control objective selection approach* [17], is shown to be effective at reducing the frequency of replanning required by the controller. The last section presents an alternate form of robustness that guarantees that the spacecraft will remain inside an error box, and explicitly accounts for the possibility of future feedback. This *closed-loop robustness* approach [17], is shown to be less conservative than requiring that the initial planned trajectory robustly remain inside the error box without replanning.

Finally, the three major developments are combined to create a hierarchical fleet-optimal control system that uses periodic terminal conditions, closed-loop robustness, and virtual-center coordination. The performance of this system is demonstrated in a fully non-linear formation flying simulation.

8.2 Dynamics of formation flight

The performance of any controller design depends on the fidelity of the embedded system dynamics model. This section presents the dynamics for the relative motion of a satellite with respect to a reference satellite (or a reference point) on an eccentric orbit. Even though these orbital dynamics are well known, its highly non-linear form complicate the development of precise control laws and online optimization algorithms. To simplify the dynamics for analysis and control, it is common to linearize with respect to the separation distance between satellites in the formation. A common set of linearized dynamics for circular orbits is Hill's Equations of Motion [18]. For elliptical orbits, linearizations are typically parameter-varying [19–21], where the parameter is the reference orbit true anomaly. This transition to linearized dynamics in case of short baseline separations for circular and eccentric Keplerian reference orbits is especially useful in modeling multiple spacecraft control and coordination problems. In the last part of this section, we summarize various extensions to these equations in order to account for disturbances. Our brief but precise development of the formation flight equations of motion follows Ref. [21], and the full details are available from classical texts such as Refs. [19, 22, 23].

The location of each spacecraft within a formation can be given by

$$\vec{R}_j = \vec{R}_{fc} + \vec{\rho}_j \tag{8.1}$$

where $\vec{R}_{fc}$ and $\vec{\rho}_j$ correspond to the location of the formation center and the relative position of the jth spacecraft with respect to that point. The choice of formation center is rather arbitrary and can either be fixed to an orbiting satellite, or just be chosen as a local point that provides a convenient reference for linearization and development of linear control laws. The reference orbit in the Earth Centered Inertial (ECI) reference frame is represented by the standard orbital elements $(a, e, i, \Omega, \omega, \theta)$, which correspond to the semi-major axis, eccentricity, inclination, right ascension of the ascending node, argument of periapsis and true anomaly.

With the assumption that $|\vec{\rho}_j| \ll |\vec{R}_{fc}|$, the equations of motion of the jth spacecraft under the gravitational attraction of a main body

$$_i\ddot{\vec{R}}_j = -\frac{\mu}{|\vec{R}_j|^3}\vec{R}_j + \vec{f}_j \tag{8.2}$$

can be linearized around the formation center to give

$$_i\ddot{\vec{\rho}}_j = -\frac{\mu}{|\vec{R}_{fc}|^3}\left(\vec{\rho}_j - \frac{3\vec{R}_{fc}\cdot\vec{\rho}_j}{|\vec{R}_{fc}|^2}\vec{R}_{fc}\right) + \vec{f}_j, \tag{8.3}$$

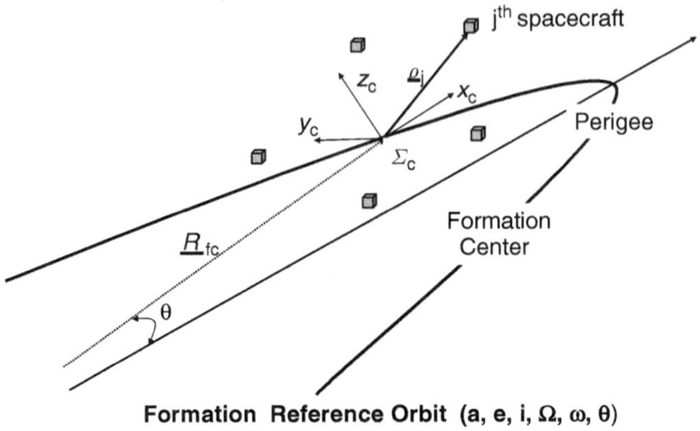

Fig. 8.1. Spacecraft formation.

where the accelerations associated with other attraction fields, disturbances or control inputs are included in $\vec{f}_j$. The derivatives in the ECI reference frame are identified by the preceding subscript i. A natural basis for inertial measurements and scientific observations is the orbiting (*non-inertial*) reference frame Σ_c, fixed to the formation center (see Figure 8.1). Using kinematics, the relative acceleration observed in the inertial reference frame ${}_i\ddot{\vec{\rho}}_j$ can be related to the measurements in the orbiting reference frame

$$
{}_i\ddot{\vec{\rho}}_j = {}_c\ddot{\vec{\rho}}_j + 2{}_i\dot{\vec{\Theta}} \times {}_c\dot{\vec{\rho}}_j + {}_i\dot{\vec{\Theta}} \times ({}_i\dot{\vec{\Theta}} \times \vec{\rho}_j) + ({}_i\ddot{\vec{\Theta}} \times \vec{\rho}_j), \tag{8.4}
$$

where ${}_i\dot{\vec{\Theta}}$ and ${}_i\ddot{\vec{\Theta}}$ correspond to the angular velocity and acceleration of this orbiting reference frame. The fundamental vectors $(\vec{\rho}_j, \vec{R}_{fc}, {}_i\dot{\vec{\Theta}})$ in Eqs. (8.3) and (8.4) can be expressed in Σ_c as

$$
\vec{\rho}_j = x_j\hat{k}_x + y_j\hat{k}_y + z_j\hat{k}_z \tag{8.5}
$$

$$
\vec{R}_{fc} = R_{fc}\hat{k}_x \tag{8.6}
$$

$$
i\dot{\vec{\Theta}} = \dot{\theta}\hat{k}_z, \tag{8.7}
$$

where the unit vector $\hat{k}_x$ points radially outward from Earth's center (anti-nadir pointing) and $\hat{k}_y$ is in the in-track direction along increasing true anomaly. This right-handed reference frame is completed with $\hat{k}_z$, pointing in the cross-track direction. All of the proceeding vectors and their time rate of changes are expressed in the orbiting reference frame Σ_c.

Combining Eqs. (8.3) and (8.4) to obtain an expression for ${}_c\ddot{\vec{\rho}}_j$, and using Eqs. (8.5)–(8.7), it is clear that the linearized relative dynamics with respect to an eccentric orbit can be expressed via a unique set of elements and their time rate of change. This set consists of the relative states $[x_j, y_j, z_j]$ of each satellite, the radius R_{fc} and the angular velocity $\dot{\theta}$ of the formation center. Using fundamental orbital mechanics describing planetary motion [24, 25], the radius and angular velocity of the formation center can be written as

$$|\vec{R}_{fc}| = \frac{a(1-e^2)}{1+e\cos\theta}, \quad \text{and} \quad \dot{\theta} = \frac{n(1+e\cos\theta)^2}{(1-e^2)^{3/2}}, \tag{8.8}$$

where $n = \sqrt{\mu/a^3}$ is the natural frequency of the reference orbit. These expressions can be substituted into the equation for ${}_c\ddot{\vec{\rho}}_j$ to obtain the relative motion of the jth satellite in the orbiting formation reference frame. The first correct formulation of the below equations in its original form is attributed to Lawden [19].

$$\frac{\mathrm{d}}{\mathrm{d}t}\begin{bmatrix}\dot{x}\\ \dot{y}\\ \dot{z}\end{bmatrix}_j = -2\begin{bmatrix}0 & -\dot{\theta} & 0\\ \dot{\theta} & 0 & 0\\ 0 & 0 & 0\end{bmatrix}\begin{bmatrix}\dot{x}\\ \dot{y}\\ \dot{z}\end{bmatrix}_j - \begin{bmatrix}-\dot{\theta}^2 & 0 & 0\\ 0 & -\dot{\theta}^2 & 0\\ 0 & 0 & 0\end{bmatrix}\begin{bmatrix}x\\ y\\ z\end{bmatrix}_j - \begin{bmatrix}0 & -\ddot{\theta} & 0\\ \ddot{\theta} & 0 & 0\\ 0 & 0 & 0\end{bmatrix}\begin{bmatrix}x\\ y\\ z\end{bmatrix}_j$$

$$+ n^2\left(\frac{1+e\cos\theta}{1-e^2}\right)^3\begin{bmatrix}2x\\ -y\\ -z\end{bmatrix}_j + \begin{bmatrix}f_x\\ f_y\\ f_z\end{bmatrix}_j. \tag{8.9}$$

The terms on the right-hand side of this equation correspond to the Coriolis acceleration, centripetal acceleration, accelerating rotation of the reference frame, and the virtual gravity gradient terms with respect to the formation reference. The right-hand side also includes the combination of other external and control accelerations in $\vec{f}_j$. These terms can be explicitly presented for each spacecraft as

$$\begin{bmatrix}f_x\\ f_y\\ f_z\end{bmatrix}_j = \begin{bmatrix}u_x\\ u_y\\ u_z\end{bmatrix}_j + \begin{bmatrix}w_x\\ w_y\\ w_z\end{bmatrix}_j \tag{8.10}$$

where $u = [u_x(t)\ u_y(t)\ u_z(t)]^T : \mathbb{R} \to \mathbb{R}^3$ represents the control inputs and

$$w = [w_x(t)\ w_y(t)\ w_z(t)]^T : \mathbb{R} \to \mathbb{R}^3$$

represents the combination of other external accelerations, such as disturbances. Note that care must be taken when interpreting and using the equations of motion and the relative states in a non-linear analysis. The difficulty results from the linearization process, which maps the curvilinear space to a rectangular one via a small curvature approximation. In this case, a relative separation in the in-track direction in the linearized equations actually corresponds to an incremental phase difference in true anomaly, θ.

For a circular reference orbit, $e = 0$, substituting $\dot{\theta} = n_o$, $\ddot{\theta} = 0$, and the well known Clohessy–Wiltshire or Hill's equations are

$$\frac{\mathrm{d}}{\mathrm{d}t}\begin{bmatrix}\dot{x}\\ x\\ \dot{y}\\ y\\ \dot{z}\\ z\end{bmatrix} = \begin{bmatrix}0 & 3n_o^2 & 2n_o & 0 & 0 & 0\\ 1 & 0 & 0 & 0 & 0 & 0\\ -2n_o & 0 & 0 & 0 & 0 & 0\\ 0 & 0 & 1 & 0 & 0 & 0\\ 0 & 0 & 0 & 0 & 0 & -n_o^2\\ 0 & 0 & 0 & 0 & 1 & 0\end{bmatrix}\begin{bmatrix}\dot{x}\\ x\\ \dot{y}\\ y\\ \dot{z}\\ z\end{bmatrix} + \begin{bmatrix}1 & 0 & 0\\ 0 & 0 & 0\\ 0 & 1 & 0\\ 0 & 0 & 0\\ 0 & 0 & 1\\ 0 & 0 & 0\end{bmatrix}\begin{bmatrix}u_x\\ u_y\\ u_z\end{bmatrix} + \begin{bmatrix}1 & 0 & 0\\ 0 & 0 & 0\\ 0 & 1 & 0\\ 0 & 0 & 0\\ 0 & 0 & 1\\ 0 & 0 & 0\end{bmatrix}\begin{bmatrix}w_x\\ w_y\\ w_z\end{bmatrix} \tag{8.11}$$

Numerical experience have shown that the applicability of Hill's equations to various formation flying maneuvers is limited to strictly circular orbits and short baseline separations on the order of 1 m to 1 km. However, transforming in-track separations to relative anomaly angles (rectangular map of curvilinear space), allows for larger baseline formation initialization and formation-keeping maneuvers.

8.2.1 Extensions to formation flight dynamic representations

Linearized relative orbital dynamics equations of motion of the type discussed in this chapter typically rely on several key assumptions: small vehicle separations ($\rho/R \ll 1$) and Keplerian motion. They ignore the non-Keplerian effects of drag and J_2[1] and some (e.g., Hill's equations) also assume no reference orbit eccentricity. In recent years, a number of new dynamics models have been published that capture the effects of these disturbances. Table 8.1 summarizes the dynamics models that have been commonly used for formation-flight control.

The inherent inability of Hill's equations to account for the eccentricity of the Keplerian reference orbit provides a dominant source of error, even for typical STS mission eccentricities such as $e = 0.005$. In comparison, Lawden's equations [19, 21] provide the flexibility to account for reference orbit eccentricities up to $e = 0.8$. Similar to Hill's equations, the applicability of these equations is also limited by the non-linearity introduced with large vehicle separations. The addition of second and third order non-linearity effects [27, 29, 31] provides better approximations for larger vehicle separations. These non-linearity corrections are, however, limited to circular reference orbits and degrade rapidly with larger eccentricity. The Ref. [26] modifies Hill's equations to include differential J_2 effects.

Differential orbital elements [22, 30] and Gauss' Variational Equations (GVE) [15, 23, 32, 33] can be used to simultaneously account for large baselines, reference eccentricity, and J_2 effects [28, 33]. The linearized GVE approach in Ref. [33] uses slowly changing differential orbital elements that are linearized about the desired orbit of each spacecraft in the spacecraft formation. Linearization error then grows with the state error of each

Table 8.1
The matrix of formation-flight dynamics used for control and the corrections introduced by the corresponding authors.

	$e = 0$ no J_2	$0 < e < 1$ no J_2	$e = 0$ with J_2	$0 < e < 1$ with J_2
Linearized dynamics	Hill's [18]	Lawden [19] Inalhan [21]	Schweighart [26]	
Long baseline capable	Karlgrad [27] Mitchell [29] Alfriend [31]	GVEs [23] Alfriend [30]	Gim [28]	Gim [28] Breger [33]

[1] J_2 is the major correction term to uniform gravitational attraction because of Earth's oblateness.

spacecraft rather than the overall size of the formation. As in the case of Lawden's equations, the linearized GVEs in Ref. [33] are linear parameter-varying (LPV), making them particularly well-suited to use within the model predictive control presented in this chapter.

The next section reviews the techniques available for formation flight control and develops a model predictive controller using one of the linearized dynamics presented in this section.

8.3 Formation flight control and the model predictive control formulation

The goal of formation flying control is for several vehicles to create and maintain a desired formation utilizing the limited resources available on each spacecraft while acting as a single system. In this context, we can divide this specific control task into two distinct parts. First is the *formation initialization and keeping* which initializes the desired formation to a passive aperture[2] and maintains it against disturbances. Second is the *formation planning* which creates trajectories for the formation to follow during maneuvers such as reshaping or retargeting of the formation. The primary focus of the formation flying control research to date has been to develop fuel efficient methods of performing these two distinct tasks. For formation planning and keeping, many formation control approaches have been suggested in recent years spanning a large range of control techniques, including PD, LQR, LMI, nonlinear, Lyapunov, impulsive, and model predictive control [12, 34–41].

For formation initialization and keeping, typically it is assumed that a formation is initialized to a passive aperture and deviations caused by disturbances such as differential drag and/or differential J_2 is corrected through feedback laws. Passive apertures, created through slight differences in eccentricity, inclination, and argument of latitude, provide a natural structure for formation initialization. These apertures inherently take advantage of the orbital dynamics of the spacecraft in the absence of disturbance forces to create a periodic relative motion through establishing no-drift conditions which set zero differential energy across the fleet [21, 42, 43]. The size, shape and the relative initial conditions for these passive apertures is designed by using the closed-form solutions for the non-linear or the linearized orbital equations. This idea can be further extended to mitigate the effects of the disturbances, such as J_2, by establishing initial conditions [43] which results in zero average relative drifts in special orbits. However, when such special orbit selection is not available, it is necessary to develop control laws for keeping spacecraft in formation.

Control law approaches, such as Lyapunov and PD controllers [37], require that control be applied continuously, a strategy both prone to high fuel use and difficult to implement when thrusting requires attitude adjustment. Other approaches, such as the impulsive thrusting scheme introduced in Ref. [44], require spacecraft to thrust at previously specified times and directions in the orbit, ensuring that some of the maneuvers will not be fuel-optimal. Many of the non-linear feedback control schemes available in literature [10]

[2] Passive apertures are typically short baseline periodic formation configurations that provide good, distributed, Earth imaging and reduce the tendency of the vehicles to drift apart.

utilize feedback linearization, wherein control commands are used to cancel the nonlinear dynamics and replace them with linear dynamics that are typically not the natural dynamics of an orbiting satellite. The linearized relative dynamics discussed in the previous section also provide many avenues for control development. For example, numerous linear quadratic regulators have been developed that force a vehicle to track a desired state [45–47]. However, these feedback control schemes require almost constant control effort which leads to high fuel costs over the length of a mission. In order to reduce fuel costs, recent research has focused on developing methods to generate fuel-time optimal control sequences over a period of time rather than just one step.

This section presents a method of determining fuel/time optimal control inputs and trajectories using linear programming (LP), which was first introduced in Ref. [12]. This LP formulation is the base of the control work presented in this chapter. Linear programming solves for the minimum fuel maneuver explicitly by minimizing a sum of the control inputs for the solution over a given planning horizon. The general formulation can include any form of linearized dynamics and disturbance models. The LP formulation also provides a general framework for including various types of state and actuator constraints. LP can be used for different types of maneuvers: formation maneuvers, individual station-keeping, formation-keeping, or general trajectory planning. Note that this type of control system, in addition to providing an optimal planning algorithm, replaces a reactive control system when used as a feedback controller. In literature, this technique is known as *model predictive control* (MPC) [48]. Figure 8.2 shows the MPC algorithm that each spacecraft in the formation implements.

To develop a model predictive controller for formation flight[3], we consider the general set of linear time-varying (LTV) continuous equations of motion

$$\dot{x}(t) = A(t)x(t) + B(t)u(t) + B_d(t)w(t), \tag{8.12}$$

which can be used to describe the relative dynamics of each satellite with respect to a reference Keplerian orbit. Here $x(t) \in \mathbb{R}^n$ are the states, $u(t) \in \mathbb{R}^m$ are control inputs, $w(t) \in \mathbb{R}^p$ are differential disturbances. Note that, in what follows, the dynamics could

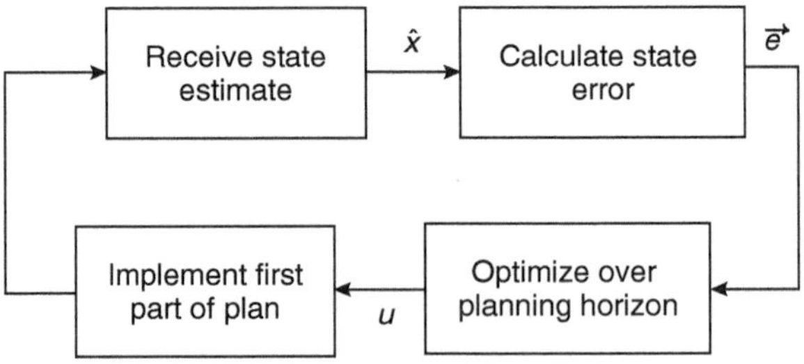

Fig. 8.2. Algorithm followed by each spacecraft in the formation.

[3] The technique developed in this section is independent of different techniques that can be used for discretization of system dynamics.

have be written as a linear parameter-varying (LPV) model, where the parameter is the true anomaly of the reference orbit, as is the case for Lawden's equations [19, 12], by using Kepler's equation to determine the true anomaly as a function of time [23]. There are a number of ways to develop discrete dynamics for optimization: approximation methods [13], direct integration [14] or closed-form solutions [15]. A straightforward, yet computationally intensive approach, is to numerically integrate the A and B matrices to obtain the discrete dynamics [49] using the following identify

$$\frac{\mathrm{d}}{\mathrm{d}t}\Phi(t, t_0) = A(t)\Phi(t, t_0)\ \forall t, \quad \Phi(t_0, t_0) = I, \tag{8.13}$$

where $\Phi(t, t_0)$ is the state transition matrix from a time t_0 to time t. In some cases, Φ can be found by solving the equations of motion directly, or by using the fundamental solution matrix [15, 50], which corresponds to the homogenous solution to Eq. (8.13). This matrix is represented by F_k for a specific time interval

$$F_k = \Phi((k+1)T_s, kT_s), \tag{8.14}$$

where T_s is the duration of a time step and k corresponds to the kth step in the discretization. The discrete input matrix G_k is given by

$$G_k = \int_{kT_s}^{(k+1)T_s} \Phi((k+1)T_s, \tau)B(\tau)\mathrm{d}\tau \tag{8.15}$$

using a zero-order hold (ZOH) assumption on the input. The ZOH assumption is appropriate for periods of thrusting that are significant relative to the duration of a time step. Equation (8.15) can be accurately computed using direct integration, but this is typically computationally expensive for online implementation. If the equations of motion are linear time invariant (LTI), or if the time step is sufficiently short that the parameter can be considered constant between discretization points, then F_k and G_k take the form [51]

$$F_k = e^{A_k T_s}, \quad G_k = \int_{kT_s}^{(k+1)T_s} e^{A_k \tau}\mathrm{d}\tau B_k, \tag{8.16}$$

where A_k and B_k are the constant state transition and input matrices for the interval $((k+1)T_s, kT_s)$. The approximate forms of F_k and G_k in Eq. (8.16) are significantly less expensive to compute than the direct integration in Eq. (8.15), but any integration errors must be carefully monitored to determine their effect on the closed-loop performance. Note that if, instead, the thrusting occurs quickly relative to the duration of a time step, then the input can be assumed to be impulsive, and $G_k = B(kT_s)$ [51]. A similar analysis must also be performed for the disturbance input matrix, M_k, which is calculated using the disturbance input matrix B_{dk}.

Using any of these discretization options, the discrete dynamics take the following form

$$x(k+1) = F_k x(k) + G_k u(k) + M_k w(k), \quad t = kT_s. \tag{8.17}$$

For any given k^{th} time step, the variables of interest $z(k) \in \mathbb{R}^l$ can be extracted from the states $x(k)$ and input variables $u(k)$ as

$$z(k) = H_k x(k) + J_k u(k), \quad t = kT_s, \tag{8.18}$$

where H_k and J_k correspond to output matrices for the optimization. The result $z(k)$ can be explicitly written as a function of the input variables $u(1), \ldots, u(k-1)$ up to the kth step, initial state $x(0)$, and the discrete system dynamics using the convolution sum [49]

$$z(k) = H_k F^{(k,k)} x(0) + J_k u(k) + \sum_{i=0}^{k-1} H_k F^{(k-i-1,k)} \left[G_i u(i) + M_i w(i)\right] \quad k \geq 1, \tag{8.19}$$

where $F^{(j,k)}$ corresponds to

$$F^{(j,k)} = \begin{cases} F_{(k-1)} \cdots F_{(k-j+1)} F_{(k-j)} & 2 \leq j \leq k \\ F_{(k-1)} & j = 1 \\ I & j = 0 \end{cases} . \tag{8.20}$$

Further, Eq. (8.19) can be structured into a compact matrix representation

$$z(k) = \Gamma(k) U_k + h(k), \tag{8.21}$$

where

$$\Gamma(k) = \left[\, H_k F^{(k-1,k)} G_0 \;\; H_k F^{(k-2,k)} G_1 \ldots H_k F^{(0,k)} G_{k-1} \;\; J_k \,\right] \tag{8.22}$$

and

$$h(k) = \left[\, H_k F^{(k-1,k)} M_0 \;\; H_k F^{(k-2,k)} M_1 \ldots H_k F^{(0,k)} M_{k-1} \,\right] \begin{bmatrix} w(0) \\ w(1) \\ \vdots \\ w(k-1) \end{bmatrix} + H_k F^{(k,k)} x(0) \tag{8.23}$$

and the input vector is

$$\mathrm{U}_k = [\, u(0)^T u(1)^T \ldots \;\; \ldots u(k-1)^T \; u(k)^T \,]^T \tag{8.24}$$

The disturbance inputs $w(k)$ are used to account for *known* dynamics that are not modeled in the A_k matrices. For example, Hill's dynamics do not model aerodynamic drag, however equations describing these effects are readily found in the literature [52]. By calculating the effects of drag on each satellite at each time step a priori, these effects can be included in the optimization. By including known disturbances on the dynamics, the controller is better able to prevent constraint violations and conserve fuel in cases where the disturbance acts as required input would have.

This affine plant description is the basis of the formation keeping control problem. The objective of each vehicle in this problem is to minimize fuel use

$$J = \min_{U_n} \sum_{j=1}^{m} c_j \| u(j) \|_1, \tag{8.25}$$

where c_j terms are scalar weights. Note that the 1-norm is used as the fuel use metric, because it correctly identifies the velocity change costs (ΔV) for spacecraft with axial thrusters. The vehicles are constrained to maintain their state to within some tolerance of

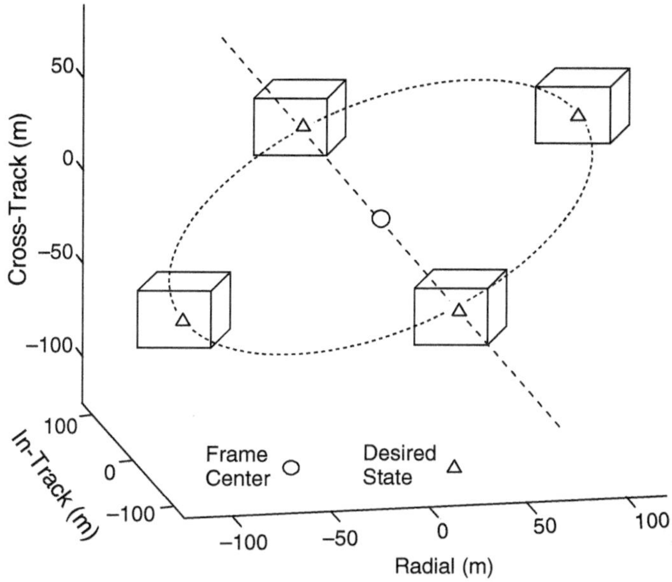

Fig. 8.3. Error boxes and desired states for a formation in a relative frame centered around a reference orbit.

a specified desired set of coordinates at each time-step k. These state constraints create an "error-box" around the desired state of the spacecraft (see Figure 8.3). Using an error box has advantages over tracking a desired point: it is more fuel efficient, better captures the mission constraints, and allows "breathing room" for the controller to account for modeling errors. The performance specification at each step k is

$$a|z_j(k) - z_{j_{\text{des}}}(k)| \leq z_{\text{tol}\,j}, \quad z_{\text{tol}\,j} \geq 0 \tag{8.26}$$

$$\Rightarrow \; -z_{\text{tol}\,j} \leq z_j(k) - z_{j_{\text{des}}}(k) \leq z_{\text{tol}\,j} \qquad \forall\, j = 1, \ldots, l \tag{8.27}$$

where $z_{\text{tol}\,j}$ is the error bound associated with each coordinate z_j. The vector $z_{\text{des}}(k)$ is the *desired state* at step k and has the same dimensions as z. The resulting MPC optimization uses the cost function from Eq. (8.25) and the constraints from Eq. (8.27).

The one-norm cost function used in the linear program (LP) is formulated by splitting the input matrix into positive, U_n^+, and negative, U_n^-, parts that are recombined after optimization

$$U_n = U_n^+ - U_n^-, \quad U_n^+ \geq 0, \; U_n^- \geq 0 \tag{8.28}$$

which gives the cost function

$$J^* = \min_{U_n^+, U_n^-} \begin{bmatrix} C^T & C^T \end{bmatrix} \begin{bmatrix} U_n^+ \\ U_n^- \end{bmatrix}, \tag{8.29}$$

where $C^T = [c_0, \ldots, c_n]$. Constraints on the maximum input can be imposed using

$$\begin{bmatrix} I & 0 \\ 0 & I \end{bmatrix} \begin{bmatrix} U_n^+ \\ U_n^- \end{bmatrix} \leq \begin{bmatrix} u_{\max} \\ u_{\max} \end{bmatrix}, \tag{8.30}$$

where $u_{\max}$ is the maximum thruster input. A variety of additional practical constraints are described in Ref. [12]. The constraints on spacecraft state at each step k are then written

$$\begin{bmatrix} \Gamma(k) & -\Gamma(k) \\ -\Gamma(k) & \Gamma(k) \end{bmatrix} \begin{bmatrix} U_k^+ \\ U_k^- \end{bmatrix} \leq \begin{bmatrix} z_{\text{des}}(k) - h(k) + z_{\text{tol}} \\ -z_{\text{des}}(k) + h(k) + z_{\text{tol}} \end{bmatrix} \quad \forall\, k \in \{1 \ldots n\} \tag{8.31}$$

$$\begin{bmatrix} -I & 0 \\ 0 & -I \end{bmatrix} \begin{bmatrix} U_n^+ \\ U_n^- \end{bmatrix} \leq \begin{bmatrix} 0 \\ 0 \end{bmatrix} \tag{8.32}$$

in order to ensure that the spacecraft remains inside an error box centered on a desired state (Eq. (8.31)) and that elements of U_n^+ and U_n^- are optimized to be positive (Eq. (8.32)). Note that the constraint matrices and vectors here can be compiled in a single large set of constraints

$$\mathbf{A}\hat{U} \leq \mathbf{b}, \tag{8.33}$$

where $\hat{U}$ is the concatenation of U_k^+ and U_k^-, and $\mathbf{A}$ and $\mathbf{b}$ capture the constraints at all steps k considered in the optimization.

Remark 8.3.1. One method to achieve robustness to unmodeled disturbance and noise terms is to identify the worst-case disturbance sequence that would force the spacecraft out of the desired error boxes. One way of approximating this is to find the disturbance sequence, w_{um}^*, that would result in total maximum variation in the spacecraft states of interest. Assuming that the unmodeled terms are set bounded (such as $|w(k)| \leq w_{\max}(k)$, for convenient LP formulation), this problem can be formulated using the convolution step in Eq. (8.22)

$$w_{um}^* = \arg \max_{[w(0),\ldots,w(m-1)]} \sum_{k=0}^{m-1} z_{um}(k) \tag{8.34}$$

$$\text{subject to } z_{um}(k) = \begin{bmatrix} H_k F^{(k-1,k)} M_0 & H_k F^{(k-2,k)} M_1 & \ldots & H_k F^{(0,k)} M_{k-1} \end{bmatrix} \begin{bmatrix} w(0) \\ w(1) \\ \vdots \\ w(k-1) \end{bmatrix}$$

$$z_{um}(k) \geq 0$$

$$|w(k)| \leq w_{\max}(k), \quad \forall k = 0, \ldots, m-1.$$

Define z_{um}^* as the value of $z_{um}(k)$ for w_{um}^*, then the error box tolerance constraints in the original LP formulation (i.e., Eq. (8.27)) can be modified through $z_{tol}^{new}(k) = z_{tol}^{old} - z_{um}^*(k)$.

Thus, at any time step, the constraints are contracted to account for the unmodeled worst-case disturbance sequence that could force them out of the errors boxes in a given time horizon, m. If the LP formulation is used as an open-loop feedback loop using replanning, then distinct parts of the time horizon can be emphasized by introducing weights to states in the cost function: $\sum_{k=0}^{m-1} c_{um}(k) z_{um}(k)$.

Remark 8.3.2. Error box constraints sometimes arise from the requirement that spacecraft achieve their desired states to within 10% of the formation separation [53]. Also, some missions would require that the error box be enforced at all times to achieve continuous observations [53], and others would only impose the constraint for brief periods during each orbit [54]. The constraints in Eq. (8.32) easily handle these cases through the addition or subtraction of steps k to the optimization, thereby enforcing the error box constraints at all times or only specific times.

Remark 8.3.3. This optimization problem is readily formulated as a linear program. Linear programming is a very fast optimization method that uses a linear, convex cost function and linear, convex constraints [55]. Linear programming was used for the simulation examples in this chapter, and in all cases required no more than a small fraction of a second to solve on a 3 GHz computer. Likewise, formulating the dynamics matrices used in the linear programs never required more than 10 seconds of computation time.

The center of each error box is referenced to the formation center, which could be a set of reference orbital elements or another spacecraft. The choices of the formation center are discussed in detail in the following section.

8.4 Distributed coordination through virtual center

Section 8.3 presented a model predictive controller that minimizes the fuel use and guarantees that the spacecraft will remain within an error box. However, a typical formation has multiple spacecraft that must all be constrained not to drift, and for some missions, form or maintain a particular shape. Thus the relative state requirements for the entire formation must be specified with respect to some reference point. This section investigates the effect of three different methods of specifying this reference point. The first is a point on a reference orbit that is propagated with the fleet. The second is a traditional *leader–follower*, where a leader is the formation reference point. The third approach involves a new method, called the *virtual center* [16, 56], which uses measurements taken by the spacecraft to calculate the location of the center. The virtual center approach is similar to the formation feedback method presented in Ref. [35], but is applied to spacecraft formation flying in low Earth orbit and explicitly uses a fuel weighting in the center calculation to equalize fuel use across the fleet. Another distinction is that the calculation of the virtual center is based on measurements available from the relative navigation estimator developed for this application [57, 58]. Using the virtual center extends the previous *formation-keeping control* in Section 8.3 to *formation flying control* by enabling extensive cooperation between the vehicles. At the end of this section, we compare these three approaches through simulation and show that the virtual center method results in fleet-wide fuel savings and fuel balance across the formation.

Remark 8.4.1. Desired states relative to the reference point are specified using passive apertures designed with the closed-form solutions to various linearized equations of relative motion [21, 42, 43]. Passive apertures are designed to result in drift-free motion, but disturbances such as differential drag cause the formation to disperse, necessitating feedback control. Our approach uses a control algorithm based on linear programming (LP) to minimize fuel cost (see Section 8.3).

8.4.1 Reference point coordination

One of the first steps in applying the LP technique for a spacecraft control system is to determine the *desired state*, which is the current state in the desired trajectory for the spacecraft. An error box is fixed to the desired state to provide a position tolerance for the satellite. A key point in this section is that the error box is specified relative to a desired point for the spacecraft. This section investigates various techniques for specifying the desired points for the formation, and demonstrates how any error in the spacecraft location relative to the current desired point can be estimated using the onboard carrier-phase differential GPS (CDGPS) measurements [58].

The formation-keeping LP algorithm in Section 8.3 is formulated to control a single spacecraft to maintain a desired state to within some tolerance specified by an error box. The formation-keeping algorithm is applied independently to each spacecraft, which enables the required computation to be distributed across the fleet. The desired state for each spacecraft is specified relative to a *reference point*, which can be chosen to enable *cooperation* between the spacecraft in the fleet, thereby enabling true formation flying.

Figure 8.4 depicts a typical scenario for a formation of three spacecraft, in which the desired trajectories of the spacecraft have been designed to create a projected ellipse in a relative frame [59]. This type of formation is known as a passive aperture and has a size determined by the formation radius. The formation angle is measured from maximum positive radial displacement. The initial conditions and closed-form solutions to the relative dynamics are used (with the drift-free constraints imposed [21]) to find the desired states at future times. The desired state is specified relative to the formation center, which is determined relative to a *formation reference point*. Three methods for determining the reference point are discussed in the following subsections. Each method is evaluated for its complexity and the amount of information flow required for its execution.

8.4.1.1 Reference orbit

A simple method of specifying the reference point is using the reference orbit. The reference orbit is a point in space that is propagated using a model that describes the average fleet orbit. The formation center is attached to the reference orbit and is used to specify the desired spacecraft states. The reference point is described by the non-linear orbit equations, requiring little communication between vehicles. Also note that the

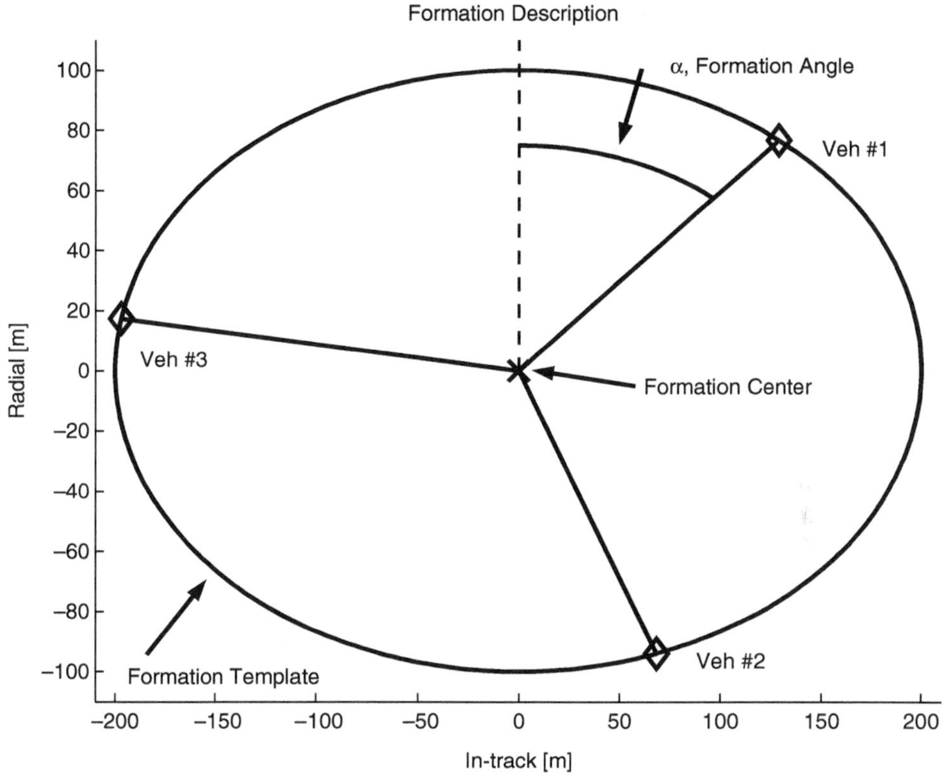

Fig. 8.4. The formation geometry is described relative to the formation center, marked by a ×. Each vehicle state is specified by a radius and an angle.

reference point is not specified using measurements, so there is no uncertainty in the state due to sensor noise. A disadvantage of this approach is that the reference point does not naturally experience the fleet disturbances. Instead, a disturbance model must be included in the propagation. If the model is inaccurate, the fleet will track a reference orbit that does not describe the fleet motion. Instead of using control effort to maintain the fleet, effort would be wasted "chasing" a mathematical point in space that does not move with the fleet.

8.4.1.2 Leader–follower

Another common method of specifying the reference point is to let a vehicle be the *leader* and fix the reference point to the leader spacecraft. The advantage is that the reference point is on a spacecraft, which eliminates the need to propagate the motion and it naturally captures the absolute disturbances. The leader–follower method requires little information flow, because the reference state is just the state of the leader spacecraft.

One leader–follower configuration places the leader at the aperture center, but this makes it prohibitively expensive to change leaders in the formation. Alternately, the leader spacecraft could be one of the vehicles on the passive aperture. The desired state for each follower spacecraft then becomes the desired state for the follower relative to the aperture center minus the desired state of the leader from the aperture center. This simplifies the transition between leaders, because no maneuvers are required. However, instantaneously switching leaders could cause a jump in the desired state of each spacecraft and must be done with care.

A disadvantage of this method is that the leader does not represent the average fleet motion, forcing some followers to overcome larger disturbances than others. Also, the leader spacecraft will use minimal fuel, because its state never experiences error. To equalize control effort across the fleet, the leader spacecraft can be alternated based on the fuel usage/status within the fleet.

8.4.1.3 Virtual center

An alternative approach to reference orbit and leader–follower tracking is to use a "virtual center" as the reference point. The reference state in this case is estimated using measurements between the spacecraft in the fleet. An advantage of the virtual center is that it represents the weighted average motion of the fleet, including an average of the actual disturbances. The weighted average enables cooperation within the fleet. The virtual center method presented here is similar to the formation feedback method for multiple vehicle control presented in Ref. [35], but our approach differs, because we show how the virtual center can be implemented using sensors planned for formation flying missions [57, 58]. The navigation algorithm presented in Ref. [57] uses decentralized estimators to filter the CDGPS measurements, precisely determining the location of each spacecraft relative to a reference vehicle. The following discussion assumes that the reference vehicle is the leader, but that is not necessary in general. Given the estimated states relative to the leader, it is possible to precisely determine the formation center.

Figure 8.5 shows a formation of three spacecraft. The thick solid lines are known or measurable distances. The thin solid lines represent the true distances to the virtual center, which are compared to the specified desired state relative to the virtual center (dashed lines). To calculate the relative position and velocity of the center, a measurement reference state must be specified. In the figure, the reference frame is attached to spacecraft #1, which will be referred to as the reference spacecraft. Inter-spacecraft states, $\vec{x}_{1i}$, are measured relative to the reference spacecraft and are represented by the solid lines in Figure 8.5. The virtual center state, $\vec{x}_c$, is also specified relative to the reference spacecraft. Each spacecraft state relative to the virtual center is

$$\vec{x}_{ci} = \vec{x}_{1i} - \vec{x}_c. \tag{8.35}$$

The error states are the difference between the state of each spacecraft relative to the center, x_{ci}, and the desired state for that spacecraft, which is also specified relative to the center. Error states in the figure are the differences between the $\Diamond$ and $\circ$ for each spacecraft.

$$\vec{x}_{ci} - \vec{x}_{i,des} = \vec{e}_i. \tag{8.36}$$

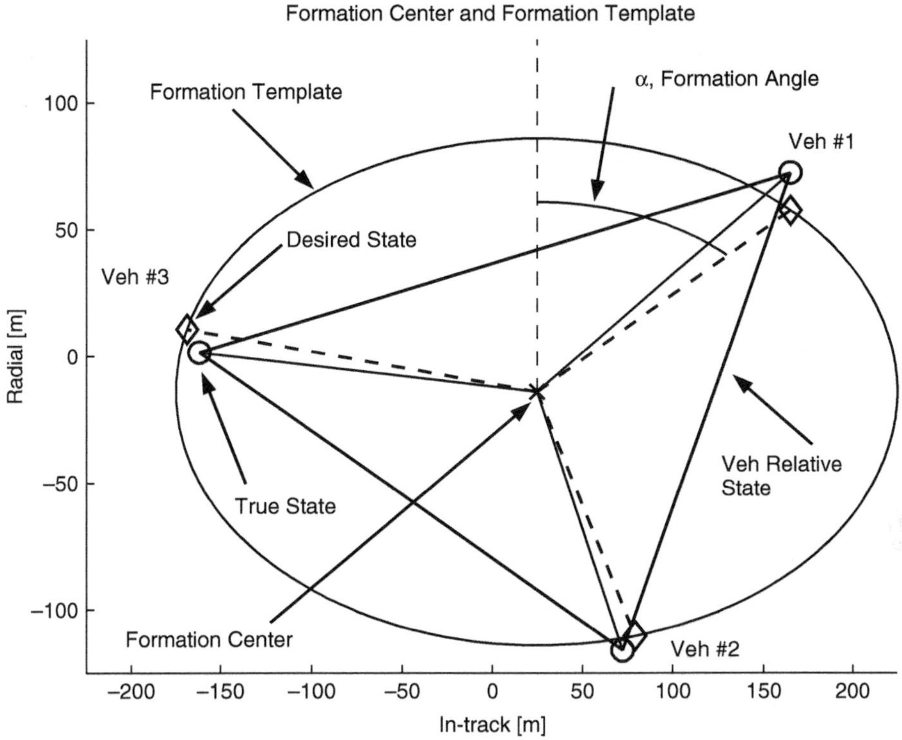

Fig. 8.5. Virtual center calculated from measured relative states (thick solid) and used to determine desired states (dashed) and actual state (thin solid) of each spacecraft.

Substituting Eq. (8.35) for x_{ci} yields the following expression for the vehicle error in terms of known quantities and the unknown virtual center, $\vec{x}_c$,

$$\vec{x}_{1i} - \vec{x}_c - \vec{x}_{i,des} = \vec{e}_i. \tag{8.37}$$

The error equation for each spacecraft becomes

$$\begin{bmatrix} \vec{x}_{11} - \vec{x}_{1,des} \\ \vec{x}_{12} - \vec{x}_{2,des} \\ \vdots \\ \vec{x}_{1N} - \vec{x}_{N,des} \end{bmatrix} - \begin{bmatrix} I \\ I \\ \vdots \\ I \end{bmatrix} [\vec{x}_c] = \begin{bmatrix} \vec{e}_1 \\ \vec{e}_2 \\ \vdots \\ \vec{e}_N \end{bmatrix} \equiv \vec{e} \tag{8.38}$$

which can be compactly rewritten as

$$b_c - A_c x = \vec{e}. \tag{8.39}$$

The virtual center location $\vec{x}_c$, is chosen to minimize the sum of the errors, $\|\vec{e}\|_2 = (b_c - A_c x)^T (b_c - A_c x)$. A weighting matrix, W, can be included to increase the importance of a vehicle or state, giving the weighted least-squares problem $\min(b_c - A_c x)^T W(b_c - A_c x)$, with solution

$$\hat{x}_c = \left(A_c^T W A_c\right)^{-1} A_c^T W b_c. \tag{8.40}$$

Note that using the normalized current fuel consumption in the weighting matrix allows fuel use across the fleet to be equalized over time.

Remark 8.4.2. Given the special form of b_c and A_c the calculation of the virtual center can be decentralized using the following algorithm

$$\hat{x}_{c_1} = b_1, \tag{8.41}$$

$$\hat{x}_{c_i} = \hat{x}_{c_{i-1}} + \frac{w_i}{\bar{w}_{i-1} + w_i}(b_i - \hat{x}_{c_{i-1}}), \tag{8.42}$$

where $b_i = \vec{x}_{1i} - \vec{x}_{i,\text{des}}$, w_i is the weight of the ith estimate, and $\bar{w}_i = \sum_{j=1}^{i} w_j$. In this formulation, spacecraft i passes its current state estimate, $\hat{x}_{c_i}$ and the scalar $\bar{w}_i$ to spacecraft $i+1$ to update the estimate of the optimal center position. The error-minimizing fuel-weighted virtual center can be computed in one cycle around a formation. The final estimate can then be shared with the rest of the fleet.

Using the virtual center method, updates can be made to the virtual center state every time step or periodically with a propagation of the virtual state between updates. A key advantage of this method is that the disturbances affecting each spacecraft become differential disturbances relative to the fleet average, which will lower fuel costs. Measuring spacecraft error relative to a regularly updating virtual center makes the absolute motion of the fleet unobservable to individual spacecraft. Thus, absolute motion will not enter into the LP, ensuring control effort is only utilized for relative geometry maintenance. A disadvantage is that the virtual center calculation must be centralized, since the current and desired states of all spacecraft must be collected in one place to find the virtual center, requiring an increase in communication throughout the fleet. Also, noise and uncertainty in measurements will lead to uncertainty in the virtual center state.

A further issue with this approach is that the virtual center is a function of the states of all the vehicles in the fleet, so any control effort by one vehicle will influence all of the other vehicles. When a vehicle uses a control input to correct an error, the control input assumes the virtual center is fixed over the plan horizon. However, the location of the center will change over time as each vehicle moves. The control inputs from the other vehicles can be included in this decentralized control algorithm by having all vehicles "publish" a list of planned control actions and then having each vehicle include the inputs of the other vehicles as *disturbance inputs* into their dynamics. The control inputs get scaled to give the motion for the virtual center in the near future. Unfortunately, there is no guarantee the published plans will get fully implemented, which may cause errors in the trajectory design.

Another way to predict the effect of external control inputs on the virtual center is to form a centralized LP to solve for all vehicles' control inputs simultaneously. The virtual center state at each time step is described in terms of the vehicle states, as in

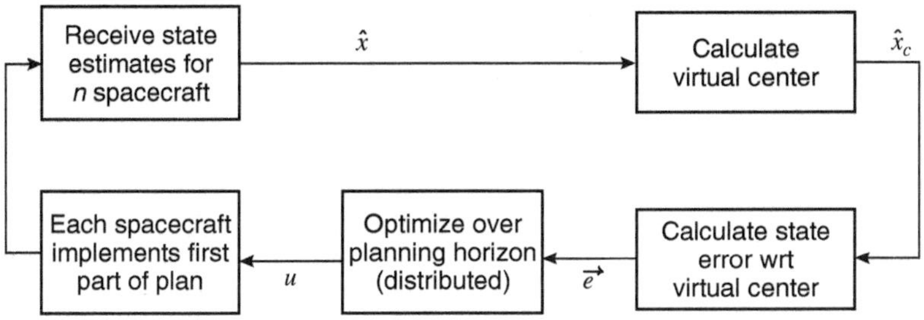

Fig. 8.6. Algorithm for model predictive control with virtual center coordination. Note that the calculation of the virtual center can be decentralized and the plan optimization step occurs concurrently for all spacecraft in the formation.

Eq. (8.40), capturing the center motion due to all control inputs. Control input solutions and trajectories would then have to be sent to each vehicle, increasing the communication load, thereby making this approach intractable for larger fleets. Figure 8.6 shows an updated version of the MPC algorithm (see Figure 8.2) that includes the virtual center calculation.

8.4.2 Simulations results

Several simulations were performed to demonstrate the effectiveness of the new coordination method. FreeFlyer™ [60] orbit simulation software is used as the nonlinear propagator for each satellite while MATLAB™ mathematical software is used to implement the controller. The entire control system is executed without human intervention.

The simulation consists of three vehicles, each modeled as a 45 kg Orion spacecraft [57] with different drag coefficients (2.36, 2.20, and 2.12). Other disturbances, such as gravity perturbations, solar radiation pressure, atmospheric lift, and third body effects are activated in the FreeFlyer™ propagator. Sensor noise is included in the simulation as a white noise component added to the true relative state. The magnitude of the sensor noise is bounded by 2 cm for position and 0.5 mm/s for velocity, based on expected CDGPS sensor noise [57]. Spacecraft thrusters provide a maximum acceleration of 0.003 m/s^2, which corresponds to continuous thrusting for a full time-step. The formation is initialized on a reference orbit (semi-major axis 6900 km, eccentricity of 0.005) similar to a space shuttle orbit. The reference orbit inclination is 35°, introducing significant differential gravity disturbances for spacecraft with inclination differences. See Ref. [56] for the full details on the simulation parameters.

When using the virtual center procedure, the reference point is updated at every time-step. The relative dynamics are discretized on a 10.8 second time interval to match the propagation step-size. Formation flying problems are planned over a half-orbit time horizon. The LP formation flying formulation restricts control inputs and applies position constraints to every sixth time-step [61], which reduces LP solution times to about 1–3 seconds. The robust LP approach in Ref. [7] is used to account for sensor noise and the always feasible approach in Ref. [61] is also used. The error box size for position

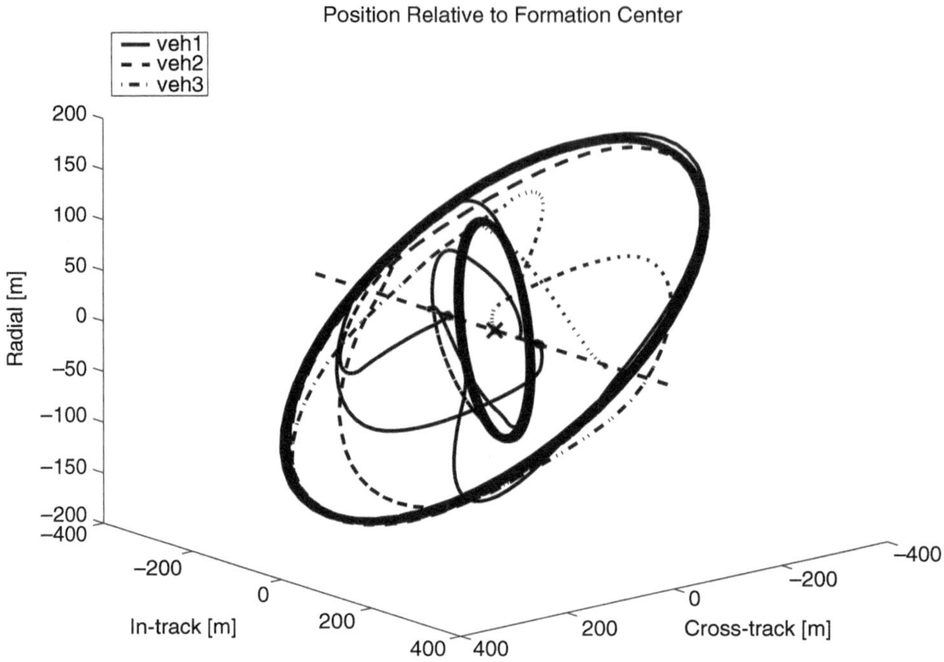

Fig. 8.7. Relative motion for three-vehicle formation. Sequence is in-track separation, small ellipse, larger ellipse, in-track separation.

tolerance is 10 m in-track, 5 m radial, and 5 m cross-track, which meets the tolerance requirement of 10% of the baseline for all formations in the simulation.

The simulation contains three formation maneuvers with formation flying at each configuration. The spacecraft paths during maneuvers are shown with respect to the virtual reference point in Figure 8.7. The formation begins and ends in similar in-track separations. Two passive aperture formations are maintained for approximately seven days each to observe any long- and short-term effects of the disturbances, particularly the gravity perturbation effects. The first aperture projects a 400×200 m ellipse in the in-track–radial plane and a circle with a 100 m radius in the radial–cross-track plane. The second aperture projects a 600×300 m ellipse in the in-track–radial plane and a 300 m radius circle in the in-track–cross-track plane. Aperture position assignment is coordinated through the procedure described in Section 8.3 with a plan horizon of one orbit.

8.4.2.1 Analysis of controller performance

Full simulation fuel costs for the leader–follower and fuel-weighted virtual center methods are shown in Figures 8.8 and 8.9. The fuel cost figures show three reconfiguration maneuvers, each of which uses a significant amount of fuel over a short period of time. The longer, constant slope segments correspond to the periods of formation-flying. Comparing the two figures, it is clear that the Leader–Follower method has a higher fuel

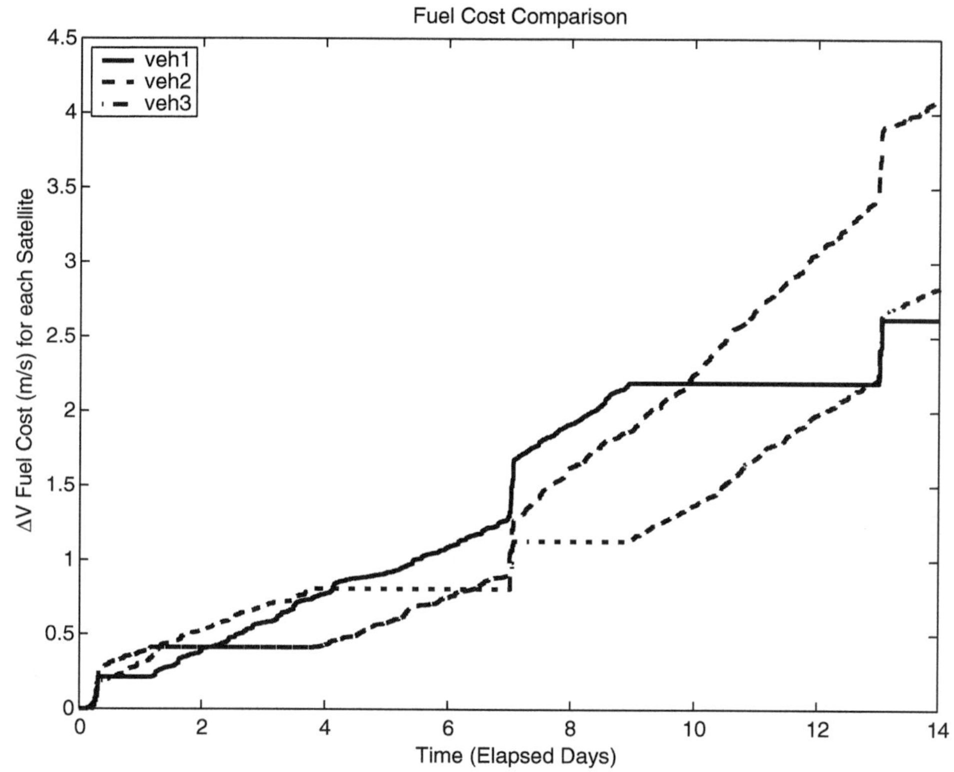

Fig. 8.8. ΔV's for each vehicle, using Leader–Follower—sharp rises indicate formation maneuvers and constant slope parts correspond to formation flying.

cost than the virtual center method throughout the mission. In the leader–follower figure, a spacecraft exerting no control (the flat lines) is presently the leader of the formation and, consequently, has no state error. When the total control effort exerted by one spacecraft significantly exceeds that of the other spacecraft, the leader is switched to balance overall fuel use. In comparison, the fuel-weighted virtual center method spreads the error out among all of the spacecraft, with the objective of placing the virtual center in such a way as to minimize global control effort across the formation. As a result, there is a non-zero fuel cost for all three spacecraft during the formation-flying mode.

The simulations using a virtual center reference point were performed for three different levels of fleet cooperation. The first simulation calculates the virtual center in the formation with equal weights on each vehicle in the fleet. The second simulation includes the control actions of other spacecraft in the control determination. The third simulation includes the external control inputs, as in the second simulation, and also adjusts the weighting of the vehicles based on fuel use. All three methods successfully achieve and maintain the specified configurations during the simulation. In the formation-flying mode, the vehicles are maintained approximately within the specified position tolerance due to the always-feasible formulation. The maximum deviation

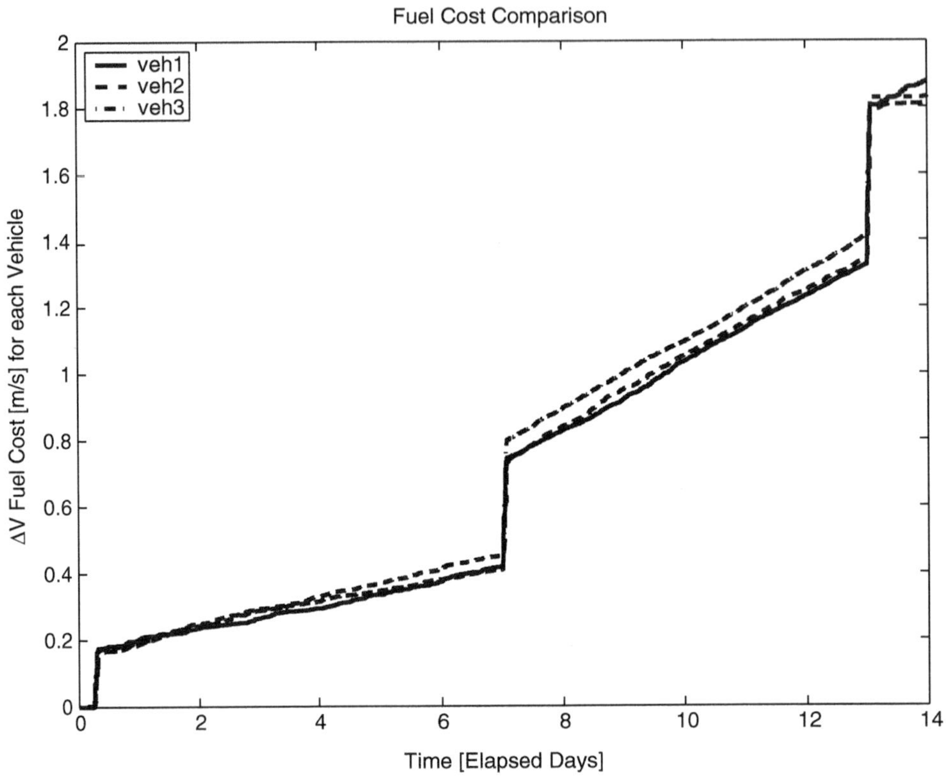

Fig. 8.9. ΔV's for each vehicle, using Virtual Center—sharp rises indicate formation maneuvers and constant slope parts correspond to formation flying. Note that the y-axis scale differs from that of Figure 8.8. The plots show that the Virtual Center method achieves both better fuel distribution across the fleet and better overall fuel minimization than the Leader–Follower method.

from the desired state for any simulation was less than 11 m in-track, 5 m radial, and 7 m cross-track. The total fuel cost data for each of the simulations is contained in Table 8.2.

8.4.2.2 Formation flying analysis—FM

The results in Table 8.2 show that there is no appreciable difference in fuel cost between the three different controllers for the formation maneuvers; however, this is not unexpected. The difference between the first two simulations is the inclusion of the control inputs of other vehicles in the low-level controller for formation-flying. Therefore, there is no expected improvement in the formation maneuvers from this change. The last simulation adds fuel weighting to the calculation of the formation center. The fuel weighting is only updated once every two orbits, whereas the formation maneuvers occur over a single orbit. Some benefit can be expected, because the fuel weighting will reduce

Table 8.2
Table of ΔV use for the virtual center simulations. Simulation number corresponds to level of coordination used. Spacecraft are indicated by number. The maneuver types are followed by the number of orbits the maneuver was performed for. FF indicates formation flying and FM represents formation maneuvers.

	Sim 1			Sim 2			Sim 3		
Maneuver type	SC 1	SC 2	SC 3	SC 1	SC 2	SC 3	SC 1	SC 2	SC 3
FF #1 (4) mm/s/orbit	0.509	0	0.523	0.652	0	0.112	0.542	0	0.341
FM #1 (1) mm/s	163	150	171	160	148	168	169	157	165
FF #2 (101) mm/s/orbit	3.07	2.82	2.43	2.70	2.54	2.26	2.39	2.48	2.51
FM #2 (1) mm/s	315	291	315	339	291	320	275	306	314
FF #3 (90) mm/s/orbit	8.14	6.90	6.18	7.42	6.57	6.59	6.59	6.73	6.82
FM #3 (1) mm/s	415	440	391	408	410	374	404	392	393
FF #4 (14) mm/s/orbit	2.64	4.69	2.30	1.71	0.696	1.45	3.90	2.25	0.541
FM Total (3) mm/s	893	881	877	907	849	862	848	855	872
FF Total (209) mm/s	1141	1040	899	1019	920	931	979	1004	964
Total Fuel (212) mm/s	2034	1921	1776	1926	1769	1793	1827	1859	1836

the differential disturbances of vehicles that have used large amounts of fuel, however, this change will be minimal over the course of one orbit.

8.4.2.3 Formation flying analysis—FF

For the two passive aperture formation flying maneuvers, the rate at which fuel is expended for each vehicle is heavily dependent on the cross-track disturbance. The in-track and radial control efforts are approximately equal for each vehicle in the formation, regardless of the spacecraft location in the aperture; however, the cross-track fuel use varies significantly for each vehicle. The cross-track disturbance results in a secular increase in the amplitude of the cross-track oscillatory motion, and the magnitude of this increase depends on the cross-track phasing. With a three vehicle formation, it is impossible to eliminate the disturbance completely for every vehicle, therefore, at least two vehicles will experience a cross-track disturbance and will expend more control effort than the other in response to the cross-track disturbance. Altering the phasing over time can equalize the average cross-track disturbance for all vehicles [45]. This method could be included in the control system in Section 8.3, but note that the coordinated virtual center equalizes the fuel cost due to the cross-track disturbance through the calculation of the fuel-weighted virtual center.

8.4.2.4 Total fuel cost analysis

The controllers can be compared by the total fuel cost for the mission. If a formation flying mission requires the entire fleet to perform the science observations, then the mission life will be limited by the vehicle with the greatest fuel use. The fuel expenditure for each vehicle during the mission is summarized in the last row of Table 8.2. The results show that the maximum fuel cost is reduced from 2.03 m/s for the first simulation to 1.93 m/s for the second simulation. However, the fuel cost for one vehicle is much larger than for the other two vehicles in both simulations. The third simulation utilizes the fuel weighting method to reduce the maximum fuel use by shifting control effort to the lower fuel cost vehicles. The result is a reduction in the maximum fuel cost to 1.86 m/s.

8.4.3 Summary

This section addresses the three main issues of the formation flying coordination problem: the reference point for the formation, the specification of the desired state, and the control to achieve or maintain the desired state across the whole formation. The new virtual center method presents a procedure for calculating the reference point for the fleet from which the desired states for each spacecraft can be readily calculated. Note that the calculation of this virtual center is closely tied to the planned formation flying sensor (CDGPS in LEO). The selection of the location of the center also includes a weighting on fuel use across the fleet, which facilitates increased coordination and cooperation within the distributed model predictive control system. The result is an efficient real-time control system using the benefits of a fuel-optimal controller to plan control actions and coordination between the fleet to further reduce fuel effort. The simulation results indicate that this control system can adequately maintain a formation at a fuel cost of 2–8 mm/s per orbit. The simulations also clearly show that the virtual center approach required significantly less fuel than the leader–follower technique. In the next two sections, two distinct approaches to achieve disturbance and sensor noise robustness are presented.

8.5 Open-loop robust control and replan frequency

The model predictive control system described in Section 8.3 has many tunable parameters, such as error box size, planning horizon length, replan frequency, terminal condition, and robustness level. The closed-loop behavior of the control system and its performance level can be significantly altered by posing the optimization problem in different ways. However, since the control inputs are determined using online optimization, the best choice of these parameters is typically not obvious. The Ref. [62] analyzes formation flying model predictive control (MPC) mission parameters when a particular form of closed-loop robustness is used. Here, robustness refers to the ability of a controller to operate in the presence of navigation error and can be closed-loop, where feedback effects are exploited, or open-loop, where full trajectories can be implemented without additional feedback. In the case of the open-loop robustness method in Ref. [12], another approach to choosing control parameters is required. This section examines the effect of terminal conditions on replan frequency of the controller described in Section 8.3.

Section 8.3 presents an MPC approach where the implementation horizon was made variable and replanning was triggered by the vehicle approaching the edge of the error box. Using a variable implementation horizon with a problem that has disturbances and noise will result in a control system with an uncertain replan rate. An investigation of

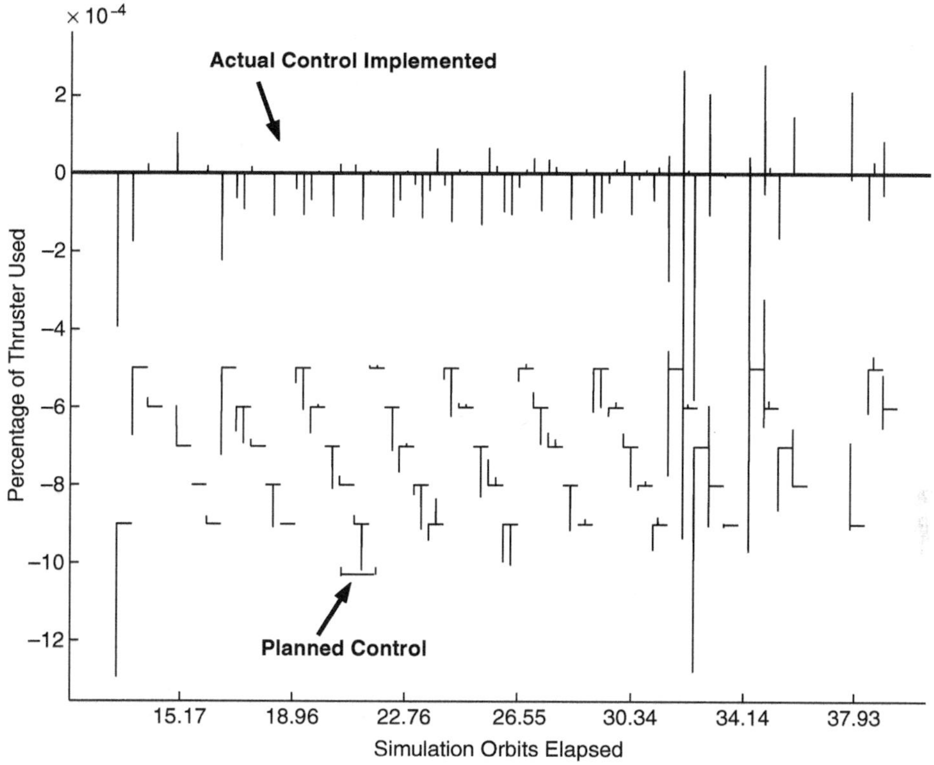

Fig. 8.10. Plans created using unmodified terminal conditions. The horizontal axis crossing the origin shows the thrusting that was implemented. The floating axes below the implementation axis indicate the planned thrusting sequence at time matching the far left side of each floating axis.

this controller showed that a replan was occurring at the end of each planning horizon. Figure 8.10 shows this phenomenon, where the actual control implemented (in thruster firings) is plotted along the x-axis and the plan being implemented at any given time is shown using floating axes in the lower half of the figure. Times when floating axes overlap indicate that a plan has been abandoned in favor of a new plan (i.e., a replan occurred). In this control formulation, the only event that can trigger a replan is a deadband violation. The figure shows that the majority of replans occur soon after previous plans have completed (as opposed to replanning before a plan has completed, or after a long period of drifting).

In a typical error-box problem with fixed terminal time, the optimal solution is known to be a "bang-off-bang," which applies control for a set period, then allows the system to evolve unperturbed, and then applies control for another set period. The solution of the online optimization problem should reproduce this optimal solution. However, a model predictive controller (MPC) that has solved for a single plan does not necessarily implement that entire plan. Instead, the controller implements only the first part of the plan (the *implementation horizon*) and then computes a new plan (re-planning). With no disturbances and no sensing noise, only one plan needs to be computed. However, when process and sensing noise is included, frequent re-planning may be required.

The cause of the specific replan rate is the noise robustness formulation. The noise robustness formulation used in Ref. [12] guarantees that a plan will be feasible (i.e., remain inside the error box) for a range of possible initial conditions h_i. Here, n_{ic} initial conditions are chosen to capture the bounds of expected initial condition error and each is used to evaluate a different set of vectors $h_i(k)\ \forall\ k \in \{1 \ldots n\}$. This formulation uses the constraint

$$\begin{bmatrix} \Gamma(k) & -\Gamma(k) \\ -\Gamma(k) & \Gamma(k) \end{bmatrix} \begin{bmatrix} U_k^+ \\ U_k^- \end{bmatrix} \leq \begin{bmatrix} z_{\text{des}}(k) - h_i(k) + z_{\text{tol}} \\ -z_{\text{des}}(k) + h_i(k) + z_{\text{tol}} \end{bmatrix} \forall \begin{array}{c} k \in \{1 \ldots n\} \\ i \in \{1 \ldots n_{\text{ic}}\} \end{array} \tag{8.43}$$

$$\begin{bmatrix} -I & 0 \\ 0 & -I \end{bmatrix} \begin{bmatrix} U_n^+ \\ U_n^- \end{bmatrix} \leq \begin{bmatrix} 0 \\ 0 \end{bmatrix}. \tag{8.44}$$

Ref. [12] presents a method of solving an identical optimization using fewer constraints. This robustness technique tends to keep the spacecraft inside the deadband for the duration of the planning horizon, thus eliminating the need for the plan to be interrupted. However, at the end of the plan, the spacecraft is often left near the edge of an error box (allowing drift as far as possible tends to minimize fuel use) and therefore requires almost immediate replanning (see Figure 8.11).

Although the motion shown in Figure 8.11 is the optimal solution to the problem that was posed, it results in very regular (several times an orbit) low level thrusting to return the spacecraft to the deadband. This is not desirable for several reasons: incorrect thrusting (not modeled here) introduces error into the state estimate and it requires regular GN&C computation and communication. A preferable system would achieve similar or better fuel use, while allowing for long periods of drifting inside the deadband. Two approaches were identified to alter the replan rate.

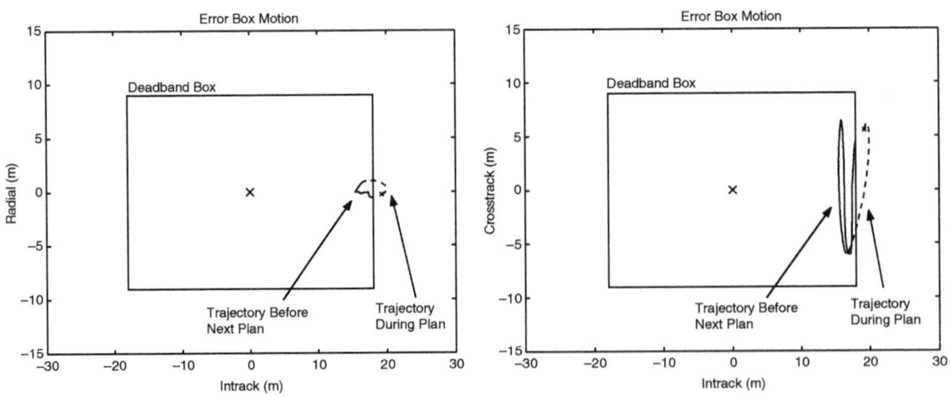

Fig. 8.11. Trajectory using unmodified terminal conditions.

1. The first imposes a fixed replan rate on the problem that is significantly shorter than the actual plan length. This is more typical of MPC algorithms than the variable replan length. Using this approach, replanning still forces the end of the planned trajectory to always be near the edge of the deadband. However, the high frequency of the replanning means that only the first few steps of a plan are ever implemented, so the spacecraft rarely reaches the edge of the box. This approach has the added advantage that it does not require as high a level of robustness, because the high feedback rate mitigates the effects of sensor noise.
2. A second approach retains the variable replan rate, but alters the terminal condition of the optimization. In the original problem, the terminal condition stated that the spacecraft should be guaranteed to finish the path inside the deadband. The new terminal condition specifies that the spacecraft should be guaranteed to end inside the deadband, but also that the nominal trajectory (i.e., the projected motion based on the nominal state estimate) should stay inside the error box for a full additional orbit, and that the terminal and initial states of that nominal motion must match. This new terminal condition is more restrictive than the original, but it can result in trajectories that drift for many (i.e., 5–12) orbits inside the deadband.

The second approach uses an alternate terminal condition and replaces the constraints in Eq. (8.44) with

$$\begin{bmatrix} \Gamma(k) & -\Gamma(k) \\ -\Gamma(k) & \Gamma(k) \end{bmatrix} \begin{bmatrix} U_k^+ \\ U_k^- \end{bmatrix} \leq \begin{bmatrix} z_{\text{des}}(k) - h_i(k) + z_{\text{tol}} \\ -z_{\text{des}}(k) + h_i(k) + z_{\text{tol}} \end{bmatrix} \forall \begin{array}{l} k \in \{1, \ldots, n\} \\ i \in \{1 \ldots n_{\text{ic}}\} \end{array}$$
$$\begin{bmatrix} \Gamma(n) & -\Gamma(n) \\ -\Gamma(n) & \Gamma(n) \end{bmatrix} \begin{bmatrix} U_n^+ \\ U_n^- \end{bmatrix} \leq \begin{bmatrix} z_{\text{des}}(j) - h_i(j) + z_{\text{tol}} \\ -z_{\text{des}}(j) + h_i(j) + z_{\text{tol}} \end{bmatrix} \forall \begin{array}{l} j \in \{n+1, \ldots, n+q\} \\ i \in \{1, \ldots, n_{\text{ic}}\} \end{array} \tag{8.45}$$

and

$$\begin{bmatrix} (\bar{\Gamma}(n) - \Gamma(q)) & (\Gamma(q) - \bar{\Gamma}(n)) \end{bmatrix} \begin{bmatrix} U_q^+ \\ U_q^- \end{bmatrix} = \begin{bmatrix} h_{\text{nom}}(q) - h_{\text{nom}}(n) \end{bmatrix} \tag{8.46}$$

$$\begin{bmatrix} -I & 0 \\ 0 & -I \end{bmatrix} \begin{bmatrix} U_n^+ \\ U_n^- \end{bmatrix} \leq \begin{bmatrix} 0 \\ 0 \end{bmatrix} \tag{8.47}$$

where h_{nom} is the vector based on the nominal initial conditions (i.e., the current best state estimate), $\bar{\Gamma}(n)$ is $[\ \Gamma(n)^T \ \ 0_{1\times q}\ ]^T$, and q is the number of time steps beyond the end of the plan that the spacecraft is required to drift inside the error box. For the altered terminal conditions, q is set to the number of time steps in an orbit, to ensure the spacecraft remains in the error box without thrusting for a full orbit beyond the end of the thrusting plan. Because relative orbital dynamics (e.g., Hill's equations) are typically cyclic over the period of an orbit, an elliptical relative trajectory that remains inside the error box and begins and terminates at the same state is a nominal invariant terminal set [63]. Attaining this ideal terminal trajectory would preclude all future error box violations. However, no single state can be attained robustly in the presence of sensing noise and hence, the objective is made to *nominally* achieve the invariant trajectory while robustly satisfying the actual error box constraints.

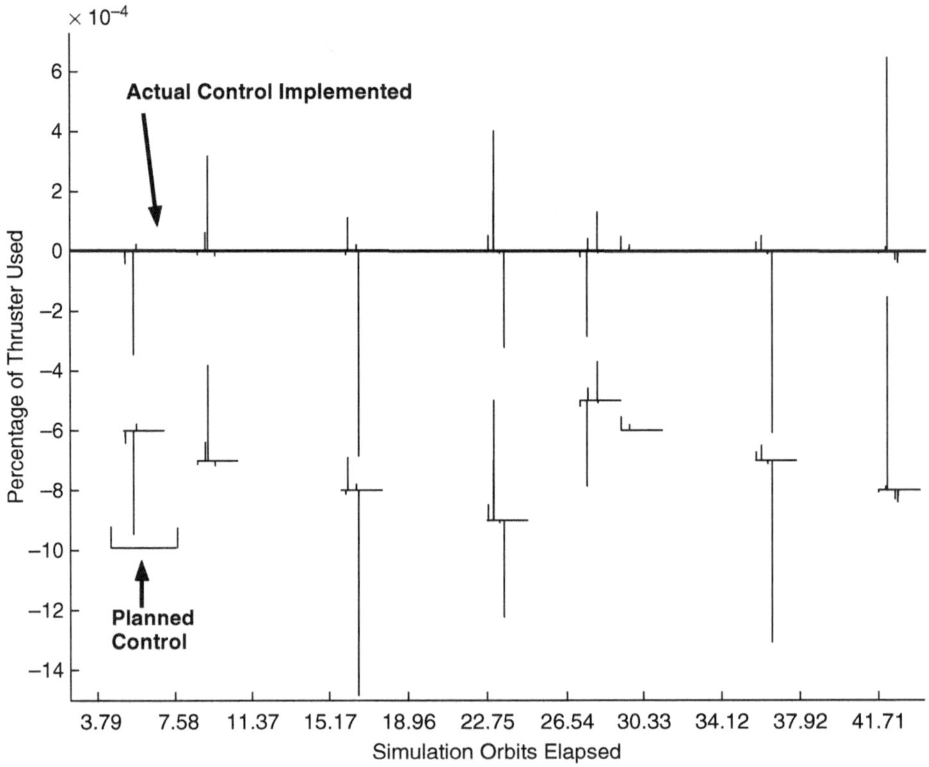

Fig. 8.12. Plans created using closed ellipse terminal conditions. The horizontal axis crossing the origin shows the thrusting that was implemented. The floating axes below the implementation axis indicate the planned thrusting sequence at time matching the far left side of each floating axis.

Figures 8.12 and 8.13 show the effect of changing the terminal condition. Both the frequent replan and altered terminal condition methods have similar fuel use rates. Interestingly, both tend to use slightly more fuel than the original problem in Ref. [59], which is likely a result of the more restrictive terminal conditions. However, when the terminal conditions were altered to create a closed-ellipse, the resulting spacecraft trajectories remained inside the error box without replanning, and consequently without control inputs, for much longer periods of time. This is a desirable condition, because the trajectory in Figure 8.13 now fills more of the error box and the spacecraft spends longer periods of time drifting. This both mitigates error introduced through thrusting and allows missions that can only collect science while drifting to utilize longer observation periods.

This section has examined the effects of terminal condition on the closed-loop behavior of a spacecraft formation with sensing noise. Ref. [12] showed that incorporating open-loop robustness into a formulation improved overall performance. Here, we demonstrated that the addition of a nominal terminal invariant set condition (i.e., the closed ellipse trajectory constraint) reduced the frequency with which replanning was required.

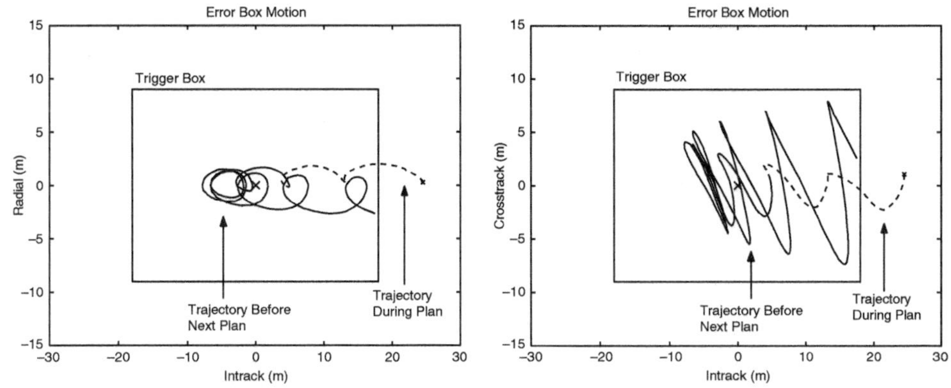

Fig. 8.13. Trajectory using closed ellipse terminal conditions.

Section 8.6, examines the comparative benefits of closed-loop robustness using a fixed replan rate and examines the application of the nominal closed ellipse condition in that formulation.

8.6 Using closed-loop robust MPC

Sensor noise in a spacecraft formation flying mission using carrier-phase differential GPS (CDGPS) will be a dominant disturbance [7]. Section 8.5 presents an approach to mitigating the effects of sensor noise on a model predictive control algorithm. This robustness was achieved by designing trajectories that would meet performance criteria for a set of possible initial conditions. The size of this set was determined by the expected sensor noise. The approach taken in that paper is characterized as "open-loop," because it generates a trajectory which is feasible for all initial conditions, without requiring replanning. An alternate approach is to design a thrusting plan which will be guaranteed to produce a feasible state and a new feasible plan at the next time step. This approach is considered "closed-loop," because each plan explicitly considers future feedback action in response to as-yet unknown information [64]. In contrast to the open-loop approach in Section 8.5, here it is desirable to use a fixed replanning rate, because the closed-loop approach exploits newly available information and must explicitly account for when that information will be available.

To demonstrate the feasibility of the robust model predictive control for actual spacecraft formation flight, a non-linear simulation with a realistic disturbance model is performed for multiple spacecraft over an extended period of time. We develop a formulation that simultaneously incorporates bounds on state error, process noise, sensor noise, and thrust availability. Simulations demonstrate the effectiveness of both the bounded models and of the closed-loop robustness technique applied to a realistic spacecraft formation control problem. The terminal conditions presented in [62] and others suggested in Section 8.5 are also examined.

8.6.1 Overview of robust MPC

For the spacecraft formation flying problem with sensor noise, robust feasibility guarantees that, provided the initial optimization is feasible and the noise is bounded, all subsequent optimizations are feasible and constraints are satisfied, e.g., the spacecraft remains inside the specified error box. This guarantee holds despite the plans being based on inaccurate information. References [17, 65] prove that the formulation reviewed in this section guarantees both robust feasibility and constraint satisfaction. Reference [63] describes an approach to transform the dynamics of the *true* state $\mathbf{x}$ to those of the *estimated* state $\hat{\mathbf{x}}$. This section assumes that an estimate of the true state is available and is equivalent to a linear combination of a bounded noise added to the true state. Robust feasibility depends on the estimate, since that is the initial condition parameter of the optimization. The dynamics of the true state are

$$\mathbf{x}(k+1) = \mathbf{F}\mathbf{x}(k) + \mathbf{G}\mathbf{u}(k) \tag{8.48}$$

and the estimation error is an additive term, applied at each time step

$$\hat{\mathbf{x}}(k) = \mathbf{x}(k) + \mathbf{n}(k) \tag{8.49}$$

$$\hat{\mathbf{x}}(k+1) = \mathbf{x}(k+1) + \mathbf{n}(k+1), \tag{8.50}$$

where $\mathbf{n}(k)$ is the navigation error at time k, which is assumed to lie in a bounded set $\mathcal{N}$. Substituting Eqs. (8.49) and (8.50) into Eq. (8.48) gives the dynamics of the estimate

$$\begin{aligned} \hat{\mathbf{x}}(k+1) &= \mathbf{F}\hat{\mathbf{x}}(k) + \mathbf{G}\mathbf{u}(k) + \mathbf{n}(k+1) - \mathbf{F}\mathbf{n}(k) \\ &= \mathbf{F}\hat{\mathbf{x}}(k) + \mathbf{G}\mathbf{u}(k) + \begin{bmatrix} -\mathbf{F} & \mathbf{I} \end{bmatrix} \begin{pmatrix} \mathbf{n}(k) \\ \mathbf{n}(k+1) \end{pmatrix}. \end{aligned} \tag{8.51}$$

With the dynamics now involving an affine disturbance, the formulation of [63] can be employed to synthesize a robustly feasible MPC algorithm. The disturbance vector is bounded using

$$\mathbf{w}(k) = \begin{bmatrix} -\mathbf{F} & \mathbf{I} \end{bmatrix} \begin{pmatrix} \mathbf{n}(k) \\ \mathbf{n}(k+1) \end{pmatrix} \in \mathcal{W} \ \ \forall k. \tag{8.52}$$

If $\mathcal{N}$ is polyhedral, the set $\mathcal{W}$ is polyhedral and can therefore be generated using a polyhedral mapping routine of the form in Ref. [66]. Output constraints take the form

$$\mathbf{y}(k) = \mathbf{H}\hat{\mathbf{x}}(k) + \mathbf{J}\mathbf{u}(k) \in \mathcal{Y} \ \ \forall k, \tag{8.53}$$

where $\mathcal{Y}$ is a bounded set which can incorporate error box and thrust constraints.

The MPC optimization is performed over a horizon of N steps and uses an arbitrary nilpotent linear control law $\mathbf{u}(j) = \mathbf{K}_{NP}\mathbf{x}(j)\ \ j \in \{0 \ldots N-1\}$. Define $\mathbf{L}(j)$ as the state transition matrix for the closed-loop system under this control law

$$\mathbf{L}(0) = \mathbf{I} \tag{8.54}$$

$$\mathbf{L}(j+1) = (\mathbf{F} + \mathbf{G}\mathbf{K}_{NP})\mathbf{L}(j) \ \ \forall j \in \{0 \ldots N\}. \tag{8.55}$$

Then the nilpotency requirement for $\mathbf{K}_{NP}$ implies

$$\mathbf{L}(N) = \mathbf{0}. \tag{8.56}$$

Define the MPC optimization problem $\mathsf{P}(\hat{\mathbf{x}}(k))$

$$J^*(\hat{\mathbf{x}}(k)) = \min_{\mathbf{u},\mathbf{x},\mathbf{y}} \sum_{j=0}^{N} \ell\left(\mathbf{u}(k+j|k), \mathbf{x}(k+j|k)\right), \tag{8.57}$$

subject to

$$\forall j \in \{0 \ldots N\}$$

$$\mathbf{x}(k+j+1|k) = \mathbf{F}\mathbf{x}(k+j|k) + \mathbf{G}\mathbf{u}(k+j|k) \tag{8.58}$$

$$\mathbf{y}(k+j|k) = \mathbf{H}\mathbf{x}(k+j|k) + \mathbf{J}\mathbf{u}(k+j|k) \tag{8.59}$$

$$\mathbf{x}(k|k) = \hat{\mathbf{x}}(k) \tag{8.60}$$

$$\mathbf{x}(k+N+1|k) \in \mathcal{X}_F \tag{8.61}$$

$$\mathbf{y}(k+j|k) \in \mathcal{Y}(j) \tag{8.62}$$

where the double subscript notation $(k+j|k)$ denotes the prediction made at time k of a value at time $k+j$. The constraint sets are chosen according to the recursion

$$\mathcal{Y}(0) = \mathcal{Y} \tag{8.63}$$

$$\mathcal{Y}(j+1) = \mathcal{Y}(j) \sim (\mathbf{H} + \mathbf{J}\mathbf{K}_{NP})\,\mathbf{L}(j)\mathcal{W} \quad \forall j \in \{0 \ldots N\} \tag{8.64}$$

where $\sim$ denotes the Pontryagin difference operation [67], defined by

$$\mathcal{X} \sim \mathcal{Y} \triangleq \{\mathbf{z} \mid \mathbf{z} + \mathbf{y} \in \mathcal{X}\ \forall \mathbf{y} \in \mathcal{Y}\} \tag{8.65}$$

and the matrix mapping of a set is defined such that

$$\mathbf{A}\mathcal{X} \triangleq \{\mathbf{z} \mid \exists \mathbf{x} \in \mathcal{X} : \mathbf{z} = \mathbf{A}\mathbf{x}\} \tag{8.66}$$

A MATLAB toolbox for performing these operations on polyhedral sets is available in Ref. [66, 68]. The choice of the terminal constraint $\mathcal{X}_F$ is typically very problem-specific. It must be a control invariant admissible set [68], i.e., there exists a control law $\kappa(\mathbf{x})$ satisfying the following

$$\forall \mathbf{x} \in \mathcal{X}_F$$

$$\mathbf{F}\mathbf{x} + \mathbf{G}\kappa(\mathbf{x}) \in \mathcal{X}_F, \tag{8.67}$$

$$\mathbf{H}\mathbf{x} + \mathbf{J}\kappa(\mathbf{x}) \in \mathcal{Y}(N). \tag{8.68}$$

The origin $\mathcal{X}_F = \{\mathbf{0}\}$ is a straightforward choice of terminal set for a linear system. However, any nominally invariant set is valid for $\mathcal{X}_F$. Note that it is demonstrated in Subsection 8.6.3 that the origin can be an overly restrictive terminal condition for this control application.

8.6.2 Bounding the process noise

Section 8.5 describes an open-loop approach to accounting for disturbances in a model predictive control algorithm. This approach uses analytic models of J_2 and drag to predict time-varying disturbances which are then added to the LP formulation in Eq. 8.32. The closed-loop approach in Refs. [62, 63, 65] uses constant dynamics and a bounded disturbance model. The method for computing this model develops polytopic bounds on the disturbance set by simulating two spacecraft in close proximity to one another and propagating both using the high fidelity nonlinear integration-based propagator (NLP) and Hill's equations. The NLP used for the bounding process includes the perturbations due to J_2, drag, third body effects, and solar pressure. At each time step in the simulation propagates the previous state of the NLP forward using both a Hill's propagator and the NLP. The bounding method stores the magnitude of the difference between the two different states for both the position and the velocity states. The simulations use different initial starting states of the second satellite within an error box ($5 \times 10 \times 5$ m in the radial, in-track, and cross-track directions, respectively) centered about the first satellite. The maximum perturbations for position and velocity are found by calculating the absolute value of the differences between the nonlinear and linear propagated states from all of the simulations.

For a low Earth orbit (LEO) reference orbit ($n = 0.001$ rad/s), the disturbance sets for a 100 second propagation time step were found to be

$$\begin{pmatrix} p_x \\ p_y \\ p_z \end{pmatrix} \leq \begin{pmatrix} 85.5 \\ 30.3 \\ 0.0168 \end{pmatrix} \quad \begin{pmatrix} v_x \\ v_y \\ v_z \end{pmatrix} \leq \begin{pmatrix} 0.635 \\ 0.323 \\ 0.00334 \end{pmatrix} \tag{8.69}$$

in units of centimeters and millimeters per second, respectively. These numbers are roughly on the same order of magnitude as the sensing noise, which is expected, given the large integration time step and the presence of many dynamic effects not modeled by Hill's equations. Another approach to developing a disturbance model of this type would be to use analytical models of the effects of J_2, drag, and nonlinearities to due separation distance and eccentricity. For a given reference orbit, the maximum perturbation predicted by each model would be combined to give the largest possible unmodeled disturbance on Hill's equations.

8.6.3 Controller implementation

To use the model predictive control formulation reviewed in Section 8.6.1, the system described in Eq. (8.48) is augmented with an additive disturbance, $\mathbf{q}(k)$, which will be used to represent process noise. The system in Eq. (8.48) then becomes

$$\mathbf{x}(k+1) = \mathbf{F}\mathbf{x}(k) + \mathbf{G}\mathbf{u}(k) + \mathbf{q}(k) \tag{8.70}$$

where $\mathbf{q}(k)$ in a vector belonging to a bounded polyhedral set $\mathcal{Q}$. Likewise, the estimated state with sensing noise in Eq. (8.51) becomes

$$\hat{\mathbf{x}}(k+1) = \mathbf{F}\hat{\mathbf{x}}(k) + \mathbf{G}\mathbf{u}(k) + [\,-\mathbf{F} \quad \mathbf{I} \quad \mathbf{I}\,]\begin{pmatrix} \mathbf{n}(k) \\ \mathbf{n}(k+1) \\ \mathbf{q}(k) \end{pmatrix} \tag{8.71}$$

This altered formulation yields the new bounded disturbance set $\mathcal{W}$

$$\mathbf{w}(k) = [\,-\mathbf{F} \quad \mathbf{I} \quad \mathbf{I}\,]\begin{pmatrix} \mathbf{n}(k) \\ \mathbf{n}(k+1) \\ \mathbf{q}(k) \end{pmatrix} \in \mathcal{W}\ \forall k. \tag{8.72}$$

The robust formulation in Section 8.6.1 can accommodate a disturbance set of this form, but its implementation is complicated by the high dimensionality of the uncertainty set and constraints. In particular, the calculation of the Pontryagin difference is the subject of on-going work. Therefore, an approximation is used, whereby two scalar noise inputs capture the dominant sensing uncertainty. This approximation is extended to represent sensing noise uncertainty in all states and a term is added to the to represent the process noise. The bounds on a single sensor noise $\mathbf{n}(k)$ are $\pm\mathbf{e}\bar{N}$. This constraint is unchanged if a new vector, $\mathbf{e}_{\text{sn}}$ is defined as

$$\mathbf{e}_{\text{sn}} = \bar{N}\mathbf{e} \tag{8.73}$$

and $\bar{n}(k)$ is now distributed over the bounded set

$$-1 \le \bar{n}(k) \le 1. \tag{8.74}$$

For the examples in this section, $\mathbf{e}_{\text{sn}}$ is defined to be the expected noise on relative spacecraft states in a CDGPS system: 0.02 meters for position sensing and 0.0005 m/s for velocity sensing.

A vector, $\mathbf{e}_{\text{pn}}$, describing the maximum process noise magnitude on each state (taken directly from Eq. (8.69)) is introduced to form an approximation for the total possible state perturbation due to noise at any step k. The disturbance vector $\mathbf{w}(k)$ is now defined to be

$$\mathbf{w}(k) = [\,(\mathbf{e}_{\text{pn}} - \mathbf{F}\mathbf{e}_{\text{sn}}) \ \ \mathbf{e}_{\text{sn}}\,]\begin{pmatrix} \bar{n}(k) \\ \bar{n}(k+1) \end{pmatrix} \in \overline{\mathcal{W}} \quad \forall\, k. \tag{8.75}$$

The new set $\overline{\mathcal{W}}$ attempts to capture the uncertainty present in the formation flying demonstrations conducted in this section. Current research is investigating computationally efficient methods of accurately bounding the actual $\mathcal{W}$.

8.6.4 Demonstration results

The demonstration of closed-loop robustness in this section incorporates the error box concept from Section 8.3 and the virtual center concept for coordination from Section 8.4. In addition, the effect of the nominal closed-ellipse constraint from Section 8.5 is also examined. The demonstration is done using a nonlinear orbit propagator [60] with realistic disturbance and sensor noise models.

In this demonstration, the output constraints on each spacecraft will be both on the spacecraft state and on the input magnitude. The spacecraft state will be constrained to error box of dimensions (in meters) of $5 \times 10 \times 5$ in the radial, in-track, and cross-track directions, respectively, of an LVLH frame. In addition, the spacecraft will be constrained to have a maximum acceleration in each direction of 0.003 m/s^2. The cost function will be the one-norm of the thrust inputs over the planning horizon. The controller's cost function and constraints are both linear, so the controller optimizations are formulated as linear programs. A two week simulation of four spacecraft on an equally spaced passive aperture formation is shown in Figure 8.14. The passive aperture formation is a drift-free in-track–cross-track projected circle with a 100 meter radius and in-track–radial 400×200 m ellipse. The "tube" of spacecraft trajectories is essentially the error boxes in Figure 8.3 moving around the relative trajectory of the formation. The fuel-weighted virtual center method described in Section 8.4, is used to minimize state error and equalize fuel use across the formation. Spacecraft error box motion throughout the duration of the simulation is shown in Figure 8.17. It can be observed from the figures that no spacecraft exceeds its state constraints at any time in the simulation. However, the trajectories of the spacecraft remained close to the center of their respective error boxes, likely a result of the requirement that the each spacecraft arrive at the origin at the end of its plan. On average over the course of the simulation, each spacecraft used 14.5 mm/s of fuel

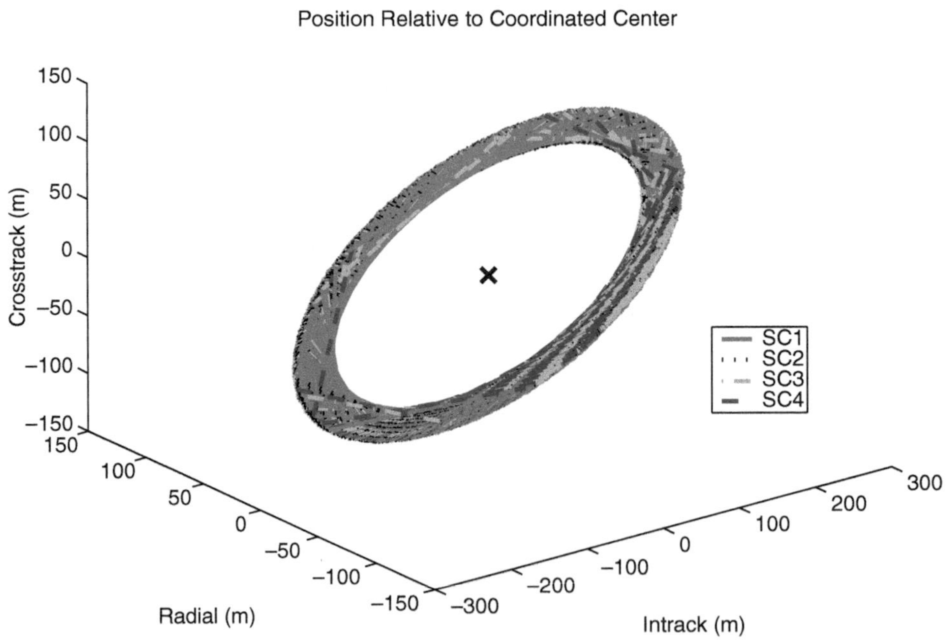

Fig. 8.14. Formation relative to the virtual center.

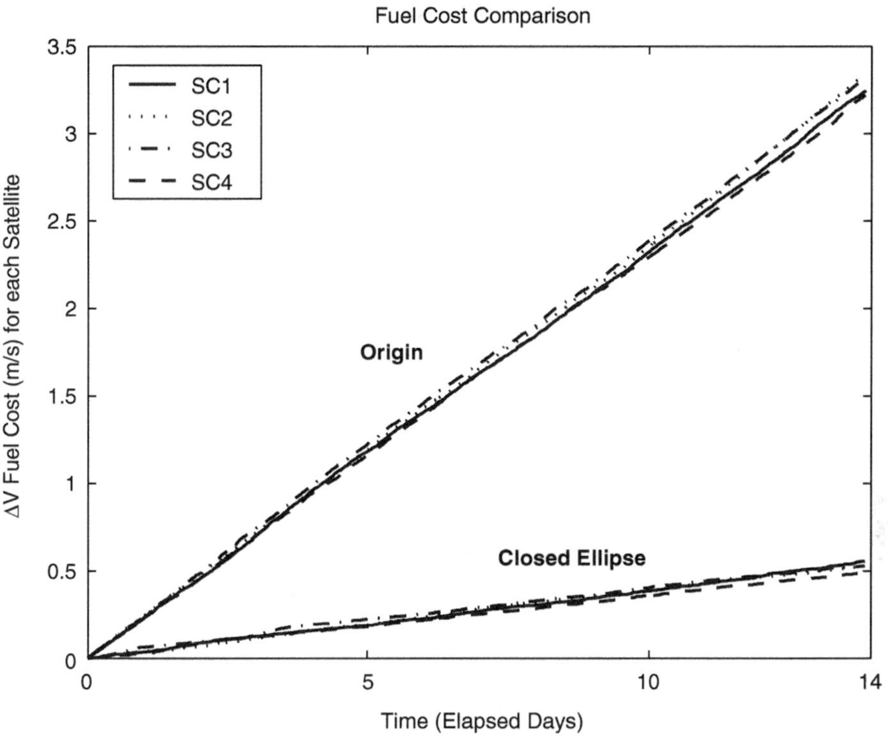

Fig. 8.15. Effect of terminal condition on fuel use rates.

per orbit, significantly more than 2.46 mm/s per orbit, the value reported for a similar simulation in Ref. [16].

These fuel expenditures can be reduced using the closed-ellipse constraint from Section 8.5 that restrict the spacecraft to terminate their plans on a closed ellipse in the LVLH frame. This requirement is enforced through two conditions:

1. The spacecraft must remain inside the error box at every time step for a full orbit after the plan ends.
2. The spacecraft state at the end of the plan is restricted to be the same as the state a full orbit after the end of the plan.

The origin terminal condition is a subset of the closed form ellipse terminal condition, because Hill's equations state that a spacecraft at the origin of an LVLH frame (i.e., zero position and zero velocity) will remain motionless in that frame. This motionless trajectory is a closed form ellipse with major and minor radii of zero meters. The difference between the terminal conditions is illustrated in Figure 8.16.

Figure 8.18 shows error box motion during a two week simulation using the closed ellipse terminal conditions. It is clear that the spacecraft motion occupies a much larger region of the error box and appears to take on the shape of an ellipse in the in-track–radial plane. As expected, the less restrictive terminal conditions led to significantly

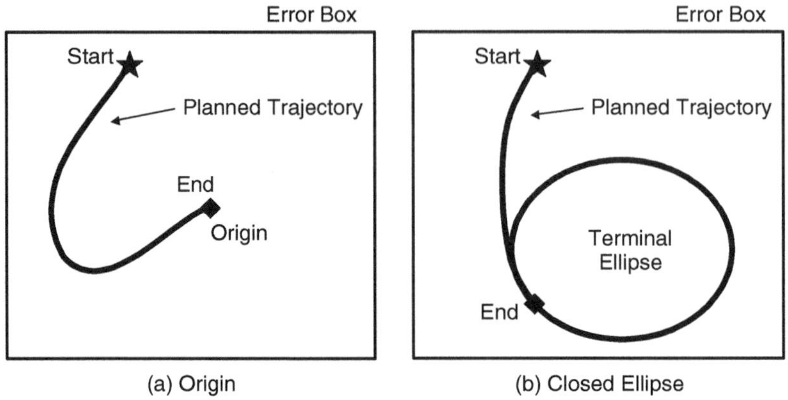

Fig. 8.16. Terminal conditions examined for closed loop MPC.

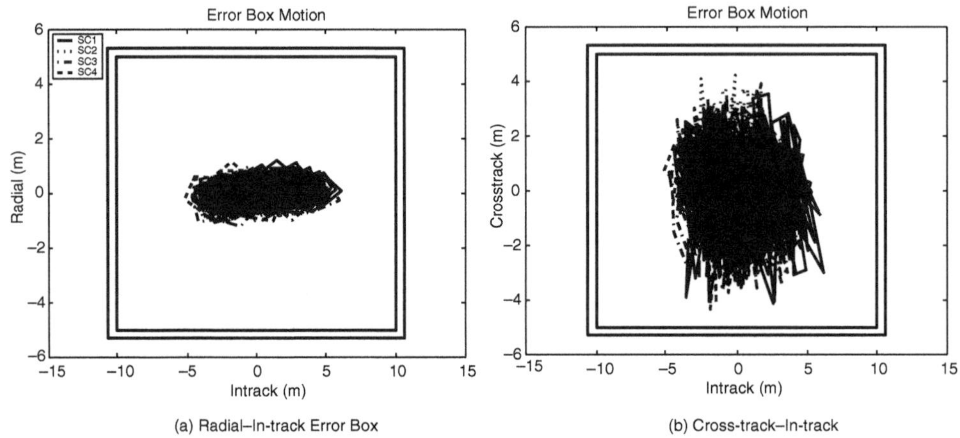

Fig. 8.17. Error box motion using Origin terminal constraint.

lower average fuel usage: 2.22 mm/s per orbit, which is a slight improvement over the results in Ref. [16]. Figure 8.15 compares the fuel use rates of a four spacecraft formation over a two week period. In contrast to the approach used in Section 8.5, the closed-loop robust method replans at all times, effectively guaranteeing that the spacecraft never drifts out of the error box. Furthermore, the spacecraft never enters an area of the error box that would be costly, from a fuel-use perspective, to prevent a constraint violation. The tradeoff for using the closed-loop method is that known time-varying disturbances must now be modeled as bounded polytopes, which does not allow the controller to fully exploit the well-known orbital dynamics. It is likely that performance can be further improved by using the LTV relative dynamics used for planning in Ref. [56] (excluding

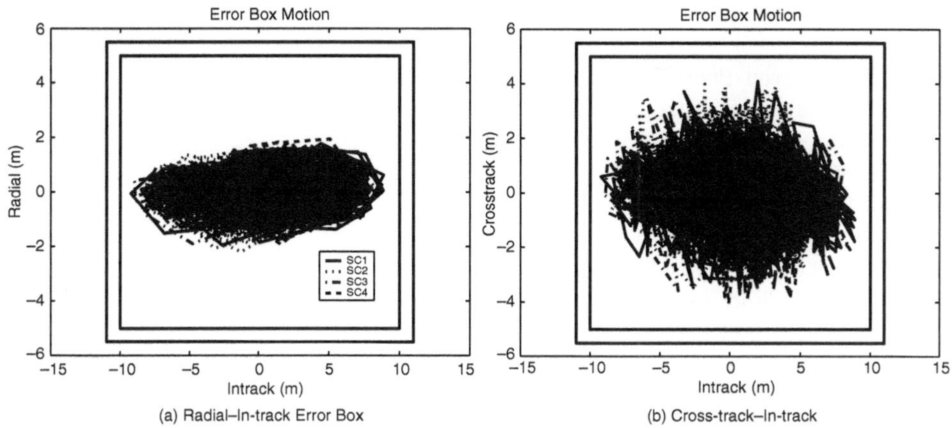

Fig. 8.18. Error box motion using Closed Ellipse terminal constraint.

cross disturbances) to capture J_2 effects and creating a new bounded process noise model. By including some of the effects of J_2 in the state transition matrix, the system should be better able to exploit natural dynamics and be capable of operating with less conservative process noise bounds. Both of these improvements should reduce overall fuel use.

Spacecraft formation flying simulations using the closed loop MPC controller were conducted in a realistic environment using realistic constraints. These simulations incorporated cooperation through the virtual center technique and a nominal terminal invariant set through the closed-ellipse technique. Results of these simulations show that the techniques can be used to control spacecraft for long periods of time reliably (i.e., no constraint violations) and with low fuel use.

8.7 Conclusions

This chapter developed three modifications to a basic formation flying model predictive control system to improve practical implementation. First, the virtual center method enabled cooperation between satellites by using a weighted optimization to find the optimal formation reference point. The addition of an feasible ellipse terminal constraint to an open-loop robust MPC formulation was demonstrated to reduce the frequency of required replanning, a desirable characteristic in an open-loop planning scheme. An alternate closed-loop robustness approach using fixed-rate replanning was extended to the formation flying control problem and was shown to be fuel efficient and capable of using virtual center fleet cooperation and the closed-ellipse terminal condition robustly. Several high-fidelity nonlinear simulations demonstrated all three modifications in use simultaneously to robustly and efficiently control a three satellite formation over a two week period in the presence of realistic disturbances and sensor noise.

8.8 Nomenclature

CDGPS—Carrier-Phase Differential GPS
GN&C—Guidance, Navigation, and Control
GPS—Global Positioning System
GVEs—Gauss' Variational Equations
LEO—Low Earth Orbit
LP—Linear Program
LPV—Linear Parameter-Varying
LTI—Linear Time Invariant
MPC—Model Predictive Control
NLP—Nonlinear Propagator
SMA—Semimajor Axis
ZOH—Zero Order Hold

Acknowledgments

The authors would like to acknowledge Dr. Arthur Richards for his insights into the design of a computationally tractable closed-loop robust MPC. This work was funded under Air Force Grant F49620-99-1-0095, NASA Grant #NAG5-10440, and Cooperative Agreement NCC5-729 through the NASA GSFC Formation Flying NASA Research Announcement. Any opinions, findings, and conclusions or recommendations expressed in this material are those of the author(s) and do not necessarily reflect the views of the National Aeronautics and Space Administration.

References

1. Bauer, F.H., Hartman, K., Bristow, J., Weidow, D., How, J.P. and Busse F. (1999). Enabling spacecraft formation flying through spaceborne GPS and enhanced autonomy technologies. *ION-GPS '99*, Proceedings of the *12th International Technical Meeting of the Satellite Division of the Institute of Navigation*, Sept. 14–17, pp. 369–383.
2. Leitner, J., Bauer, F., Folta, D., Moreau, M., Carpenter, R. and How, J. (2002). Distributed Spacecraft Systems Develop New GPS Capabilities, in *GPS World: Formation Flight in Space*, Feb. 2002.
3. The MAXIM Mission (2005). Online at http://maxim.gsfc.nasa.gov, last accessed Nov. 2005.
4. The Stellar Imager Mission (2005). Online at http://hires.gsfc.nasa.gov/si/ last accessed Nov. 2005.
5. The Terrestrial Planet Finder mission (2005). Online at http://tpf.jpl.nasa.gov last accessed Nov. 2005.
6. Carpenter, J.R., Leitner, J.A., Folta, D.C. and Burns, R.D. (2003). Benchmark Problems for Spacecraft Formation Flying Missions. AIAA *Guidance, Navigation and Control Conference*, Austin, TX, AIAA-5364.
7. How, J.P. and Tillerson, M. (2001). Analysis of the Impact of Sensor Noise on Formation Flying Control. *Proc. of the American Control Conf.*, June, pp. 3986–3991.
8. Carpenter, R. and Alfriend, K. (2003). Navigation accuracy guidelines for orbital formation flying. *Proceedings of the AIAA Guidance Navigation and Control Conference*, Austin, TX, August 11–14.
9. Alfriend, K.T. and Lovell, T.A. (2003). Error Analysis of Satellite Formations in Near-Circular Low-Earth Orbits. (AAS 03-651), Vol. 116, Advances in the Astronautical Sciences.
10. Scharf, D., Ploen, S. and Hadaegh, F. (2003). A Survey of Spacecraft Formation Flying Guidance and Control (Part I): Guidance. *IEEE American Control Conference*, June.

11. Scharf, D., Hadaegh, F. and Ploen, S. (2004). A Survey of Spacecraft Formation Flying Guidance and Control (Part II): Control. *IEEE American Control Conference*, June, 2976–2985.
12. Tillerson, M., Inalhan, G. and How, J. (2002). Coordination and Control of Distributed Spacecraft Systems Using Convex Optimization Techniques. *International Journal of Robust and Nonlinear Control*, **12**(2–3), pp. 207–242.
13. Melton, R.G. (2002). Time explicit representation of relative motion between elliptical orbits. *Journal of Guidance, Control, and Dynamics*, **23**(4), pp. 604–610.
14. Breger, L.S., Ferguson, P., How, J.P., Thomas, S., McLoughlin, T. and Campbell, M. (2003). Distributed control of formation flying spacecraft built on OA. Presented at the AIAA *Guidance, Navigation and Control Conference*, August.
15. Gim, D.W. and Alfriend, K.T. (2003). State transition matrix of relative motion for the perturbed noncircular reference orbit. *AIAA Journal of Guidance, Control, and Dynamics*, **26**(6), pp. 956–971.
16. Tillerson, M., Breger, L.S. and How, J.P. (2003). Multiple spacecraft coordination & control. Presented at the *American Control Conference*, June, pp. 1740–1745.
17. Richards, A.G. and How, J.P. (2005). A computationally-efficient technique for robust model predictive control, submitted to IEEE *Transactions on Automatic Control*, Feb. 2005.
18. Hill, G.W. (1878). Researches in Lunar Theory, *American Journal of Mathematics*, **1**, pp. 5–26, 129–147, 24–260.
19. Lawden, D. (1963). *Optimal Trajectories for Space Navigation*, Butterworths, London.
20. Carter, T. (1990). New form for the optimal rendezvous equations near a keplerian orbit. *AIAA Journal of Guidance, Control, and Dynamics*, **13**, pp. 183–186.
21. Inalhan, G., Tillerson, M. and How, J.P. (2002). Relative dynamics and control of spacecraft formations in eccentric orbits. *AIAA JGCD*, **25**(1), pp. 48–59.
22. Marec, J. (1979). *Optimal Space Trajectories* Elsevier Scientific, NY.
23. Battin, R.H. (1987). *An Introduction to the Mathematics and Methods of Astrodynamics*, AIAA Education Series, New York.
24. Bate, R., Mueller, D. and White, J. (1971). *Fundamentals of Astrodynamics*, Dover Publications Inc., NY.
25. Chobotov, V. (1996). *Orbital Mechanics*. Second Edition, AIAA Educational Series.
26. Schweighart, S. (2001). *Developement and Analysis of a High Fidelity Linearized J2 Model for Satellite Formation Flying*, S.M. Thesis, Massachusetts Institute of Technology, Dept. Aeronautics and Astronautics, June.
27. Karlgrad, C.D. and Lutze, F.H. (2002). Second order relative motion equations. *Advances in the Astronautical Sciences*, **109**, Pt. 3, 202, pp. 2429–2448.
28. Gim, D.T. and Alfriend, K.T. (2003). State Transition Matrix of Relative Motion for the Perturbed Non-circular Reference Orbit, *Journal of Guidance, Control and Dynamics*, **26**(6), pp. 956–971.
29. Mitchell, J.W. and Richardson, D.L. (2003). A third order analytic solution for relative motion with a circular reference orbit. *Journal of the Astronautical Sciences*, **51**(1), pp. 1–12.
30. Alfriend, K.T. and Yan, H. (2005). Evaluation and comparison of relative motion theories. *Journal of Guidance, Control and Dynamics*, **28**(2), pp. 254–261.
31. Alfriend, K.T., Schaub, H. and Gim, D.W. (2002). Formation flying: accomodating non-linearity and eccentricity perturbations. Presented at the 12th *AAS/AIAA Space Flight Mechanics Meeting*, January 27–30.
32. Breger, L.S. and How, J.P. (2004). GVE-based dynamics and control for formation flying spacecraft. Presented at the *2nd International Formation Flying Symposium*, September.
33. Breger, L.S. and How, J.P. (2005). J_2-modified GVE-based MPC for formation flying spacecraft. Presented at the *AIAA Guidance, Navigation and Control Conference*, August.
34. Kapila, V., Sparks, A.G. Buffington, J.M. and Yan, Q. (1999). Spacecraft formation flying: dynamics and control. *American Control Conference*, San Diego, CA, June 2–4, Institute of Electrical and Electronic Engineers, pp. 4137–4141.
35. Ren, W. and Beard, R. (2002). Virtual structure based spacecraft formation control with formation feedback. *AIAA GN&C Conference*, August.
36. Mishne, D. (2002). Formation control of LEO satellites subject ot drag variations and J_2 perturbations. *AAS/AIAA Astrodynamics Specialist Conference*, Monterey, CA, August.
37. Gurfil, P. (2003). Control-theoretic analysis of low-thrust orbital transfer using orbital elements. *AIAA Journal of Guidance, Control, and Dynamics*, **26**(6), November–December, pp. 979–983.

38. Schaub, H. and Junkins, J.L. (2003). *Analytical Mechanics of Space Systems*, AIAA Education Series, Reston, VA.
39. Naasz, B. (2002). *Classical Element Feedback Control for Spacecraft Orbital Maneuvers*, S.M. Thesis, Dept. of Aerospace Engineering, Virginia Polytechnic Institute and State University, May.
40. Yan, Q., Yang, G., Kapila, V. and de Queiroz, M. (2000). Nonlinear Dynamics and Output Feedback Control of Multiple Spacecraft in Elliptical Orbits. *Proceedings of 2000 American Control Conference*, Chicago, IL, June 28–30, **2** (A01-12740 01-63), Piscataway, NJ, Institute of Electrical and Electronics Engineers, pp. 839–843.
41. de Queiroz, M., Yan, Q., Yang, G. and Kapila, V. (1999). Global output feedback tracking control of spacecraft formation flying with parametric uncertainty. IEEE Conference on Decision and Control, 38th, Phoenix, AZ, Dec. 7–10, 1999, Proceedings. **1** (A00-4816014-63), Piscataway, NJ, Institute of Electrical and Electronics Engineers, Inc., pp. 584–589.
42. Sedwick, R., Miller, D. and Kong, E. (1999). Mitigation of differential perturbations in clusters of formation flying satellites, *Proceedings of the AAS/AIAA Space Flight Mechanics Meeting*, Breckenridge, CO, Feb. 7–10, 1999. Pt. 1 (A99-39751 10-12), San Diego, CA, Univelt, Inc. (Advances in the Astronautical Sciences. Vol. 102, pt. 1), pp. 323–342.
43. Schaub, H. and Alfriend, K. (1999). J_2 Invariant relative orbits for spacecraft formations. In *Goddard Flight Mechanics Symposium*, May 18–20, Paper No. 11.
44. Schaub, H. and Alfriend, K. (2001). Impulsive feedback control to establish specific mean orbit elements of spacecraft formations. *AIAA Journal of Guidance, Control, and Dynamics*, **24**(4), July–August, pp. 739–745.
45. Vadali, S., Vaddi, S., Naik, K. and Alfriend, K.T. (2001). Control of Satellite Formations, *Proceedings of the AIAA Guidance, Navigation, and Control Conference*, Montreal, Canada, August 6–9. AIAA Paper 2001-4028.
46. Sparks, A. (2000). Satellite formationkeeping control in the presence of gravity perturbations, *Proceedings of the 2000 American Control Conference*, Chicago, IL, June 28–30, 2000, **2** (A01-12740 01-63), Piscataway, NJ, Institute of Electrical and Electronics Engineers, pp. 844–848.
47. Redding, D., Adams, N. and Kubiak, E. (1989). Linear-quadratic stationkeeping for STS orbiter. AIAA *Journal of Guidance, Control, and Dynamics*, **12**, March–April, pp. 248–255.
48. Maciejowski, J.M. (2002). *Predictive Control with Constraints*, Prentice Hall.
49. Kwakernaak, H. and Sivan, R. (1972). *Linear Optimal Control Systems*, Wiley-Interscience.
50. Broucke, R.A. (2003). Solution of the elliptic rendezvous problem with the time as an independent variable. *AIAA Journal of Guidance, Control, and Dynamics*, **26**, July–August, pp. 615–621.
51. Franklin, G., Powell, J. and Workman, M. (1998). *Digital Control of Dynamic Systems*. Third Edition, Addison-Wesley.
52. Vallado, D. (1997). *Fundamentals of Astrodynamics and Applications*. McGraw-Hill.
53. Das, A. and Cobb, R. (1998). TechSat 21—Space missions using collaborating constellations of satellites. *Proceedings of AIAA/USU Annual Conference on Small Satellites*, 12th, Utah State University, Logan, August 31–September 3, 1998, Proceedings (A99-10826 01-20), Logan, UT, Utah State University.
54. Curtis, S. (1999). The Magnetospheric Multiscale Mission Resolving Fundamental Processes in Space Plasmas, NASA GSFC, Greenbelt, MD, December, NASA/TM2000-209883.
55. Bertsimas, D. and Tsitsiklas, J.N. (1997). Introduction to Linear Optimization, Athena Scientific, Belmont.
56. Tillerson, M. (2002). *Coordination and Control of Multiple Spacecraft using Convex Optimization Techniques*, S.M. Thesis, Dept. of Aeronautics and Astronautics, MIT, June.
57. Busse, F.D. and How, J.P. (2002). Real-time experimental demonstration of precise decentralized relative navigation for formation-flying spacecraft. *AIAA GNC*, August, Paper 2002–5003.
58. Busse, F.D. and How, J.P. (2002). Four-vehicle formation flying hardware simulation results. Presented at the *ION-GPS Conference*, September.
59. Inalhan, G., Tillerson, M. and How, J.P. (2002). Relative Dynamics & Control of Spacecraft Formations in Eccentric Orbits. *AIAA Journal of Guidance, Navigation, and Control*, **25**(1), pp. 43–53.
60. Solutions, A.I. (1999). *FreeFlyer User's Guide*. Version 4.0, March.
61. Tillerson, M. and How, J.P. (2002). Advanced guidance algorithms for spacecraft formation flying. Presented at the *American Control Conference*, May 2002, pp. 2830–2835.
62. Breger, L.S., Richards, A. and How, J.P. (2005). Model predictive control of spacecraft formations with sensing noise. Presented at the IEEE *American Control Conf.*, June, pp. 2385–2391.

63. Richards, A.G. and How, J.P. (2003). Model predictive control of vehicle maneuvers with guaranteed completion time and robust feasibility. *Proceedings of the American Control Conference*, June, pp. 4034–4040.
64. Scokaert, P.O.M. and Mayne, D.Q. (1998). Min-max feedback model predictive control for constrained linear systems. *IEEE Transactions on Automatic Control*, **43**(8), August, p. 1136.
65. Richards, A.G. and How, J.P. (2004). Robust constrained model predictive control with analytical performance prediction. Presented at the *AIAA Guidance, Navigation and Control Conf.*, Aug 2004. AIAA-2004-5110.
66. Kerrigan, E.C. (2003). Invariant Set Toolbox for Matlab, available at http://www-control.eng.cam.ac.uk/eck21, July.
67. Kolmanovsky, I. and Gilbert, E.G. (1995). Maximal Output Admissible Sets for Discrete-Time Systems with Disturbance Inputs. IEEE *American Control Conference*.
68. Kerrigan, E.C. (2000). Robust Constraint Satisfaction: Invariant Sets and Predictive Control. PhD Thesis, Cambridge University, November.

INDEX

Color plate 1. Covering of part of the global unstable manifold of an unstable periodic orbit in the circular restricted three-body problem (projection onto configuration space). The blue body depicts the Earth, the black trajectory is a sample orbit which leaves the periodic orbit in the direction of the Earth. The coloring indicates the temporal distance from the periodic orbit. (see Fig. 5.5)

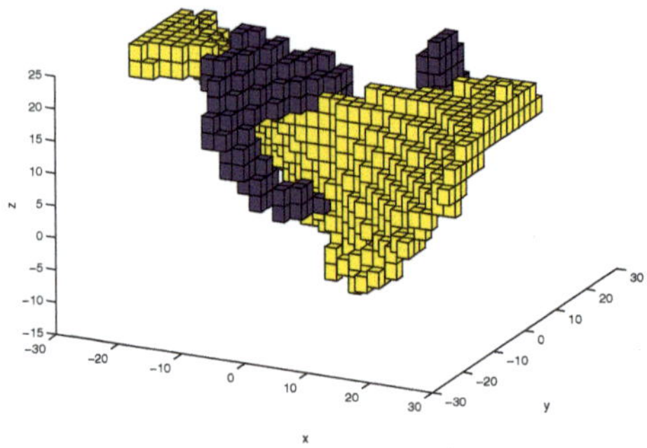

(c) $k = 15$

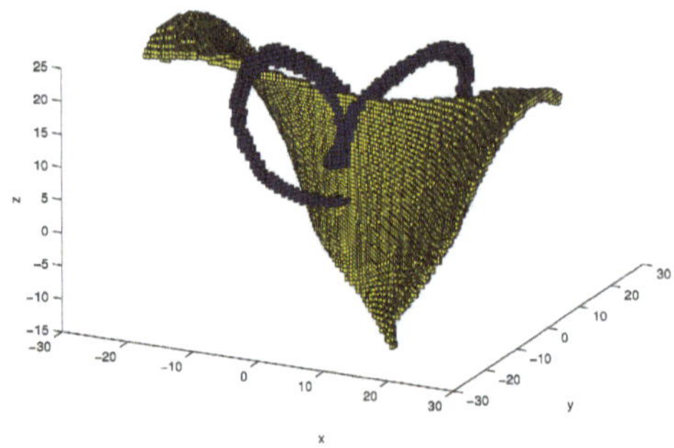

(b) $k = 21$

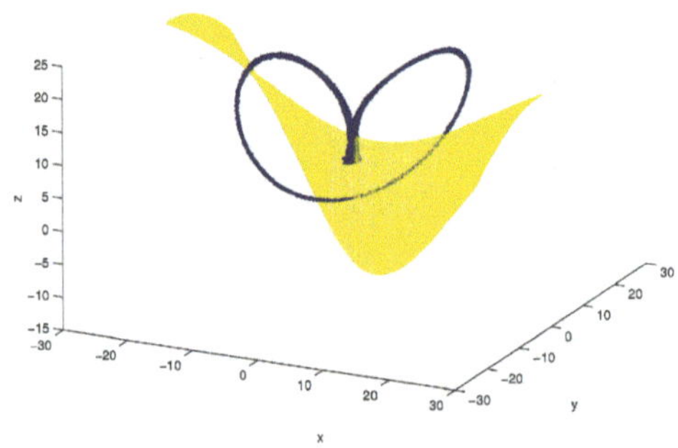

(c) $k = 27$

Color plate 2. Lorenz system: The coverings $\mathcal{U}_{(k)}(14)$ (blue) and $\mathcal{S}_{(k)}(14)$ (yellow) for different k. (see Fig. 5.11)

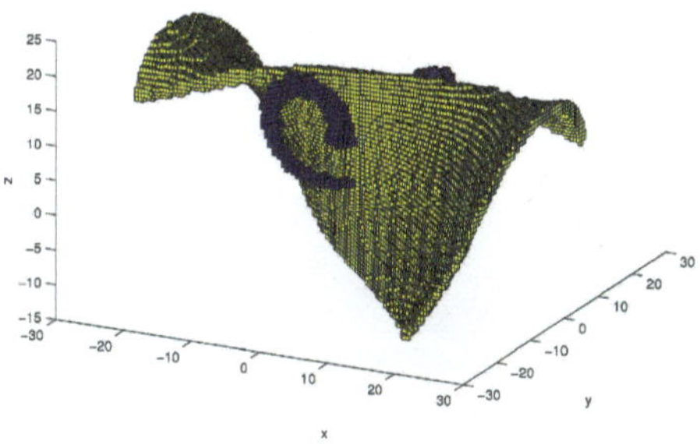

(a) $\rho = 10$

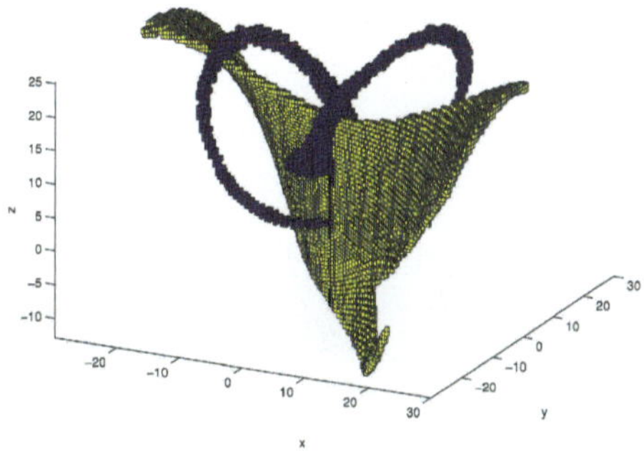

(b) $\rho = 14$

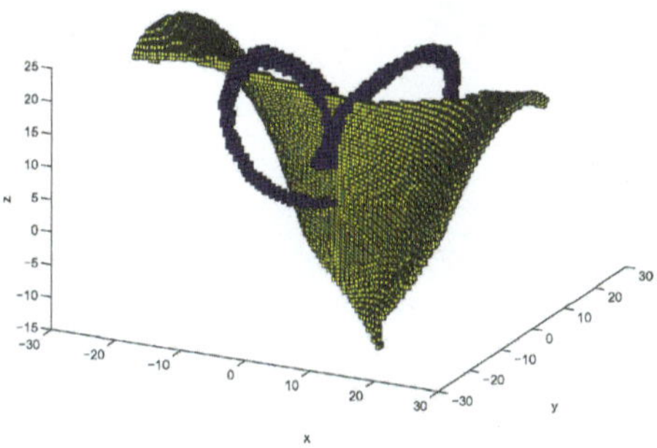

(c) $\rho = 17$

Color plate 3. Lorenz system: The coverings $\mathcal{U}^{(21)}(\rho)$ (blue) and $\mathcal{S}^{(21)}(\rho)$ (yellow) in dependence of ρ. (see Fig. 5.12)

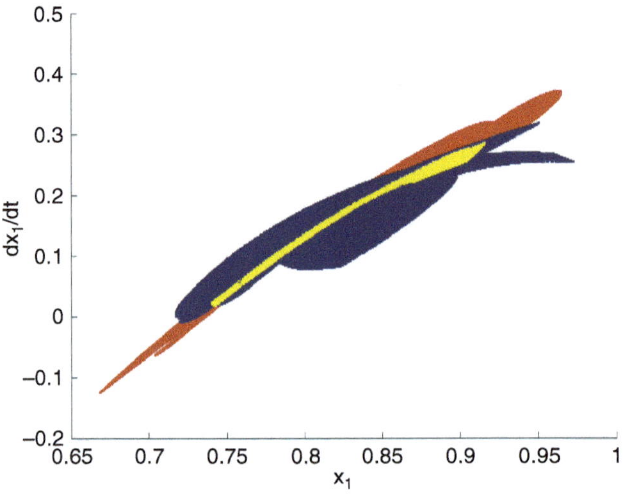

Color plate 4. Intersection of two reachable sets in a common intersection plane. Red: reachable set of the gateway set of Earth, blue: reachable set of the gateway set of Venus; yellow: their intersection. Shown is a projection of the covering in three space onto the $(x_1, \dot{x}_1)$-plane (normalized units). (see Fig. 5.14)

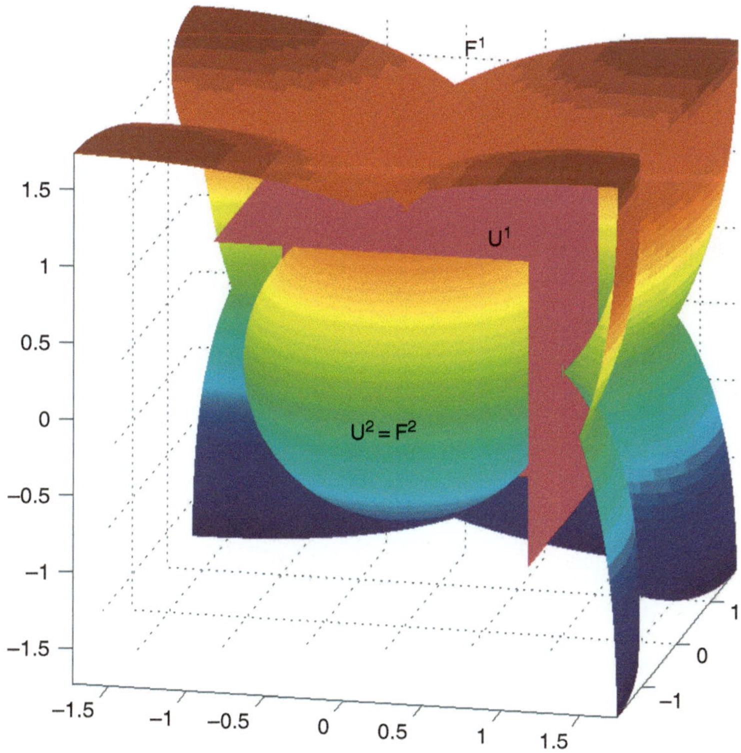

Color plate 5. Cutaway views of the geometries of the control space and their corresponding mass-flow rates. (see Fig. 6.3)

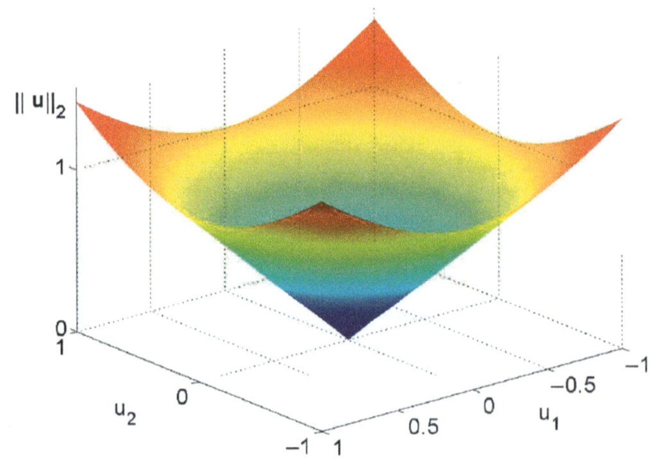

Color plate 6. Illustrating the nonsmooth structure of $\mathbf{u} \mapsto \|\mathbf{u}\|_2$. (see Fig. 6.9)

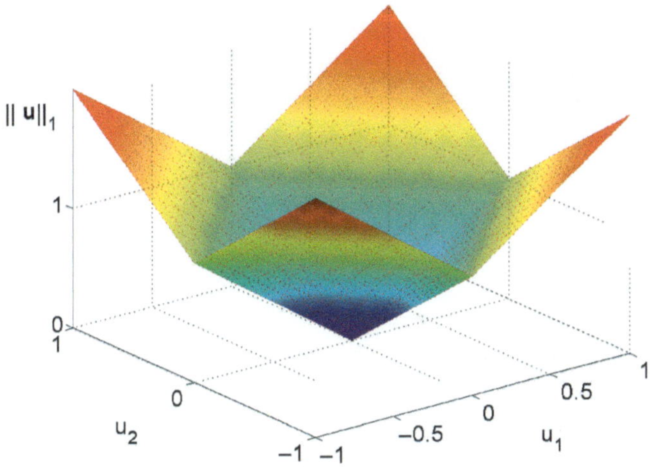

Color plate 7. Illustrating the nonsmooth structure of $\mathbf{u} \mapsto \|\mathbf{u}\|_1$. (see Fig. 6.10)

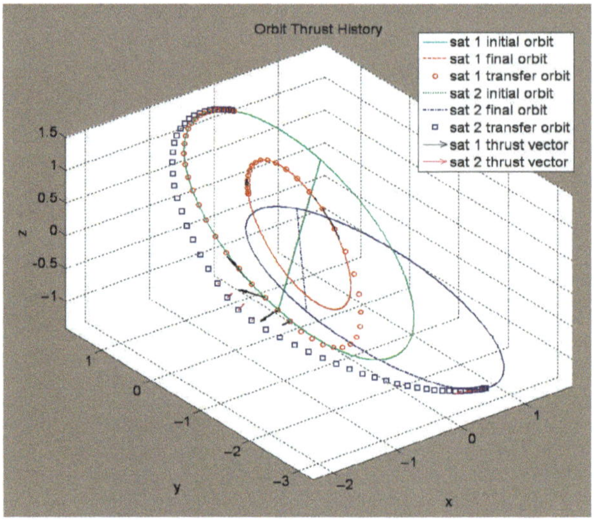

Color plate 8. A two-agent optimal space trajectory; from Ref. [15]. (see Fig. 6.14)

FSC
www.fsc.org
MIX
Papier | Fördert
gute Waldnutzung
FSC® C083411

Kolibri 360 GmbH
Ferdinand-Jühlke-Straße 7
99095 Erfurt, Deutschland
produktsicherheit@kolibri360.de